Computational Intelligence Methods for Super-Resolution in Image Processing Applications

Anand Deshpande • Vania V. Estrela
Navid Razmjooy
Editors

Computational Intelligence Methods for Super-Resolution in Image Processing Applications

Springer

Editors
Anand Deshpande
SAEF's Angadi Institute of Technology and Management
Belagavi, Karnataka, India

Vania V. Estrela
Electrical & Computer Engineering
Universidade Federal Fluminese
Duque de Caxias, Rio de Janeiro, Brazil

Navid Razmjooy
Amirkabir University of Technology
Ghent, Belgium

ISBN 978-3-030-67920-0 ISBN 978-3-030-67921-7 (eBook)
https://doi.org/10.1007/978-3-030-67921-7

This Springer imprint is published by the registered company Springer Nature Switzerland AG
The registered company address is: Gewerbestrasse 11, 6330 Cham, Switzerland

Preface

Super-resolution (SR) evolves day by day, given the importance of computer vision and computational intelligence (CI) to increase resolution as well as enhance visualization and analysis of images. Computer vision applications include but are not limited to surveillance, target tracking/identification, remote sensing, microscopy, astronomy, medical imaging, biometrics, and forensics. SR can enhance low-resolution (LR) images in several straightforward, average-complexity, and advanced uses.

This book emphasizes introductory definitions, applications, and deployment strategies in hardware/software with algorithmic approaches. CI allows better SR models, architectures, and outcomes highlighted by various authors' research surveys.

Imageries captured by different sensing modalities and treated by computerized systems through various techniques to visualize several phenomena is the fundamental underlining foundation of computer vision. The image acquisition hardware then processes the captured scene into an image rendition. The acquired image may have disturbances, noise, or blur that will require up-sampling, down-sampling, and possibly various other mechanisms to deliver the desired output. Better performance in computer vision applications is frequently associated with high-resolution images. This book's chapters also highlight SR usage in innumerable domains.

The list of chapters below illustrates the significance of the combined use of SR and CI:

Part I: A Panorama of Computational Intelligence in Super-Resolution Imaging

1. Introduction to Computational Intelligence and Super-Resolution
2. Review on Fuzzy Logic Systems with Super-Resolved Imaging and Metaheuristics for Medical Applications
3. Super-Resolution with Deep Learning Techniques: A Review
4. A Comprehensive Review of CAD Systems in Ultrasound and Elastography for Breast Cancer Diagnosis

Part II: State-of-the-Art Computational Intelligence in Super-Resolution Imaging

5. Pictorial Image Synthesis from Text and Its Super-Resolution Using Generative Adversarial Networks
6. Analysis of Lossy and Lossless Compression Algorithms for Computed Tomography Medical Images Based on Bat and Simulated Annealing Optimization Techniques
7. Super-Resolution-Based Human-Computer Interaction System for Speech and Hearing Impaired Using Real-Time Hand Gesture Recognition System
8. Lossy Compression of Noisy Images Using Autoencoders for Computer Vision Applications
9. Recognition of Handwritten Nandinagari Palm Leaf Manuscript Text
10. Deep Image Prior and Structural Variation-Based Super-Resolution Network for Fluorescein Fundus Angiography Images
11. Lightweight Spatial Geometric Models Assisting Shape Description and Retrieval and Relative Global Optimum-Based Measure for Fusion
12. Dual-Tree Complex Wavelet Transform and Deep CNN-Based Super-Resolution for Video Inpainting with Application to Object Removal and Error Concealment
13. Super-Resolution Imaging and Intelligent Solution for Classification, Monitoring, and Diagnosis of Alzheimer's Disease
14. Image Enhancement Using Nonlocal Prior and Gradient Residual Minimization for Improved Visualization of Deep Underwater Image
15. Relative Global Optimum-Based Measure for Fusion Technique in Shearlet Transform Domain for Prognosis of Alzheimer's Disease

Prospective readers will experience several facets of SR and CI besides noticing the target topics' evolving nature. This book offers different alternatives and methods to expand existing implementations with successful results in several realms, for example, graduate course classrooms, research facilities, healthcare services, non-destructive investigations, ambient intelligence, cultural heritage preservation, and industrial plants. This book also made it possible to gather a diverse and interesting group of international authors, which put forward a different understanding of the SR and CI within their respective chosen research fields.

Belagavi, Karnataka, India — Anand Deshpande
Duque de Caxias, Rio de Janeiro, Brazil — Vania V. Estrela
Ghent, Belgium — Navid Razmjooy

Acknowledgment

There are a number of people we would like to thank for helping us bring this book to completion. First of all, we would like to thank God. In the process of putting this book together, we realized how true this gift of writing is for us. He has given the power to believe in our passion and pursue our dreams. We could never have done this without the faith we have in Him, the Almighty.

We would like to express our sincere thanks to all authors for their excellent contributions. It was appreciable that we did not have to send too many reminders to the contributors for their submissions. We thank the reviewers for agreeing to our request to review chapters and their valuable contributions for the improvement of quality and content presentation of chapters.

We thank the Springer Nature team for their constant support. They allowed us liberal extensions of deadlines whenever required. We also thank Late Shri Suresh Angadi, Chairman of Suresh Angadi Education Foundation, and the management and staff members at Angadi Institute of Technology & Management, Belagavi, for their constant support. We thank all those who have supported directly or indirectly in helping us bring this book to completion.

Dr. Anand Deshpande
Dr. Vania V. Estrela
Dr. Navid Razmjooy

About the Book

Super-resolution (SR) techniques can be used in general image processing, microscopy, security, biomedical imaging, automation/robotics, and biometrics, among other areas, to handle the dimensionality conundrum posed by the conflicts caused by the necessity to balance image acquisition, image modality/resolution/representation, subspace decomposition, compressed sensing, and communications constraints. Lighter computational implementations are needed to circumvent the heavy computational burden brought in by SR image processing applications. Soft computing and, specifically, deep learning (DL) ascend as possible solutions to SR efficient deployment. The amount of multi-resolution and multimodal images has been augmenting the need for more efficient and intelligent analyses, for example, computer-aided diagnosis via computational intelligence (CI) techniques.

To facilitate this to the research community working in various fields, we and Springer Nature have come up with this book consolidating the work done in the subject of computational intelligence methods for super resolution in image processing applications. The intent behind publishing this book is to serve researchers, technology professionals, academicians, and students working in the area of latest advances and upcoming technologies employing CI methods for SR in image-processing applications. This book explores the application of deep learning techniques within a particularly difficult computational type of computer vision problem: SR. This book aspires to provide an assortment of novel research works that focuses on broad challenges of CI approaches for SR in image-processing applications.

Contents

About the Editors

Anand Deshpande is currently serving as the Principal and Director of the Angadi Institute of Technology and Management (AITM), India. He received his PhD in Electronics and Communication and a Master of Technology degree in Digital Communication and Networking from Visvesvaraya Technological University, and a Bachelor of Engineering degree in Electronics and Communication Engineering from Karnatak University, Dharwad. His research work has been published in international journals, international conferences, and books, and he has filed patents in several areas. Dr. Deshpande is a reviewer for a number of journals published by the IEEE, The Institution of Engineering and Technology (IET), and Springer. His research interests include artificial intelligence, image and video analytics, data analytics, and machine vision.

Vania V. Estrela is currently a member of the faculty in the Department of Telecommunications at Federal Fluminense University (UFF), Brazil. Professor Estrela obtained her BSc degree in Electrical and Computer Engineering (ECE) from Federal University of Rio de Janeiro (UFRJ), an MSc in ECE from the Technological Institute of Aeronautics (ITA) and Northwestern University, and her PhD in ECE from the Illinois Institute of Technology (IIT). She has taught at DeVry University, State University of Northern Rio de Janeiro (UENF), and the West Zone State University, Brazil. Her research interests include signal/image/video processing, inverse problems, computational and mathematical modeling, stochastic models, multimedia, electronic instrumentation, computational intelligence, automated vehicles, machine learning, and remote sensing. She is an Editor for the International Journal of Ambient Computing and Intelligence, International Journal on Computational Science & Applications, and the EURASIP Journal on Advances in Signal Processing, and a member of the IEEE and the Association for Computing Machinery (ACM).

Navid Razmjooy holds a PhD in Electrical Engineering (Control and Automation) from Tafresh University, an MSc with honors in Mechatronics Engineering from the Isfahan Branch of Islamic Azad University (IAU), and a BSc from the Ardabil Branch of IAU. His research interests include renewable energies, control, interval analysis, optimization, image processing, machine vision, soft computing, data mining, evolutionary algorithms, and system control. He is a senior member of the IEEE and Young Researchers Club of IAU. Dr. Razmjooy has published five books and more than 120 papers in English and Farsi in peer-reviewed journals and conferences. He is a reviewer for several national and international journals and conferences.

Part I
A Panorama of Computational Intelligence in Super-Resolution Imaging

Introduction to Computational Intelligence and Super-Resolution

Anand Deshpande, Navid Razmjooy, and Vania V. Estrela

1 Introduction

The world has been progressing toward an unimaginable direction in both hardware and software technologies. Evolution has allowed companies to take off advanced technology to produce electronic gadgets and equipment cost-effectively. The camera companies, such as Nikon, Canon, and Sony, have also advanced their R&D and manufacturing capabilities in high dense pixel properties and quality HR digital cameras. Even though technology progresses with HR, multimodality cameras, there is a further need for computer vision applications. Computer vision applications include but are not limited to surveillance, target identification, satellite imaging, medical imaging, and forensic. The need for HR images is a requisite for further computer vision (CV). Thus, to match HR images' requirement for the applications, image-processing techniques evolved to improve HR images' quality. It is often paramount to have a clear vision. Unclear image capture could lead to catastrophic or disastrous aftermath.

Vision captured by hardware and computer-intensive structures afford a more precise visualization, which is the necessary underlining foundation of super-resolved CV that extracts information from still pictures or video (in frames per second) via assorted types of cameras and imaging modalities. The camera, then, processes the acquired image that could be disturbed, noisy, or blurred. This image could undergo up-sampling, down sampling, and possibly various other mechanisms to get the desired output. The whole CV process of capturing and processing images, in a nutshell, called for better performance and improved HR image versions of a scene [1–3].

A. Deshpande (✉)
Angadi Institute of Technology and Management, Belagavi, Karnataka, India

N. Razmjooy · V. V. Estrela
Telecommunications Department, Federal Fluminense University, Rio de Janeiro, Brazil

A. Deshpande et al. (eds.), *Computational Intelligence Methods for Super-Resolution in Image Processing Applications*, https://doi.org/10.1007/978-3-030-67921-7_1

2 Computational Intelligence

"The measure of intelligence is the ability to change," is stated, Albert Einstein. Change is a continuous process in human life. Thus, the evolution and progression toward different theories and paradigms in the field of intelligence arose.

Artificial intelligence (AI) is a practically impossible topic to talk about before the invention of computers. After the advent of computers in human life, the discussion of AI became a hot topic. AI is a part of computer science from a scientific point of view. AI relies mostly on hard computing techniques as it inclines to favor designs with more robust theoretical warranties and still has a significant community focused on purely deductive reasoning. Today, it is studied and researched as a fundamental and applied topic in various fields of basic sciences and engineering. Loosely speaking, that is, in colloquial terms, AI is a set of algorithms and other types of intelligence that attempts to reproduce human intelligence and can handle the following kinds of systems or combination of types [4–7]:

1. Fully observable or partially observable
2. Single-agent, multi-agent or adversarial agent
3. Deterministic or stochastic
4. Static or dynamic
5. Discrete or continuous

Computational intelligence (CI) or soft computing (SC) has been in vigorous development with classical well-established schemes and algorithms like machine learning approaches, classification procedures, cluster techniques, and data mining practices playing essential roles. Persistent improvements and refinements appeared, but these CI methods became truly popular when computer resources experienced a steady increase in computational speed with affordable prices.

"Computational intelligence is the theory, design, application, and development of biologically and linguistically motivated computational paradigms" [4], and it relies mainly on fuzzy systems, neural networks, and lastly on evolutionary computation. However, due to evolution, many nature-inspired computing paradigms have been continuously developing. Hence, CI incorporates artificial life, ambient intelligence, artificial endocrine networks, cultural learning, artificial hormone networks, and social reasoning because of evolution.

Computation intelligence has imprinted significant developments in the development of intelligent systems. The intelligent systems include the popular cognitive developmental systems and games. With the evolution, products, and effects, there is an explosion of interest toward deep convolutional neural networks. The building blocks of computational intelligence appear in Fig. 1 [16, 17].

The three main CI foundations (i.e., fuzzy systems, neural networks, and metaheuristics) further escalated the interest in the research sector.

Nowadays, CI permeates many applications, directly or indirectly. Nature-inspired metaheuristic algorithms begin to demonstrate auspicious power in many other areas [8]. As such, CI started evolving at a tremendous pace along with the

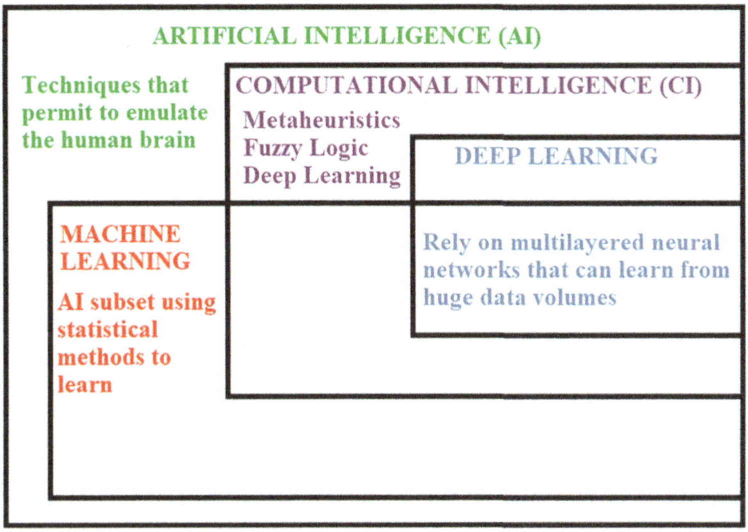

Fig. 1 Building blocks of computational intelligence

other subordinates. CI is one of the most essential and practical subsections of AI, in which various tools implement the AI. CI often employs mathematical tools inspired by nature (aka metaheuristics) or deep learning (DL) and knowledge about the world where interactions happen.

Concisely, CI comprehends and learns a particular task from either experimental observations or data.

2.1 Metaheuristics

A metaheuristic is a higher-level scheme or heuristic that can be a CI subset, intended to discover, generate, or choose a heuristic (aka partial search algorithm). The metaheuristic may offer a sufficiently excellent solution to an optimization problem, principally with incomplete or imperfect data or inadequate computation effort [9, 10]. These algorithms usually sample a solution subset, otherwise too vast to be fully itemized or explored. Metaheuristics may handle reasonably few assumptions regarding the optimization problem being serviceable for an assortment of issues.

Unlike optimization and iterative algorithms, metaheuristics do not warranty a globally optimal solution on some categories of problems [9, 10]. Many metaheuristics contain stochastic optimization, thus making the solution contingent on the generated set of random variables. Characteristically, combinatorial optimization searches over an enormous set of feasible solutions. Metaheuristics, on the other hand, can frequently discover decent solutions with less computational work than optimization algorithms, iterative approaches, or unpretentious heuristics [10–15].

Most literature on metaheuristics describes empirical results from computer experiments. Nevertheless, some formal theoretical outcomes are also available, often on convergence and the possibility of finding the global optimum [10].

There are many crucial issues related to the context of CI and metaheuristic algorithms as follows:

(i) Lack of a general mathematical (systematic) framework to analyze metaheuristic algorithms' convergence and stability.
(ii) Parameter tuning is still a timewasting process. The best way to tune an algorithm for an extensive assortment of problems remains an unsolved problem.
(iii) The best form to solve high-dimensional problems effectively is unclear. Efficient scalability to deal with large-scale issues is an area that requires investigation.
(iv) Discrete problems and combinatorial optimization, particularly those involving NP-hard problems, remain very puzzling to solve.

These challenges also pose great opportunities for researchers. Any evolvement in the above areas will help comprehend metaheuristic algorithms. Another useful byproduct is to increase capabilities in solving a diverse range of problems in real-world applications.

Evolutionary computation (EC) algorithms are a family of problem-solving population-based methods based on trials and errors with stochastic optimization mechanisms or metaheuristic optimization. They use the optimal solution to converge to the global or approximate optimal solution.

In evolutionary calculations, a basic set of "candidate solutions" is first formed. Evolutionary computational algorithms manipulate and update the population of candidate answers to move this population to the region containing the "global optimum" solution during the evolutionary process. In each iteration of evolutionary computational algorithms (also called "generation"), an evolutionary process will take place by eliminating undesirable responses in the population and making very small, albeit random, changes in candidate responses [18].

By modeling natural evolutionary processes in biology, evolutionary processes, for example, "natural selection" and "mutation," influence the population of candidate responses (in evolutionary computational algorithms). Natural selection is also called artificial selection process [19] in the terminology of evolutionary computational algorithms. As a result, populations of candidate responses (in evolutionary computational algorithms) evolve based on the mechanisms defined in the evolutionary process, in such a way that their "fitness" increases with the passage of time and successive generations. The "fitness function" is used to calculate the fitness of candidate solutions in evolutionary computational algorithms.

Evolutionary computational methods can produce a set of optimized solutions for a particular problem in different conditions. Such an essential feature in evolutionary computing systems and evolutionary algorithms puts them in contrast to conventional optimization methods that can produce only one deterministic solution or a limited number of approximated solutions to the problem. Besides, the ability to generate a set of candidate solutions to a given problem has made evolutionary computational algorithms the most popular problem-solving methods in the field of computer science.

So far, various versions of evolutionary computational algorithms have surfaced. Many current evolutionary computational algorithms are "domain-independent" and "problem-independent," making them a perfect option for solving a wide range of problems, primarily converts specific data structures. Evolutionary computational algorithms sometimes work as a computer-aided testing method (in silico experimental procedure) to study common aspects of general evolutionary processes.

The scope of evolutionary computation includes but is not limited to evolutionary programming, genetic algorithms, genetic programming, evolution strategies, evolvable hardware, differential evolution, multiobjective optimization, and many more.

2.2 Fuzzy Systems

The Oxford dictionary expresses the term "fuzzy" as vagueness. The theory of fuzzy sets acts in conditions of uncertainty. It can mathematically formulate many concepts, variables, and systems that are inaccurate and provide the basis for reasoning, inference, control, and decision-making in ambiguous conditions. Fuzzy systems employ knowledge or rules, and a variable assumes many logic values.

The heart of a fuzzy system is a knowledge base made up of fuzzy `if-then` rules. In other words, a fuzzy inference system is a tool for formulating a process using fuzzy `if-then` rules. Consider the fuzzy rule "If it rains a lot, then the humidity is high." Suppose

x is the linguistic variable corresponding to the amount of rainfall,
H is the set describing high rainfall,
Y is the linguistic variable of the amount of humidity, and
HN is the fuzzy set high.

In this case, the `if-then` rule becomes

If x is H then y is HN

The system that formulates an input-to-output mapping using phase logic is known as the Fuzzy Inference System (FIS). Fuzzy systems are used today in a wide range of sciences and technologies, from signal processing control, integrated circuit building communications, and expert systems to business, medicine, and social sciences. The fuzzy system includes but is not limited to fuzzy systems and sets, classification and clustering, fuzzy neural networks, linguistic summarization, and many more.

An Adaptive Neuro-Fuzzy Inference System (aka Adaptive Network-Based Fuzzy Inference System) (ANFIS) comprises a kind of Artificial Neural Network (ANN) that relying on the Takagi–Sugeno FIS (the next subsection discusses

ANNs). The procedure dates from the early 1990s [20, 21] and integrates both ANNs and fuzzy logic principles. It can capture the benefits of both schemes in a single framework. Its inference system corresponds to a set of fuzzy `if-then` rules with learning competence and can approximate nonlinear functions [22]. Therefore, ANFIS is considered a universal estimator. One can utilize the best parameters attained by genetic algorithm [23–25], with the ANFIS more efficiently and optimally.

2.3 Neural Networks (NNs) and Deep Learning (DL)

In line with many scientists, the human brain is the most complex system ever observed and studied throughout the universe. But this complex system has neither the dimensions of a galaxy nor the number of components it has more than today's supercomputer processors. This unique system's mysterious complexity goes back to the many connections between its parts, which sets the 1400 gram human brain apart from all other structures [26].

The conscious and unconscious processes within the human body's geographical boundaries are all controlled by the brain. Some of these processes are so complex that no computer or supercomputer in the world can process and perform them.

The very high speed and processing power of the human brain go back to the very massive connections between the cells that make up the brain. The human brain would be reduced to a standard system without these links and certainly did not have the current capabilities.

After all, humans have made the simulation of the brain's and its capabilities. The most critical hardware and software architects' primary goal is the brain's excellent performance in solving several kinds of problems and its high efficiency. If the day comes (which, of course, apparently is not too far away) when we can build a computer the size of the human brain, there will undoubtedly be a great revolution in science, industry, and of course, human life [27].

In the last few decades, when computers have made it possible to implement computational algorithms to simulate the human brain's computational behavior, many computer scientists, engineers, and mathematicians started research works, whose results are in a branch of artificial intelligence. ANNs fall under the computational intelligence heading and encompass several mathematical and software models inspired by the human brain. These paradigms solve a wide range of scientific, engineering, and applied problems in various fields. The ANNs include but are not limited to recurrent NNs. Self-organizing NNs, feedforward NNs, convolutional neural networks, deep learning, and many more. Some self-driving cars are examples of this system [28, 29].

2.4 *Hybrid Methods*

However, some efficient approaches can stem from combining the different algorithmic classes described above, and the eagle strategy is a two-stage strategy combining a coarse explorative stage and an intensive exploitative stage in an iterative manner [16].

3 Super-Resolution

Recently, companies like Tesla, Volvo, BMW, Audi, and Mercedes Benz utilized multiple cameras, ultrasonic sensors, and radars in their self-driving cars to allow them to detect the lane markings and identify objects (human or vehicles), traffic signals and signs. The captured images by them are for the safety of the people inside a vehicle. The importance of the quality of the image captured for further processing is a vital task. As if the input is not appropriate, then the output will be unsatisfactory as well. Thus, a low-quality output will lead to difficulty in pattern recognition and analysis of images. Hence, to overcome poor outcomes, a process called super-resolution (SR) arises. SR is the process of recovering HR images from low-resolution (LR) imageries employing a single image or a set of images (aka multiple images that utilizing frames from sequences or different scenes views) [30–33]. SR can boost forensic, surveillance, satellite imaging, and so on.

The typical approach accounts for the LR images from there sampling of HR images. The HR pictures result from resampling the LR observed images based on knowledge about the input images and the imaging model. Hence, the need for accurate modeling is crucial for the SR to avoid the image's degradation. The images originate from single or multiple cameras. SR also supports frames of video sequences. The process of registration defines a common reference frame to which images are mapped. Image registration consists of two parts, namely, geometric and photometric. The SR procedure can be within an image's specific region in the composite image after the registration process. The registration and formulation of an accurate image model is the key to a successful SR. Figure 2 states briefly the steps involved in the SR process.

Thus, to follow the SR process, the image could go through geometric and photometric registration processes. Multiple LR images possibly have different viewpoints of the same scene. Image registration involves mapping the actual points in the original scene and to the views. Then further to it, transform the data into one system. Thus, data transformation requires geometric and photometric components. Geometric registration consists of feature-based registration and maximum likelihood registration. Many other image registration transformations like planar homographic, biquadratic, and affine transformation may be necessary. Figure 3 shows the raw frame, SR frame, and images with percentage fill factors.

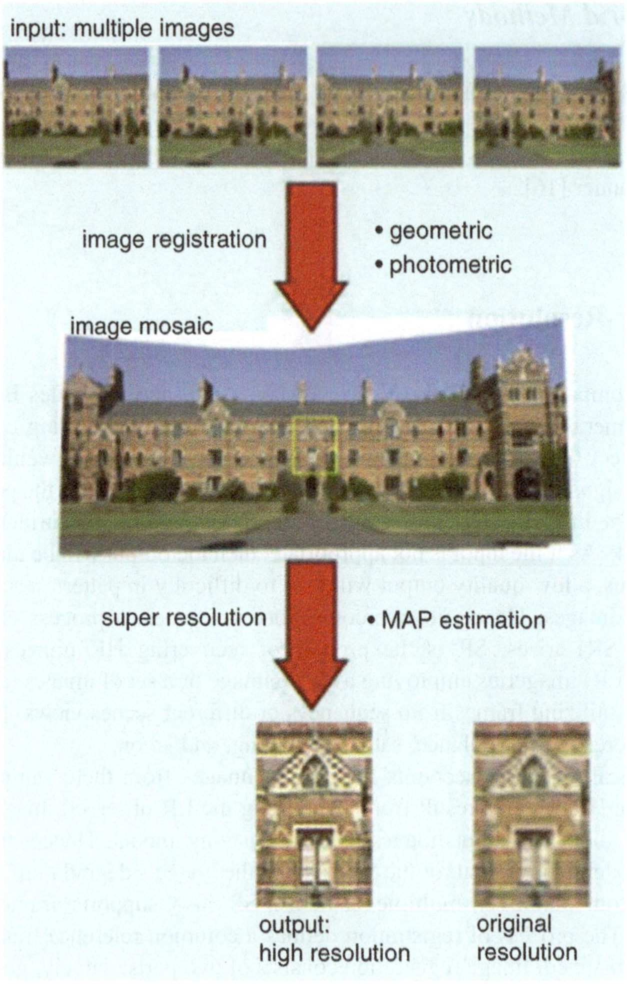

Fig. 2 Steps in the super-resolution process [2]

4 Super-Resolution Methods

Although many applications can be tapped onto SR, the existing methods include spatial domain-based, frequency domain-based, and deep learning-based SR.

In the frequency domain-based approach, SR algorithms first transform the LR images (input) to the frequency domain. Then, an HR image in this domain is estimated. In the end, the reconstructed HR image is transformed to the spatial domain [34]. In changing the images to the frequency domain, the algorithms are broken into two entities. Firstly, the Fourier transform-based and secondly, wavelet transform-based methods. The main disadvantage related to frequency domain based

Fig. 3 Comparison between the raw frame and super-resolution frame

SR is that it is insufficient to handle real-world applications. Secondly, prior knowledge is difficult to express, which helps to regularize the SR problem. Whereas the benefits of the frequency-domain-based SR approach are as follows:

- Low-computational complexity
- The high-frequency information stems from extrapolating the high-frequency content available in the LR images

4.1 Spatial Domain-Based Super-Resolution

The approaches for the spatial domain-based SR method are three fold: interpolation-based, reconstruction-based, and example-based.

4.1.1 Interpolation-Based Approach

In image processing, image interpolation is an essential operational process. It is a process of resolving the function values at a point between its samples, achieved by fitting a function throughout the discrete samples. The HR images are interpolated from the LR input using a smooth kernel function. By using either parametric or nonparametric methods, the interpolation-based SR methods upscale the LR images' size. Thus, to estimate the HR grids' pixels, a base function or an interpolation kernel is applied to obtain an HR image. This technique is, however, further divided into two segments, namely, adaptive and nonadaptive. The main advantage of the interpolation approach is that it is simple and easy to implement. Also, the interpolation approach is fast in computation. However, the interpolation approach's major disadvantage is that the SR method performs satisfactorily in low-frequency areas. However, they perform poorly in high-frequency areas. This approach's main reason for poor performance in the high-frequency regions is that they are prone to blur and produce jaggy artifacts along the edges [34].

4.1.2 Reconstruction-Based Approach

Deterministic and stochastic approaches are two categories classified in the reconstruction-based algorithm. The deterministic methodology encodes knowledge based on what the HR image should look. This evidence is built in as priors and regularizes the solution using the constrained least-squares method. The typical approach is to impose a smoothness prior via regularization on top of a least-squares optimization [38, 39]. The presence of the regularization term guarantees a convex and differentiable optimization function. Thus, a unique optimal solution can be computed using several standard methods like gradient descent. This approach's results certainly improve the LR image, but enforcing smoothness is not always the best option, especially if other priors preserve high-frequency details better.

Stochastic methods that treat SR reconstruction as a statistical estimation problem have rapidly gained prominence since they provide a robust theoretical framework for the inclusion of a priori constraints necessary for a satisfactory solution of the ill-posed SR inverse problem [40–44]. The statistical techniques explicitly handle prior information and noise [45]. The inclusion of prior knowledge is usually more natural using a stochastic approach. The stochastic SR reconstruction using the Bayesian approach provides a flexible and convenient way to model a priori knowledge about the final solution. One of this method's main advantages is its easy integration with other common image processing tasks, such as denoising, deconvolution, and enhancement. However, because of the SR problem's ill-posed nature, the reconstruction-based approach may not produce a unique solution [34].

4.1.3 Example-Based Approach

An example-based approach for video SR is proposed by Brandi et al. [37, 38]. The high-frequency information of an interpolated block is restored by searching for a similar block in a database. Adding the high-frequency of the chosen block is added to the interpolated image. The performance [47] is analyzed by testing on Foreman, Mobile, and Coast Guard CIF video sequences and achieved PSNR of 32, 23, and 28.5 dB. Shin [48] proposed a multiple multilayer perceptrons (MLPs)-based SR technique for iris recognition. This method reconstructs an HR image based on the output values of the three MLPs. Example-based approach has a high magnification factor, which is advantageous in this approach. However, this approach requires a big image database. As a result, there would be an increase in the execution time. The main reason for the rise in the execution time is that the image patches of the test images should be searched [34].

5 Deep Learning-Based Super-Resolution

The latest state-of-the-art performance attained with deep learning algorithms has facilitated super-resolving LR images. Deep learning is a wing of the machine learning algorithm. The main aim of deep learning is to learn the hierarchical representation of data [46, 49]. Compared to other machine learning algorithms, deep learning has displayed good superiority. Knowledge domains include but are not limited to speech recognition, computer vision, and natural language processing [50]. Unmanned aerial vehicles (UAV) have contributed immensely to prevent disasters and safeguard nations' borders in the field of surveillance and security. Computer vision and SR play a vital role in identifying threats or assets to the stations. The framework from [51] relies on deep learning to super-resolve LR images from UAVs. Deep learning algorithms automatically learn informative hierarchical representations by producing and leveraging the entire learning process [52].There are various types of approaches that complement the deep learning-based SR, as the next section explains.

5.1 Super-Resolution Convolutional Neural Network (SRCNN)

SR convolutional neural network (SRCNN) for the single image SR is the first deep learning method. Thus, it can directly learn LR or HR images end to end mapping.

Figure 4 shows the layout of the network structure. The network contains three layers, and whereas each layer has a convolution layer with the addition of activation function. The network's input image is a bicubic interpolation of an LR image of the same size as the HR, SR output images [53, 54].

5.2 Very Deep Super-Resolution Convolutional Networks (VDSR)

VDSR discovers the possibility of SR performance's improvement concerning the increase of the depth of the network. Because of higher depth, convergence is

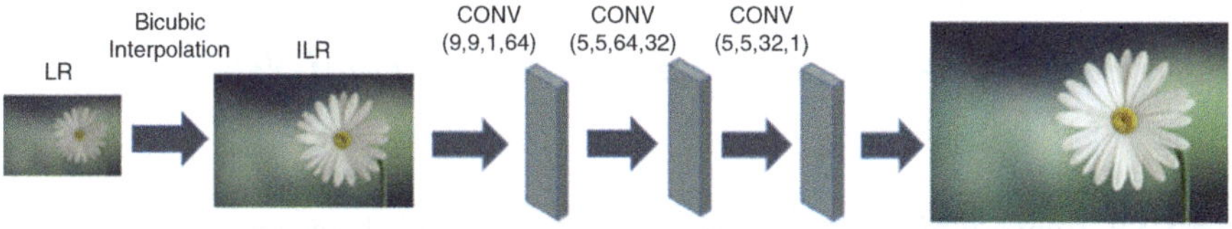

Fig. 4 Network structure layout of SRCNN [32]

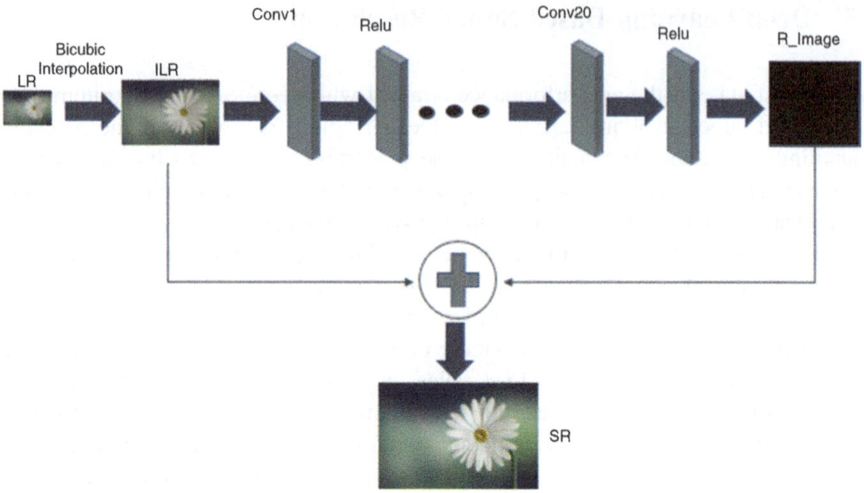

Fig. 5 Network structure of VDSR [32]

affected significantly. Thus, learning residuals is a better option because the LR and HR images share the same information to a more considerable extent to achieve better speed and performance [53, 54].

The residuals between HR and LR images learned using a too high learning rate combine LR images to generate final HR images. Figure 5 shows the network structure of the VDSR.

5.3 Deeply Recursive Convolutional Networks (DRCN)

In the field of SR reconstruction, DRCN introduces deep recursive layers. If their depths increase, then the performance may get better [53, 54]. However, there would not be an increase in the number of parameters due to convolution layers being contrary.

The reconstruction results stem from weighing the average of the results of the recursive convolution layers. Figure 6 shows the network structure of the DRCN.

5.4 Fast Super-Resolution Convolutional Neural Network (FSRCNN)

Fast SR Convolutional Neural Network is an upgraded version of SRCNN. It focuses on the acceleration of the speed of HR reconstruction. The structure of

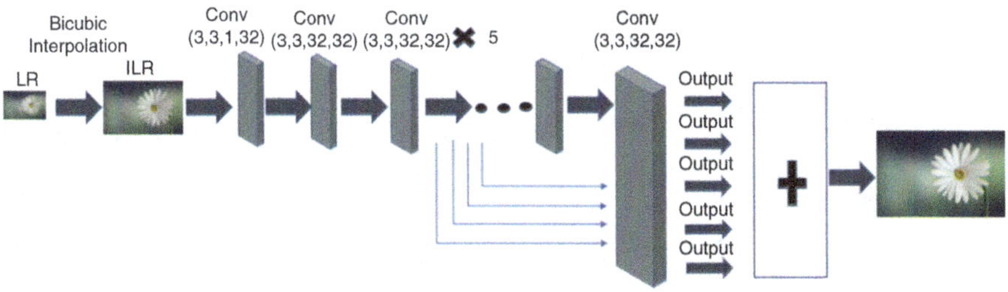

Fig. 6 Network structure of DRCN [53, 54]

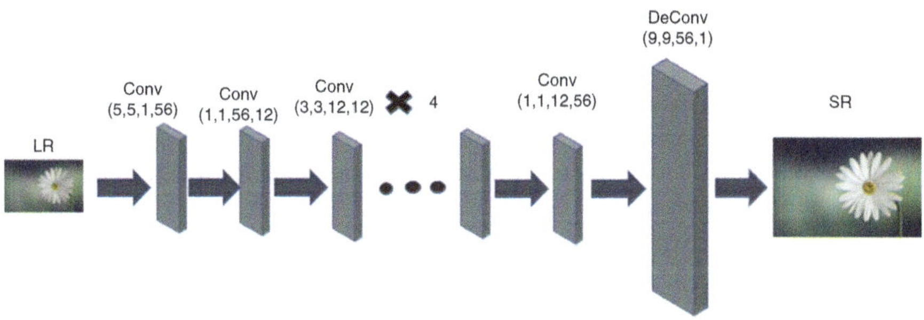

Fig. 7 Network structure of FSRCNN [32]

FSRCNN comprises five parts, namely, feature extraction, shrinking, mapping, expanding, and deconvolution.

The structure of the FSRCNN shows up in Fig. 7. The computation is reduced, and the speed is improved when the input image size is not enlarged in this network [53, 54].

6 Applications of CI and SR

6.1 Super-Resolution in Medical Imaging

During crucial situations in the health and medicine sector, a clear image of the organs or body parts plays an important role. As such, Fig. 8 shows a medical imaging scenario. Many imaging modalities exist, which possibly provide anatomical information and reveal information about the human body's structure. However, some modalities provide functional information, activity locations for specific activities, and specified tasks [55–58]. Each imaging system has a characteristic resolution, which is determined based on the system detectors' physical constraints that are tuned to signal-to-noise and timing considerations. A common goal across systems is to increase the resolution, and as much as possible, achieve true isotropic 3-D imaging. SR technology can serve to advance this goal. Research on SR in key

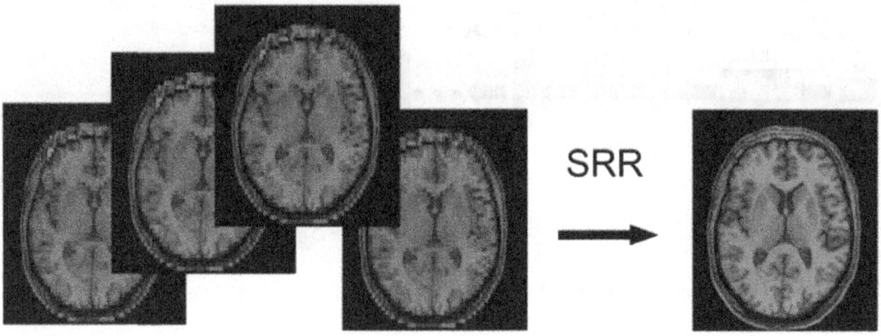

Fig. 8 Super-resolution in medical imaging [55]

medical imaging modalities, including MRI, fMRI, and PET, has emerged in recent years and reviewed herein. The algorithms used are mostly based on standard SR algorithms. Results demonstrate the potential in introducing SR techniques into practical medical applications.

6.2 Super-Resolution in Satellite Image Processing

The higher resolution of images prevails in the case of satellite imaging. SR plays a vital role in achieving higher resolution images. Satellite image processing includes the rectification, enhancement, restoration, and information extraction of the images [35, 36]. The visual interpretation of the image increases when the number of pixels grows. This could eventually eradicate the distortions and enhance the information related to geographical scenarios, location, and could aid land map constructions [59, 60]. Figure 9 shows SR in satellite imagery.

6.3 Super-Resolution in Microscopy Image Processing

SR plays a crucial role in processing microscopic images and has a substantial and useful part during the visualization of biological structures such as cells, pathways, tissues, and biological molecules. Figure 10 shows the SR in microscopy image processing.

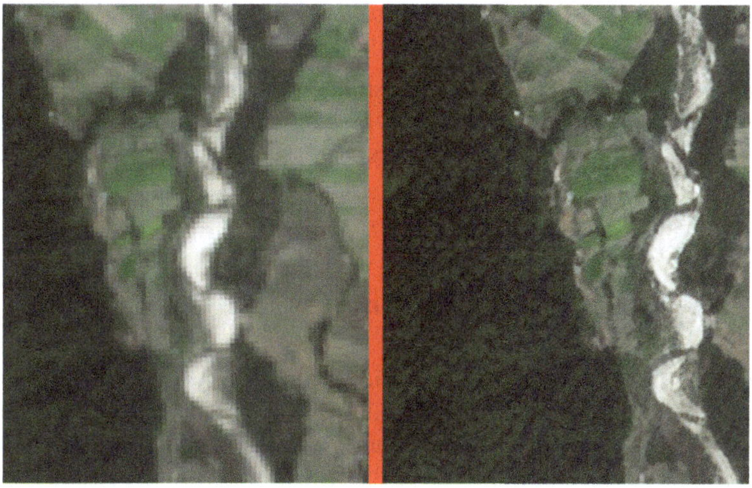

Fig. 9 Super-resolution in satellite imagery [60]

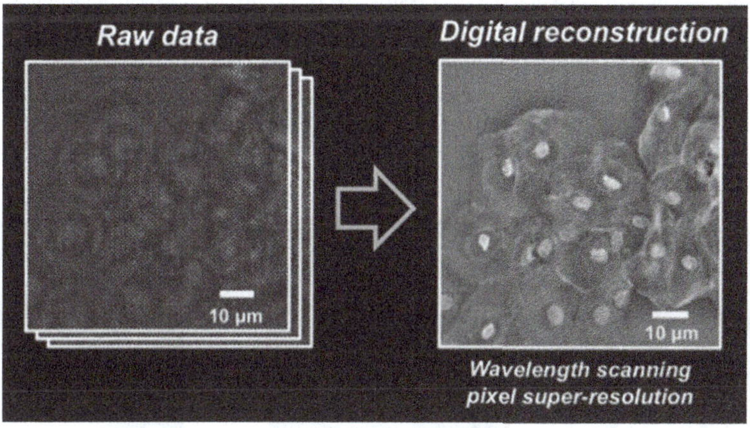

Fig. 10 Super-resolution in microscopy image processing [53]

6.4 Super-Resolution in the Multimedia Industry and Video Enhancement

In recent trends, there is an increase in the usage of multimedia-based applications. Multimedia applications involve movies, visual effects, and animations. SR enhances videos and images to improve their quality in multimedia-based applications [61].

6.5 Super-Resolution in Astronomical Studies

SR is a useful technique in the field of astrological studies. For better and efficient computation, higher resolution images of astronomy are desired. For the quality improvement of the astronomical images, SR has been used [62, 63]. Figure 11 shows the SR in astronomical studies.

6.6 Super-Resolution in Biometrics

In biometric systems, the lack of resolution of imaging systems results in adverse impacts. The adverse impact is on surveillance and long-range biometrics such as iris, face, and gait recognition [64]. To super-resolve and recognize the long-range captured polar image could consist of best frame selection algorithm, modified diamond search algorithm, Gaussian Process Regression (GPR)-based and enhanced iterated back projection (EIBP)-based SR approach, fuzzy entropy-based feature selector, and neural network (NN) classifier [65]. Figure 12 shows the application of SR in biometrics.

In an alternative to iris, ear recognition has lately picked up as an alternative to biometric recognition in recent years [68]. Thus, for detection and identification, super-resolution plays a vital role. Such detection and identification are crucial for surveillance and security in airports, secured premises, and other classified sectors.

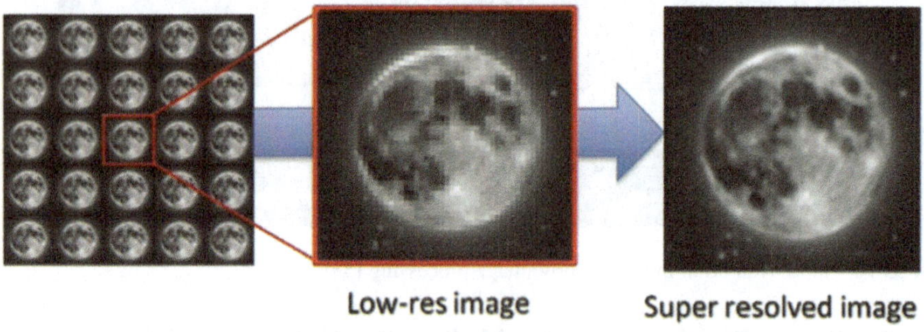

Fig. 11 The super-resolution in astronomical studies

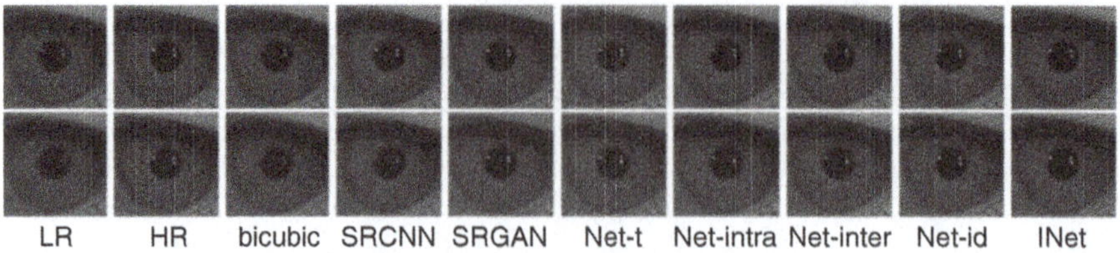

Fig. 12 Iris image super-resolution using various techniques [66]

Thus, with regards to ears, a framework was proposed to recognize the images without the external data set by super resolving ear images [67].

SR is also used in the automotive industry, surveillance, object detection, real-time processing, forensic, military, and scanning. Further, CI and SR are utilized in the following areas and domains, for example, AI in renewable energy, livable and sustainable cities, object detection and recognition, smart agriculture, ocean protection, sustainable transport system, climate change predictions, and so on.

7 Conclusion

The importance of computer vision and computational intelligence and the need for SR are on the rise as the world is evolving day by day. Computer vision applications include but are not limited to surveillance, target identification, satellite imaging, medical imaging, and forensic. For every abovementioned application, SR is required to amplify the clarity of the LR images. In this chapter, an introduction, applications, approaches, and methods of computational intelligence, SR was highlighted. In addition, this chapter discussed the different types of SR methods with an underlining of advantages and disadvantages. This chapter also highlighted SR usage in various domains and applications such as in medical, satellite, biometrics, and others [69–74].

References

1. Sung, C. P., Min, K. P., & Moon, G. K. (2003). Super-resolution image reconstruction: A technical Overview. In *IEEE Signal Processing Magazine* (Vol. 1053–5888, p. 23). IEEE.
2. Gaidhani, P. *Super-resolution*. http://homepages.inf.ed.ac.uk/rbf/CVonline/LOCAL_COPIES/AV1011/Super_Resolution_CVonline.pdf. Retrieved on 13 Jan 2020.
3. Hardie, R. C., Barnard, K. J., Bognar, J. G., Armstrong, E. E., & Watson, E. A. (1998). High-resolution image reconstruction from a sequence of rotated and translated frames and its application to an infrared imaging system. *Optical Engineering, 37*(1), 247–260.
4. Partridge, D., & Hussain, K. M. (1991). *Artificial intelligence and business management*. Norwood, NJ: Ablex Publishing.
5. Rich, E., & Knight, K. (2009). *Artificial intelligence* (2nd ed.). New York, NY: McGraw-Hill.
6. Rich, E., Knight, K., & Nair, S. B. (2009). *Artificial intelligence* (3rd ed.). New Delhi: Tata McGraw-Hill.
7. Russell, S. J., & Norvig, P. (2010). *Artificial intelligence: A modern approach* (3rd ed.). Upper Saddle River, NJ: Prentice Hall.
8. Yang, X. S. (2014). *Nature-inspired optimization algorithms*. London, UK: Elsevier.
9. Bianchi, L., Dorigo, M., Gambardella, L., & Gutjahr, W. (2008). A survey on metaheuristics for stochastic combinatorial optimization. *Natural Computing, 8*, 239–287.
10. Blum, C., & Roli, A. (2003). Metaheuristics in combinatorial optimization: Overview and conceptual comparison. *ACM Computing Surveys, 35*(3), 268–308.
11. Goldberg, D. E. (1989). *Genetic algorithms in search, optimization and machine learning*. Kluwer Academic Publishers. ISBN 978-0-201-15767-3.

12. Glover, F., & Kochenberger, G. A. (2003). *Handbook of metaheuristics* (p. 57). Springer, International Series in Operations Research & Management Science. ISBN 978-1-4020-7263-5.
13. Talbi, E.-G. (2009). *Metaheuristics: From design to implementation*. Wiley. ISBN 978-0-470-27858-1.
14. Sörensen, K. (2015). Metaheuristics—The metaphor exposed. *International Transactions in Operational Research., 22*, 3–18.
15. Kirkpatrick, S., Gelatt, C. D., Jr., & Vecchi, M. P. (1983). Optimization by simulated annealing. *Science, 220*(4598), 671–680.
16. Razmjooy, N., Estrela, V. V., & Loschi, H. J. (2019). A study on metaheuristic-based neural networks for image segmentation purposes. In *Data Science* (pp. 25–49). CRC Press.
17. Fotovatikhah, F., Herrera, M., Shamshirband, S., Chau, K.-W., Ardabili, S. F., & Piran Md, J. (2018). Survey of computational intelligence as basis to big flood management: Challenges, research directions and future work. *Engineering Applications of Computational Fluid Mechanics, 12*(1), 411–437. https://doi.org/10.1080/19942060.2018.1448896.
18. Gui-Ju, Z., Xiao, C., & Razmjooy, N. (2020). Optimal parameter extraction of PEM fuel cells by meta-heuristics. *International Journal of Ambient Energy*, 1–10.
19. Razmjooy, N., Mousavi, B. S., Khalilpour, M., & Hosseini, H. (2014). Automatic selection and fusion of color spaces for image thresholding. *Signal, Image and Video Processing, 8*(4), 603–614.
20. Jang, J.-S. R. (1991). Fuzzy modeling using generalized neural networks and Kalman filter algorithm. In *Proceedings of the 9th National Conference on Artificial Intelligence*, Anaheim, CA, USA, July 14–19. (Vol. 2, pp. 762–767).
21. Jang, J.-S. R. (1993). ANFIS: Adaptive-network-based fuzzy inference system. *IEEE Transactions on Systems, Man and Cybernetics., 23*(3), 665–685. https://doi.org/10.1109/21.256541. S2CID14345934.
22. Abraham, A. Adaptation of fuzzy inference system using neural learning, In Nedjah, N., de MacedoMourelle, L. (eds.), Fuzzy systems engineering: Theory and practice, studies in fuzziness and soft computing, 181, Germany: Springer, pp. 53–83, (2005). doi:https://doi.org/10.1007/11339366_3. ISBN 978-3-540-25322-8.
23. Jang, S. M. (1997). *Neuro-fuzzy and soft computing* (pp. 335–368). Prentice-Hall. ISBN 0-13-261066-3.
24. Tahmasebi, P. (2012). A hybrid neural networks-fuzzy logic-genetic algorithm for grade estimation. *Computers & Geosciences., 42*, 18–27. https://doi.org/10.1016/j.cageo.2012.02.004.
25. Tahmasebi, P. (2010). Comparison of optimized neural network with fuzzy logic for ore grade estimation. *Australian Journal of Basic and Applied Sciences., 4*, 764–772.
26. Yu, D., Wang, Y., Liu, H., Jermsittiparsert, K., & Razmjooy, N. (2019). System identification of PEM fuel cells using an improved Elman neural network and a new hybrid optimization algorithm. *Energy Reports, 5*, 1365–1374.
27. Yuan, Z., Wang, W., Wang, H., & Razmjooy, N. (2020). A new technique for optimal estimation of the circuit-based PEMFCs using developed Sunflower Optimization Algorithm. *Energy Reports, 6*, 662–671.
28. Hemanth, J., & Estrela, V. V. (2017). *Deep learning for image processing applications. Advances in parallel computing* (Vol. 31). Amsterdam, Netherlands: IOS Press. ISSN: 978-1-61499-822-8.
29. Razmjooy, N., & Estrela, V. V. (2019). *Applications of image processing and soft computing systems in agriculture* (pp. 1–300). Hershey, PA: IGI Global. https://doi.org/10.4018/978-1-5225-8027-0.
30. Ramavat, K., Joshi, M., & Swadas, P. B. (2016). A survey of super-resolution techniques. *International Research Journal of Engineering and Technology (IRJET), 03*(12)., e-ISSN: 2395-0056.
31. Deshpande, A., & Patavardhan, P. (2019). Survey of super-resolution techniques. *ICTACT Journal on Image & Video Processing, 9*(3).

32. Deshpande, A., & Patavardhan, P. (2016). Single frame super resolution of noncooperative iris imageS. *ICTACT Journal on Image and Video Processing, 7*, 1362–1365.
33. Deshpande, A., & Patavardhan, P. (2017). Multi-frame super-resolution for long range captured iris polar image. *IET Biometrics, 6*, 108–116.
34. Gehani, A., & Reif, J. (2007). Super-resolution video analysis for forensic investigations. In P. Craiger & S. Shenoi (Eds.), *Advances in digital forensics III. Digital forensics 2007. IFIP — The International Federation for Information Processing* (Vol. 242). New York, NY: Springer. https://doi.org/10.1007/978-0-387-73742-3_20.
35. Shermeyer, J., & Van Etten, A. *CosmiQ Works, In-Q-Tel, The effects of super-resolution on object detection performance in satellite imagery*. IEEE Xplore.
36. Tan, J. *Enhancing satellite imagery through super-resolution, case studies & projects, Economic Development*. Accessed on: https://omdena.com/blog/super-resolution/. Retrieved on 13 July 2020.
37. http://developer.amd.com/wordpress/media/2013/06/2153_final.pdf. Retrieved on 13 July 2020.
38. Guerra, M. A. J., & Estrela, V. V. (2014). Motion detection applied to microtectonics modeling. *International Journal on Computational Sciences & Applications (IJCSA), 4*(6), 47. https://doi.org/10.5121/ijcsa.2014.4604.
39. de Jesus, M. A., & Estrela, V. V. (2017). Optical flow estimation using total least squares variants. *Oriental Journal of Computer Science and Technology (OJCST), 10*, 563–579. https://doi.org/10.13005/ojcst/10.03.03.
40. Estrela, V. V., & Rivera, L. A. (2004). Pel-recursive motion estimation using the expectation-maximization technique and spatial adaptation. In *Proceedings of the 12-th International Conference in Central Europe on Computer Graphics, Visualization and Computer Vision'2004, WSCG 2004, University of West Bohemia, Campus Bory, Plzen-Bory, Czech Republic, February 2-6, 2004 (Short Papers) 2004* (pp. 47–54).
41. Estrela, V. V., & Galatsanos, N. P. (2000). Spatially adaptive regularized pel-recursive motion estimation based on the EM algorithm. In *Proceedings of the SPIE 3974, Image and Video Communications and Processing 2000, (19 April 2000)*. https://doi.org/10.1117/12.382969.
42. Estrela, V. V., Franz, M. O., Lopes, R. T., & Araujo, G. P. (2005). Adaptive mixed norm optical flow estimation. In *Proceedings Volume 5960, Visual Communications and Image Processing 2005* (Vol. 59603W). Beijing, China: SPIE. https://doi.org/10.1117/12.632674.
43. Hoze, N., & Holcman, D. (2017). Statistical methods for large ensembles of super-resolution stochastic single particle trajectories in cell biology. *Annual Review of Statistics and Its Application, 4*, 189–223.
44. Ge, W., Gong, B., & Yu, Y. (2018). Image super-resolution via deterministic-stochastic synthesis and local statistical rectification. *ACM Transactions on Graphics (TOG), 37*, 1–14.
45. Panagiotopoulou, A., & Anastassopoulos, V. (2012). Super-resolution image reconstruction techniques: Trade-offs between the data-fidelity and regularization terms. *Information Fusion, 13*, 185–195.
46. Umer, R. M., Foresti, G., & Micheloni, C. (2020). Deep generative adversarial residual convolutional networks for real-world super-resolution. *2020 IEEE/CVF Conference on Computer Vision and Pattern Recognition Workshops (CVPRW)*, 1769–1777.
47. Brandi, F., Queiroz, R., & Mukherjee, D. (2008). Super-resolution of video using key frames and motion estimation. In *2008 15th IEEE International Conference on Image Processing* (pp. 321–324).
48. Shin, K.-Y., Kang, B.-J., Park, K.-R., & Shin, J.-H., (2010). A Study on the Restoration of a Low-Resoltuion Iris Image into a High-Resolution One Based on Multiple Multi-Layered Perceptrons. Journal of Korea Multimedia Society
49. Bengio, Y., Courville, A., & Vincent, P. (2013). Representation learning: A review and new perspectives. *IEEE Transactions on Pattern Analysis and Machine Intelligence, 35*(8), 1798–1828.

50. Yang W., Zhang X., Tian Y., Wang W., Xue J.-H., Liao Q. (2019) *Deep learning for single image super-resolution: A brief review*, arXiv:1808.03344v3 [cs.CV].
51. Deshpande, A., Patavardhan, P., Estrela, V. V., & Razmjooy, N. (2020). Deep learning as an alternative to super-resolution imaging in UAV systems. In V. V. Estrela, J. Hemanth, O. Saotome, G. Nikolakopoulos, & R. Sabatini (Eds.), *Imaging and sensing for unmanned aircraft systems* (Vol. 2, 9, pp. 177–212). London, UK: IET. https://doi.org/10.1049/PBCE120G_ch9.
52. Song, H. A., & Lee, S.-Y. (2013). Hierarchical representation using NMF. *Proceedings of the International Conference on Neural Information Processing*, 466–473.
53. https://www.microscopyu.com/references/3d-superresolution-techniques. Retrieved on 13 July 2020.
54. Zhang, H., Wang, P., Zhang, C., & Jiang, Z. (2019). A comparable study of CNN-based single image super-resolution for space-based imaging sensors. *Sensors, 19*(14), 3234.
55. http://medicalimaging.spiedigitallibrary.org/article.aspx?articleid=2088621. Retrieved on 13 July 2020.
56. Kaji, S., & Kida, S. (2019). Overview of image-to-image translation by use of deep neural networks: Denoising, super-resolution, modality conversion, and reconstruction in medical imaging. *Radiological Physics and Technology*, 1–14.
57. Gupta, R., Sharma, A., & Kumar, A. (2020). Super-resolution using GANs for medical imaging. *Procedia Computer Science, 173*, 28–35.
58. Yamashita, K., & Markov, K. (2020). Medical image enhancement using super resolution methods. *Computational Science – ICCS, 2020*(12141), 496–508.
59. Sreenivas, B., & Chary, B. N. (2011). Processing of satellite image using digital image processing. *Geospatial World Forum*, 18–21 Jan. Hyderabad, India.
60. https://omdena.com/blog/super-resolution/. Retrieved on 13 July 2020.
61. Malczewski, K., & Stasiński, R. Super-resolution for multimedia, image, and video processing applications. In *Recent Advances in Multimedia Signal Processing and Communications, Volume 231 of the series Studies in Computational Intelligence* (pp. 171–208).
62. Ochsenbein, F., Allen, M. G., & Egret, D. (2004). Astronomical data analysis software and systems. In *XIII ASP Conference Series* (Vol. 314) F.
63. https://www.gla.ac.uk/schools/physics/research/groups/imagingconcepts/research_areas/computationalimaging/. Retrieved on 13 July 2020.
64. Nguyen, K., Fookes, C., Sridha, S., Tistarelli, M., & Nixon, M. (2018). Super-resolution for biometrics: A comprehensive survey. *Pattern Recognition., 78*, 23–42. https://doi.org/10.1016/j.patcog.2018.01.002.
65. Deshpande, A., & Patavardhan, P. (2017). Super-resolution and recognition of long range captured multi-frame iris images. *IET Biometrics, 6*(5), 360–368. https://doi.org/10.1049/iet-bmt.2016.0075, IET Digital Library.
66. Wang, X., Zhang, H., Liu, J., Xiao, L., He, Z., Liu, L., & Duan, P. (2019). Iris image super-resolution based on GANs with adversarial triplets. In Z. Sun, R. He, J. Feng, S. Shan, & Z. Guo (Eds.), *Biometric recognition. CCBR 2019. Lecture notes in computer science* (Vol. 11818). Cham: Springer. https://doi.org/10.1007/978-3-030-31456-9_39.
67. Deshpande, A., Patavardhan, P., & Estrela, V. V. (2020). Super-resolution and recognition of unconstrained ear image. *International Journal of Biometrics, 12*(4), 396–410.
68. Hoang, V. T. *EarVN1.0: aA new large-scale ear images dataset in the wild.* Vietnam: Ho Chi Minh City Open University.
69. Feng, X., Foody, G., Aplin, P., & Gosling, S. (2015). Enhancing the spatial resolution of satellite-derived land surface temperature mapping for urban areas. *Sustainable Cities and Society, 19*, 341–348.
70. Pereira, O., Melfi, A., Montes, C. R., & Lucas, Y. (2018). Downscaling of ASTER thermal images based on geographically weighted regression kriging. *Remote Sensing, 10*, 633.

71. Atitallah, S. B., Driss, M., Boulila, W., & Ghezala, H. B. (2020). Leveraging deep learning and IoT big data analytics to support the smart cities development: Review and future directions. *Computer Science Review, 38*, 100303.
72. Chamoso, P., González-Briones, A., Rodríguez, S., & Corchado, J. (2018). Tendencies of technologies and platforms in smart cities: A state-of-the-art review. *Wireless Communications and Mobile Computing, 3086854*, 1–3086854:17.
73. Pincetl, S., Graham, R., Murphy, S., & Sivaraman, D. (2016). Analysis of high-resolution utility data for understanding energy use in uban systems: The Case of Los Angeles, California. *Journal of Industrial Ecology, 20*, 166–178.
74. Klapp, I., Yafin, P., Oz, N., Brand, O., Bahat, I., Goldshtein, E., Cohen, Y., Alchanatis, V., & Sochen, N. (2020). Computational end-to-end and super-resolution methods to improve thermal infrared remote sensing for agriculture. *Precision Agriculture*, 1–23.

Review on Fuzzy Logic Systems with Super-Resolved Imaging and Metaheuristics for Medical Applications

Abhishek Choubey, Shruti Bhargava Choubey, and C. S. N. Koushik

1 Introduction

Fuzzy logic is the logic whose intensity level is used to determine the intensity of the truth or the closeness to logic 1. Its values range between 0 and 1 with the possibility to take infinite range values, unlike the digital Boolean values which are either 0 or 1 [81]. The development of mathematical models can be done to process the inputs and generate the outputs accordingly for the inputs given to it. The logic levels can be represented by various membership distribution functions [82]. It can be used as a self-correcting system that can govern the rate of flow of the output generation. There are various tools such as MATLAB and NETLAB that can be used to produce the models in an efficient manner that can meet the end-user requirements. The use of the fuzzy inference scheme makes the use of the fuzzy logic to be effective and by reducing the errors present in the data [1–4].

The general fuzzy inference system has the components such as the controller element and the executing unit and a feedback path for the sake of the rectification of errors. The input to the system is raw crisp data, which is generally an image whose quality must be assessed to optimize the resolution of the images. Based on the required task, the output fuzzified logic can be defuzzied or used as it is. The error in the entire task generated is handled by the controller unit to reduce them. In general, a PI controller is quite sufficient enough to reduce the effects of the error over the data [1–4]. Fortunately, some authors [24, 25] tried to offer methodical review due to which it has helped us to comprehend what is the procedure opted for healthcare and how the desingable structure must look like at the end. In [24], the author offers an overview to clarify the focus of modern-day healthcare organizations on diverse intervals. In [25], the authors point out that the frame extent networks, private area networks, gateway to the extensive area networks, extensive area networks, and

A. Choubey (✉) · S. B. Choubey · C. S. N. Koushik
Sreenidhi Institute of Science and Technology, Hyderabad, Telangana, India

A. Deshpande et al. (eds.), *Computational Intelligence Methods for Super-Resolution in Image Processing Applications*, https://doi.org/10.1007/978-3-030-67921-7_2

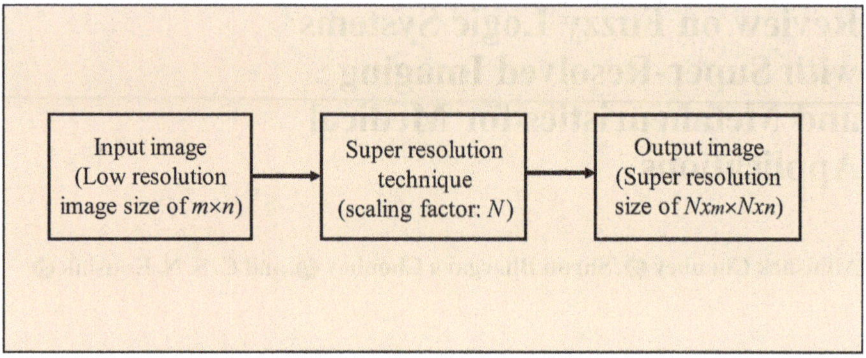

Fig. 1 Function of image super resolution

end-user healthcare intensive care applications are the main reflections of universal healthcare intensive care system. These discussions enlighten how to collect data from sensors, how to assimilate and examine the gathered data, and how to show the evidence of a healthcare system will develop important investigation matters for healthcare. A characteristic feature of healthcare is that the maximum of the training would like to apply advanced skills or systems, such as wireless sensor networks, machine to machine, and Internet of Things [26].

For the sake of the assessment of the image taken for analysis, the use of the super-resolved imaging (SR) is done along with the use of the metaheuristic algorithm. Super-resolution (SR) is the method of converting low-resolution images to a high-resolution image. As shown in Fig. 1, it is done by updating the number of pixels in the input low resolution (LR) image to a larger value as per a suitable scaling factor, thereby resulting in a high resolution at the end of the entire process as shown in Fig. 1.

As the images are captured and taken for analysis, the quality of the image needs to be assessed, especially in the case of medical applications for the sake of prevention of improper diagnosis. All the images are assessed with the help of the SR and metaheuristics in association with the fuzzy inference system. The system gives the output at a certain value whose value will range in between 0 and 1 infinitely and accordingly the image's quality can be assessed at an easier rate [4–7].

Image super resolving has many applications in various areas like the usage in satellite-imaging, medical imaging devices surveillance systems, a high-definition television (HDTV), and entertainment industry. Consider the example of medical imaging, images are obtained using a camera or by the use of MRI or CT scans for the sake of the investigative purpose of a certain kind of ailment to understand the person's anatomy and physiology [71]. Therefore, increasing the resolution of the image significantly improves the chances of proper diagnosis and detection. If the determination of the input is very low, that may lead to an incorrect diagnosis, which may cost the life of a person. Hence the highly sophisticated system is used to capture the images along with the help of certain kinds of sensors. Imaging systems

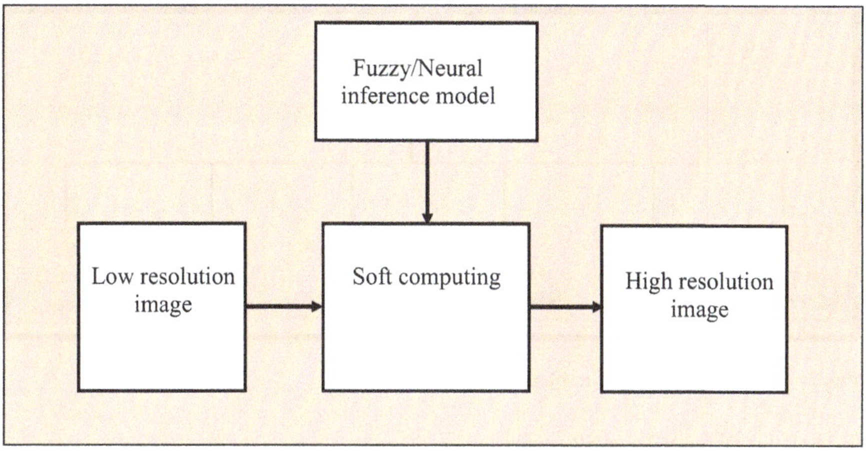

Fig. 2 Block diagram of soft-computing-based SR

such as magnetic resonance imaging (MRI), positron emission tomography (PET), and computerized tomography (CT) are used for the sake of the image capture for diagnosis with great precision in the quality of the image. In the current period, it has been pragmatic that there is a good development of methods for image dealing out based on soft computing such as neural networks and fuzzy systems. Figure 2 shows basic implementation steps of soft-computing-based SR. Despite the evolvement of the skill and in the presentation of enhanced modernization algorithms, it is hard to get an image of required resolution values due to the imaging environments and the restrictions of physical imaging schemes and quality-limiting issues such as noise and blur. Hence to solve the issue of the image quality, super-resolution (SR) methods are to be used for the handing out of such images and refining the superiority of the images.

Super resolving of the image may lead to various sorts of challenges in practicality where one faces during the implementation acting as a limiting factor for the implementation due to the complexity in terms of implementation. Certain major issues related to super-resolution are reflected in Fig. 3.

2 Review of Literature

There are different approaches offered to produce super-resolution (SR) images from low-resolution (LR) retaining the essential image specifics as shown in Table 1. In earlier methods, the SR image was generally produced by including multiple LR images to determine the specifics. Different approaches convert the single LR image to a high-resolution (HR) image. The single LR image methods are more reliable and essentially needed methods [47, 48].

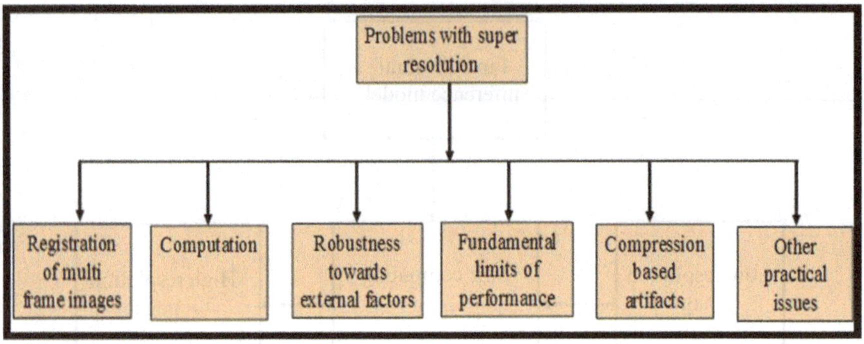

Fig. 3 Issues related to super-resolution

Table 1 Single image-based SR approaches

S. No.	Approaches	Description of approaches
1	Interpolation-based SR	Artificial increasing number of pixels
2	Learning-based super-resolution	Creating a high-frequency image form low-frequency details
3	Edge-directed super-resolution	Extension of interpolation technique
4	Sparsity-based super-resolution	Use of sparse property for calculation
5	Soft-computing-based SR	Based on imprecision

Image interruption is the collective increase of the pixel number in a firm region inside an image. It is one of the outmoded approaches used in SR. Conventional super-resolution systems used bilinear and bicubic interpolation techniques where these have the good real-time computational ability [59–50]. There are different techniques of SR dependent on an example-based learning approach [51–56].

In those methods, high-frequency details are created from LR images by using example-based learning giving HR images based on edge-focused calculations [57–61]. These are created on edge details of objects in images which are being used in the calculation of HR image from LR images. In the sparsity-based SR model, by the sparsity-based dictionaries, image specifics of HR are evaluated from sparse [62–65].

If these techniques of SR used along with the fuzzy system-based image processing system, the outputs being generated is quite fruitful in image processing due to the fuzzification/defuzzification of data along with the use of the fuzzy inference rules. Figure 4 shows the basic fuzzy interface system that can be used for image up-scaling, image filtering, exclusion of image noise, image interpretation, and image segmentation. Due to the increased range of operations by the use of the fuzzy system provides more accuracy as the values range from 0 to 1 and by these huge variations, all the error probabilities can be reduced to a large extent [66, 67, 69].

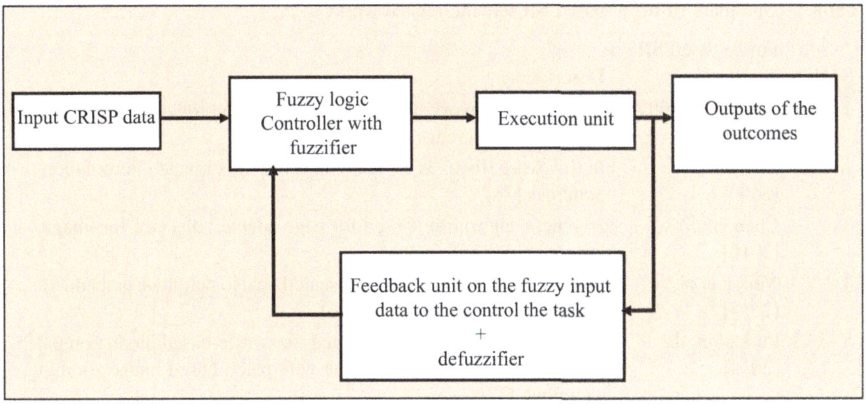

Fig. 4 Basic fuzzy inference system

Super-resolution (SR) in medical image processing has a lot of significance. Various techniques were used to expand the image quality [27]. As in CT scan or an MRI or any other medical imaging technique, a high-quality image is with good resolution and clarity and contrast for the sake of the detailing, and it is possible efficiently by the SR methods.

Medical images are of low resolution in general if captured along with geometric deformations and low contrast, for example, an X-ray image, which is of low contrast and ultrasound image having noise, etc. The image can be blurred due to the moving of the patients as well causing the quality of the image to be disrupted, and hence these problems can be reduced by the help of the super-resolution [28]. All the medical imaging applications must be accurate and fast enough with less delay in producing the outputs of the tests for easy and quick diagnosis and for that to happen, the only good way is to do the super-resolution. However, with the advent of the technology, the early stages of imaging usually have low contrast and have to be converted to a higher resolution for early diagnosis by using those novel imaging methods. Super-resolution does not require much hardware system changes, as SR is implemented in software, hence the updates can be done to the implementation, reducing the cost of the change of the hardware setup for a larger duration of time. Functional MRI (FMRI) [29–32], positron emission tomography imaging system (PET) [29, 33, 34], X-ray digital mammography, and optical coherence tomography (OCT) implement super-resolution, thereby it is the most-trusted method being selected by the medical professional experts for the sake of the diagnosis. For the sake of proper diagnosis, having higher resolution images significantly enhances the chances of the correct treatment. Table 2 reflects a descriptive analysis of other fuzzy-based SR algorithms. The higher resolution images improve the chances for the automatic detection of the diseases as well [9] in various instances.

Table 2 Summary of fuzzy-based SR scheme-related works

S. No	Fuzzy-based SR designs	Description
1	Ting et al. [1997]	The presented work has developed a method for edge preservation using fuzzy inference [73]
2	Chen et al. [2000]	In this work, focus is on fuzzy inference for image interpolation Schemes [74]
3	Chen et al. [2010]	A genetic algorithm is used for edge-adapted distance for image interpolation [75]
4	Nejiya et al. [2013]	In this paper, authors have presented sparse neighbor embedded method for fuzzy clustering [76]
5	Purkait et al. [2014]	The research paper has developed fuzzy rule-based prediction patch-based image zooming technique with patch-based image zooming technique [77]
6	Bhagya et al. [2015]	A novel approach is used for multispectral image compression using SPIHT [78]
7	Li et al. [2019]	A novel super-resolution restoration algorithm is used with fuzzy similarity fusion [79]
8	Greeshma et al. [2020]	In this work, the fuzzy deep-learning-based technique is suitable for single-image super-resolution [80]

3 SR-Based Medical Imaging

The necessities for improved resolution in all medical imaging modalities now characterize a significant and undeveloped task. Precise extent and the imagining of structure in alive tissues is essentially inadequate by the imaging organization's topographies. Imaging outside these boundaries in medical imaging is stated as super-resolution. We are providing a summary of super-resolution imaging examples of medical images.

3.1 SR-Based Digital X-Ray Mammography

Digital mammography efficiently captures numeral images of the breast, while the patient is visible to the minimal amount of radiation. The sensors cannot counselor pixel sizes to rise determination without trading off signal-to-noise ratio (SNR). Figure 5 relates a high-dosage X-ray image using the very-low revelation images used in a multiframe Scheme [41, 68]. To maximize image resolution, all multiple low-resolution images must be combined digitally considering the spatial shifts due to the patient movement or disturbances from the detector or due to the vibrations in the imaging system [83]. Applying SR processing to imaging, it requires to overcome two challenges namely the handling of the huge amount of data of LR digital mammogram images with the resolution of 10 megapixels, having a negative impact on the computation, and the other is the radiation exposure, which may affect the

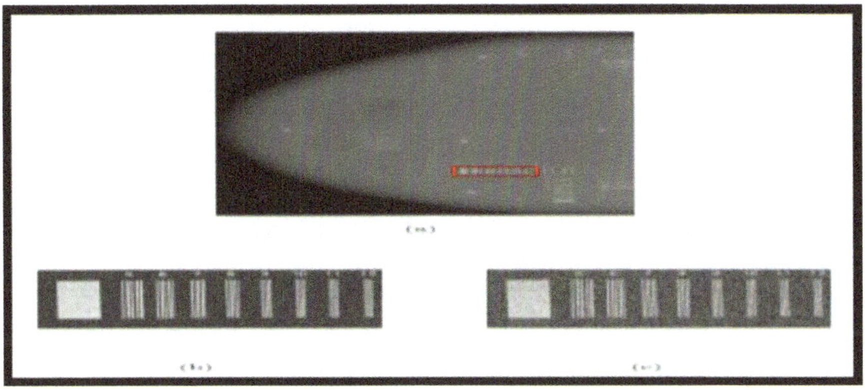

Fig. 5 Mammogram X-ray images from the phantom breast in (**a**). The red rectangular section in (**a**), zoomed in (**b**) and (**c**) [68]

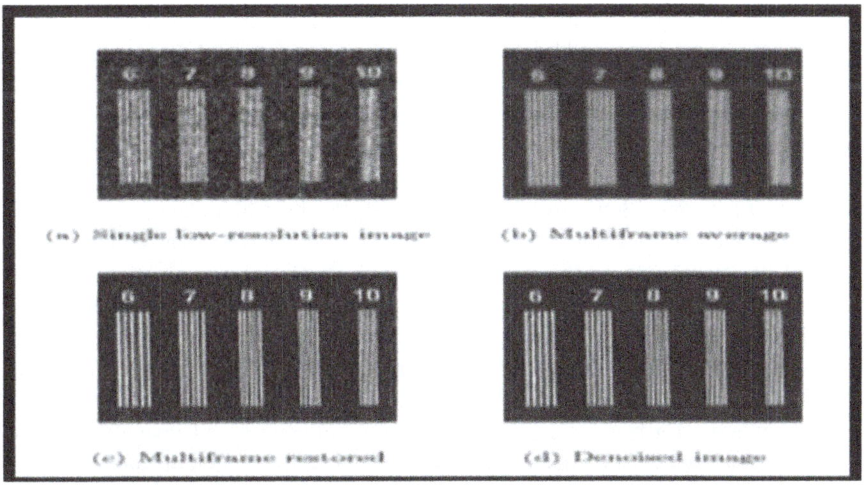

Fig. 6 Different restoration techniques on LR. (**a**) Low-resolution image, (**b**) multiframe image, (**c**) multiframe restored, (**d**) denoised SR image [68]

patient and the images causing a tremendous impact on the SNR making it to have a low peak. Figure 6 demonstrates the images throughout the SR process.

3.2 Super-Resolution in Optical Coherence Tomography (OCT)

Accurate characterization of the retinal images or any pathological aberrations is only conceivable with high-resolution 3-D ocular images. Various resolutions like

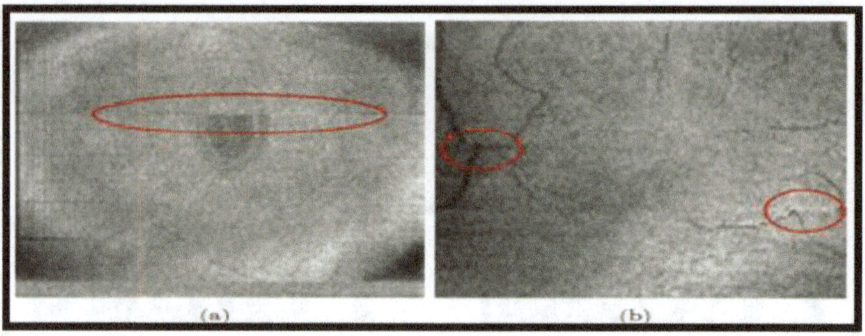

Fig. 7 SDOCT scans of two subjects [68]

the adjacent, axial, and azimuthal resolution which play a vital role for the image arrangement, solely depends on the image brightness level and bandwidth, and the light traversal in case of deflections at boundaries due to physical abnormalities if any. Using the efficient CCD detectors provided the creation of 3-D images of the anatomical structure with no aliasing [35, 36].

However, in vivo imaging submissions, the SDOCT attainment period is more where it cannot avoid abrupt motions such as blinking in the compactly experimented volumetric measurements as shown in Fig. 7. Hence, these systems are used at a nominal low-resolution value [37]. In SDOCT imaging, resolution in the azimuthal axis is subject to one amount of B-scans tested at comparatively equivalent detachments in the volumetric glance. Valid measurable measurements for retinal images depend on B-scans with the known azimuthal displacement. If one tries improving the hardware design, it gives an issue of diminishing returns where the optical components are to capture very high quality, where the dense scans develop sensitive and expensive [38, 39] hardware like in case of eye-tracking systems is not an effective alternative; hence, a software-based image processing solution is the better choice in terms of complexity in implementation and in terms of costs as well. In the handheld SDOCT systems, the motion errors are to be reduced to a large extent [40, 41].

3.3 SR in MRI, FMRI, and PET Imaging

In the case of various medical imaging methods, one can get both anatomical information, as well as functional data. However, resolution confines continuously damage the class of medical images for diagnosis [70]. SR is used through these crucial medical imaging approaches such as positron emission tomography (PET), magnetic resonance imaging (MRI), and functional MRI (fMRI), and growing the resolution of these medical images and recollecting isotropic 3-D imaging. Medical

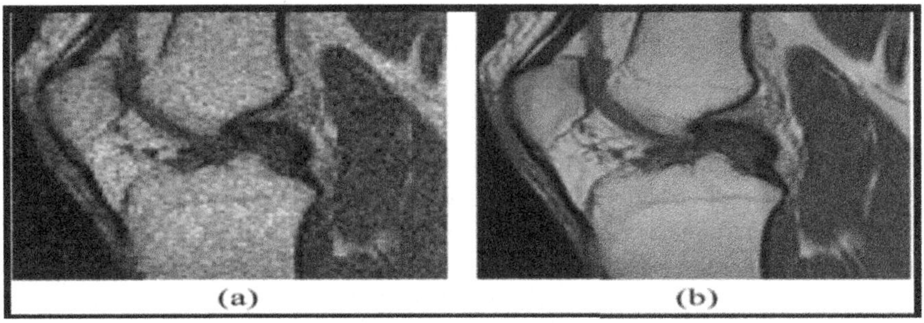

Fig. 8 MRI knee image with input and processed using single frame [45]

imaging schemes are worked under highly skillful surroundings due to which constant and multiview images are got effortlessly [42–44].

Example-based learned SR for single frames has also been used on the medical images wherein similar images are collected to establish a database. The training database has five sets of ordinary images, including computed tomography (CT) and MRI images from numerous portions of the human physique. Figure 8 shows an example of the reconstructed image of the single MRI image of the knee in [45, 46].

The quality assessment of the images is essential for the sake of the proper analysis to be done from the image captured. The images are captured by the camera or by the MRI or CT scan, etc., [72] for the sake of the analysis whose quality must be on par for diagnosis. The analysis of the image captured will help the medical experts to diagnose the problems with ease rather than to misdiagnose the image and give the wrong treatment to the patients. The data collected is given to the fuzzy inference system irrespective of the quality of the image, but one must take care the image to be clear so that the results will be more accurate. To have a good analysis of the image, the data are given to a system, which by the use of the metaheuristics and the super-resolved imaging make the analysis easier in association with the fuzzy implication system. For the proper analysis of the image to be done, the data taken from the image need to be as error-free or noise-free as possible, and hence it is possible with the help of the filters used. The filters used were linear filters such as the wiener filter or the nonlinear filter such as the median filters which were not quite efficient for the sake of noise elimination. For efficient noise removal, 2D wavelet filters were to be used such that all the noise components can be removed and then given for the sake of processing into the next stages [84]. The filters could be IIR or FIR filters with a finite sharp frequency response to remove the Gaussian or any noise elements. The noise could be estimated by the use of the peak signal-to-noise ratio and mean square error (MSE), [4–9] MSE is given by the formula:

$$\mathrm{MSE} = \frac{\sum_{j=0}^{P} \sum_{k-0}^{Q} \left(A(j,k) - \overline{A(j,k)} \right)^2}{PQ} \quad (1)$$

Hence, by the use of super-resolved imaging, a low-resolution image can be converted to a high-resolution image and as a result, the diagnosis can be done. For the super-resolved imaging, the problem of the subpixel mapping is to be handled when the resolution of the image has been either interpolated or decimated. These are in general the sampling techniques being applied on the sampled data such that the resolution of the image can either enhanced or deprived. When the super-resolved imaging is implemented, the subpixels are made to be shifted based on the multirate sampling technique applied, especially in the frequency domain. The problem that may arise in this is the aliasing, wherein the under-sampled signal may lead to it and its removal becomes a tedious task [4–9].

4 Metaheuristics for Medical Applications

Healthcare systems have a strong impact on the economy of the country and the everyday life of the individuals. However, there is a shortage in the consensus taxonomy, and terminology of devices in the country now, for example, consider the present pandemic of 2020, due to the coronavirus pandemic, there is a shortage of medical equipment, hence proper diagnosis is not possible if the people cannot have access to those. In Table 3, analysis of some metaheuristics-based algorithms are discussed. Metaheuristics [16, 33] takes input and gives the output, along with the transition, by evaluating by the help of the determination operators is done frequently until the quest procedure converges up to the predefined stopping condition [65]. Metaheuristics are used in solving the data mining issues, for example, clustering for unidentified data, an organization for a portion of unknown data, and subordinate rule for readable patterns. Metaheuristic algorithms have been extensively implemented for attempting the image registration issue, thus becoming a consistent alternate for optimization purposes.

For that to be done in an efficient and optimized manner, the data must also be given to the metaheuristic algorithms which are population-based computational algorithms. These algorithms work based on the assumptions being done on the data, and analysis is done on how the population reacts to the data given to it as the input [3, 5–10]. There are various algorithms of the metaheuristics like the above stated, but the main basic algorithms are shown in Fig. 9.

These are used to understand the data being given by considering assumptions based on the data. The basic procedure in the execution of the algorithms is that the data are initially taken from the source and then evaluated for the sake of the velocities/nectar amount/crossovers for every source and the algorithm continues till the termination condition is satisfied. At every stage, the updation or the mutation

Table 3 Survey of metaheuristic algorithms

S. No	Metaheuristic algorithms	Description
1	Das and Bhattacharya's et al. (2010)	The presented work has developed extended evolutionary self-organizing map for MR and CT modality images [16]
2	Santamaría et al. (2012)	In this paper, authors have proposed path relinking (PR) with greedy randomized adaptive search procedure (GRASP) [17]
3	Alderliesten et al. (2013)	The work is focused on a new method by sampling the estimated distribution using incremental multi-objective adapted maximum-likelihood Gaussian model mixture [18]
4	De Falco et al. (2014)	In this paper, authors have presented an adaptive invasion–based model (AIM) for searching for the optimum control parameters [19]
5	Pirpinia et al. (2015)	The authors have proposed various hybridizations based on the genetic local search (GLS) algorithms and combined objectives repeated line-search (CORL) [20]
6	Costin et al. (2016)	In this presented work, canonical multi-modal is used for rigid image restoration scenarios [21]
7	Bermejo et al. (2018)	Coral reef optimization (CRO) algorithm with substrate layers (CRO-SL) is used for medical image restoration [22]
8	Cocianu, Stan's et al. (2019)	Hybridization of a self-adaptive ES algorithm has been implemented with accelerated PSO algorithm [23]

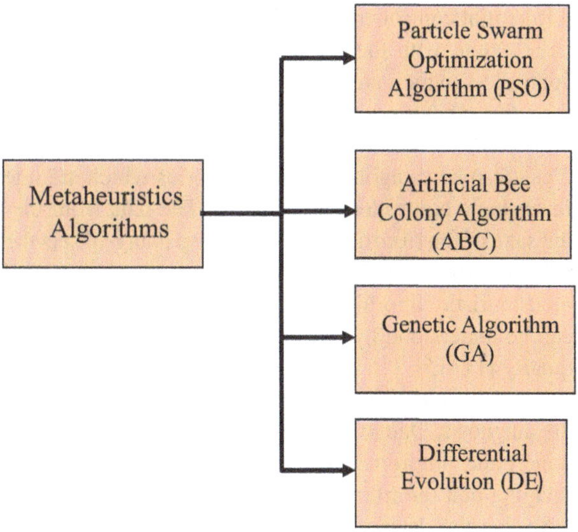

Fig. 9 Types of metaheuristics algorithms

of the source takes place. The source of data must be producing the data of the least error rate and noise. Another algorithm, namely, the Grey Wolf optimization algorithm is used similar to the hunting pattern of the wolves, wherein the reconstruction error rate is reduced to a large extent by vectors [5–11].

Table 4 The Comparison of metaheuristic algorithms

Parameter	ABC	PSO	DE	GA
Mean values	0.0991	0.0992	0.0991	0.0993
PSNR values (dB)	Around 19.45–19.50			
Execution time (sec)	18.05–18.15	17.60–17.85	18.50–18.75	14.80–15.00

Table 4 shows all the described algorithms; the most efficient and robust algorithm was found to be the ABC followed by the DE and the PSO and GA, respectively, even though the execution time for the GA is the smallest of all. Table 4 shows that the robustness of the algorithms is that, rather than improving the hardware complexity, the algorithms are made efficient enough such that the resolution of the image can be assessed for the quality at the software level only. So, this is better to use rather than using the neural networks which need more hardware complexity than it and the quality of the image cannot be determined until and unless the network is trained for the assessment. Hence, metaheuristics are advantageous than the neural networks, especially for the sake of the medical images, as the expenditure also reduces to a large extent [5–11].

The use of a fuzzy inference system is done for the improvement of the performance of the entire system. When the data are given after the super-resolved imaging and the metaheuristics stage to the fuzzy inference system, it estimates the errors and reduces them to a large extent. The controller used is a proportional integrative differential controller, wherein the error or the noise in the data can be reduced to a large extent with the support of the feedback path. There are two types of fuzzy controllers, which are type-1 and type-2, wherein the type-2 performance improved than the type-1 as it overcomes the drawbacks in the type-1 where there were limitations on the membership functions. Based on the membership function, type-2 can also be categorized into categories which are interval and grade-based type-2 fuzzy logic controller, respectively. The data from the previous stage are taken for the sake of the fuzzification of the crisp data and sets the fuzzy rules in accordance to the membership function. All the errors are removed along with the use of the metaheuristic algorithms associated with it such that the intensity of the accuracy can be understood by the medical experts for the diagnosis by assessing the image quality [11–15].

For example, consider the eye image whose image needs to be used for the sake of the diagnosis. Due to the presence of the noise or errors in the image, the diagnosis is done either improper or cannot be done. Hence, by the use of the super-resolved imaging and the filters, the resolution of the image can be increased tremendously such that the details can be understood much clearly. The use of the best suitable metaheuristic algorithm is done based on the data and the computational hardware, as more details can be extracted upon the execution completely and the quality of the image can be assessed easily and effectively with less inaccuracy value [11–15].

5 Conclusion

The fuzzy logic plays a main part to select the alternative, which is the best from the set of inputs given to it in various kinds of applications, especially in the medical field. The crisp data which have been given as input to the fuzzy logic system will be fuzzified for the data to be associated with the metaheuristic algorithm and increase its range of operation at a certain value of accuracy for the sake of the super-resolved images to be analyzed. The efficiency of the entire scheme depends on the errors in the image and the efficiency of the computing hardware. By the use of the metaheuristic algorithm for the super-resolved image, the quality of the image can be assessed easily due to the association with the fuzzy inference system. Whenever the medical images are captured, the images are assessed for the sake of the proper diagnosis where they can be captured either by the camera or by the MRI, etc. These new skills offer a technique for a healthcare organization to gather much more comprehensive and precise data so that the finding and estimation outcomes of healthcare can be enhanced. The existing trends indicate that the preference is for the metaheuristics and the accuracy can enhanced more by the use of the sensors, appliances, WSN, and IoT. With the increased performance, the output depends mainly on the picking of the appropriate features from the raw data for the attractive performance of the machine or extending the period of the sensor network of a system.

References

1. Pandey, P., Dewangan, K. K., & Dewangan, D. K. (2017, August). Enhancing the quality of satellite images using fuzzy inference system. In *2017 International Conference on Energy, Communication, Data Analytics and Soft Computing (ICECDS)* (pp. 3087–3092). IEEE.
2. Sadaei, H. J., e Silva, P. C. D. L., Guimarães, F. G., & Lee, M. H. (2019). Short-term load forecasting by using a combined method of convolutional neural networks and fuzzy time series. *Energy, 175*, 365–377.
3. Hamza, M. F., Yap, H. J., & Choudhury, I. A. (2017). Recent advances on the use of meta-heuristic optimization algorithms to optimize the type-2 fuzzy logic systems in intelligent control. *Neural Computing and Applications, 28*(5), 979–999.
4. Field, J. J., Wernsing, K. A., Domingue, S. R., Motz, A. M. A., DeLuca, K. F., Levi, D. H., et al. (2016). Superresolved multiphoton microscopy with spatial frequency-modulated imaging. *Proceedings of the National Academy of Sciences, 113*(24), 6605–6610.
5. Wang, W., Yadav, N. P., Cao, Y., Liu, J., & Liu, X. (2019). Finger skin super-resolved imaging based on extracting polarized light field. *Optik, 180*, 215–219.
6. Grußmayer, K., Geissbuehler, S., Descloux, A., Lukes, T., Leutenegger, M., Radenovic, A., & Lasser, T. (2019). Spectral cross-cumulants for multicolor super-resolved SOFI imaging. *arXiv preprint arXiv:1907.07007*.
7. Olivas, F., Valdez, F., Castillo, O., Gonzalez, C. I., Martinez, G., & Melin, P. (2013). Ant colony optimization with dynamic parameter adaptation based on interval type-2 fuzzy logic systems. *Applied Soft Computing Journal*. https://doi.org/10.1016/j.asoc.2016.12.015.

8. Zounemat-Kermani, M., Kisi, O., Piri, J., & Mahdavi-Meymand, A. (2019). Assessment of artificial intelligence–based models and metaheuristic algorithms in modeling evaporation. *Journal of Hydrologic Engineering, 24*(10), 04019033.
9. Santamaría, J., Rivero-Cejudo, M. L., Martos-Fernández, M. A., & Roca, F. (2020). An overview on the latest nature-inspired and metaheuristics-based image registration algorithms. *Applied Sciences, 10*(6), 1928.
10. Kockanat, S., & Karaboga, N. (2017). Medical image denoising using metaheuristics. In *Metaheuristics for medicine and biology* (pp. 155–169). Berlin, Heidelberg: Springer.
11. Bhattacharjee, K., & Pant, M. (2020). Applications of metaheuristics in hyperspectral imaging: A review. In *Soft computing: Theories and applications* (pp. 1005–1015). Singapore: Springer.
12. Singh, A., & Singh, J. (2020). Survey on single image based super-resolution—Implementation challenges and solutions. *Multimedia Tools and Applications, 79*(3), 1641–1672.
13. Yang, J., Shang, C., Li, Y., & Shen, Q. (2017, July). Single frame image super resolution via learning multiple ANFIS mappings. In *2017 IEEE International Conference on Fuzzy Systems (FUZZ-IEEE)* (pp. 1–6). IEEE.
14. Rajput, S. S., Bohat, V. K., & Arya, K. V. (2019). Grey wolf optimization algorithm for facial image super-resolution. *Applied Intelligence, 49*(4), 1324–1338.
15. de Jesus, M. A., Estrela, V. V., Saotome, O., & Stutz, D. (2018). Super-resolution via particle swarm optimization variants. In *Biologically rationalized computing techniques for image processing applications* (pp. 317–337). Cham: Springer.
16. Das, A., & Bhattacharya, M. (2011). Affine-based registration of CT and MR modality images of human brain using multiresolution approaches: Comparative study on genetic algorithm and particle swarm optimization. *Neural Computing & Applications, 20*, 223–237.
17. Santamaría, J., Cordón, O., Damas, S., Martí, R., & Palma, R. J. (2012). GRASP and path relinking hybridizations for the point matching-based image registration problem. *Journal of Heuristics, 18*, 169–192.
18. Alderliesten, T., Sonke, J., Bosman, P., Ourselin, S., & Haynor, D. (2013). Deformable image registration by multi-objective optimization using a dual-dynamic transformation model to account for large anatomical differences. In *Medical Imaging 2013: Image Processing*. International Society for Optics and Photonics.
19. Falco, I. D., Cioppa, A. D., Maisto, D., Scafuri, U., & Tarantino, E. (2014, July 12–16). Using an Adaptive Invasion-based Model for Fast Range Image Registration. In *Proceedings of the GECCO'14—2014 Genetic and Evolutionary Computation Conference* (pp. 1095–1102). Vancouver, BC, Canada.
20. Pirpinia, K., Alderliesten, T., Sonke, J., Bosman, M. V. H. P., & Silva, S. (2015). Diversifying multi-objective gradient techniques and their role in hybrid multi-objective evolutionary algorithms for deformable medical image registration. In *Proceedings of the GECCO'15—2015 Genetic and Evolutionary Computation Conference, Madrid, Spain, 11–15 July 2015* (pp. 1255–1262).
21. Costin, H., Bejinariu, S., & Costin, D. (2016). Biomedical image registration by means of bacterial foraging paradigm. *International Journal of Computers, Communications & Control, 11*, 331–347.
22. Bermejo, E., Cordon, O., Damas, S., & Santamaría, J. (2015). A comparative study on the application of advanced bacterial foraging models to image registration. *Information Sciences, 295*, 160–181.
23. Cocianu, C., & Stan, A. (2019). New evolutionary-based techniques for image registration. *Applied Sciences, 9*, 176.
24. Alemdar, H., & Ersoy, C. (2010). Wireless sensor networks for healthcare: A survey. *Computer Networks, 54*(15), 2688–2710.
25. Koch, S. (2006). Home telehealth–current state and future trends. *International Journal of Medical Informatics, 75*(8), 565–576.

26. Chen, M., Wan, J., González-Valenzuela, S., Liao, X., & Leung, V. C. M. (2014). A survey of recent developments in home M2M networks. *IEEE Communications Surveys and Tutorials, 16* (1), 98–114.
27. Dirk Robinson, M., Chiu, S. J., Toth, C. A., Izatt, J. A., Lo, J. Y., & Farsiu, S. *New applications of super-resolution in medical imaging, digital imaging and computer vision*. CRC Press.
28. Sable, G. S., & Gaikwad, A. (2012, November). A Novel Approach for Super Resolution in Medical Imaging. *International Journal of Emerging Technology and Advanced Engineering, 2* (11). Website: www.ijetae.com. ISSN 22502459.
29. Greenspan, H. (2009, January). Super-resolution in medical imaging. *The Computer Journal*. Oxford University Press Oxford, UK, *52*(1), 43–63.
30. Greenspan, H., Oz, G., Kiryati, N., & Peled, S. (2002). MRI inter-slice reconstruction using super-resolution. *Magnetic Resonance Imaging, 20*(5), 437–446.
31. Peled, S., & Yeshurun, Y. (2001). Superresolution in MRI: Application to human white matter fiber tract visualization by diffusion tensor imaging. *Magnetic Resonance in Medicine, 45*(1), 29–35.
32. Peeters, R. R., Kornprobst, P., Nikolova, M., Sunaert, S., Vieville, T., Malandain, G., Deriche, R., Faugeras, O., Ng, M., & Van Hecke, P. (2004). The use of super-resolution techniques to reduce slice thickness in functional MRI. *International Journal of Imaging Systems and Technology, 14*(3), 131–138.
33. Kennedy, J. A., Israel, O., Frenkel, A., Bar-Shalom, R., & Azhari, H. (2006). Super-resolution in PET imaging. *IEEE Transactions on Medical Imaging, 25*(2), 137–147.
34. Kennedy, J. A., Israel, O., Frenkel, A., Bar-Shalom, R., & Azhari, H. (2007). Improved image fusion in PET/CT using hybrid image reconstruction and super-resolution. *International Journal of Biomedical Imaging, 46846*.
35. Toth, C. A., Farsiu, S., Khanifar, A. A., & Chong, G. T. (2009). Optical coherence tomography in age-related macular degeneration. In G. Coscas (Ed.), *Application of spectral domain OCT in AMD* (pp. 15–34). Springer Medizin Verlag Heidelberg.
36. Hammer, D., Ferguson, R. D., Iftimia, N., Ustun, T., Wollstein, G., Ishikawa, H., Gariele, M., Dilworth, W., Kagemann, L., & Schuman, J. (2005). Advanced scanning methods with tracking optical coherence tomography. *Optics Express, 13*(20), 7937–7947.
37. Hammer, D., Ferguson, R. D., Iftimia, N., Ustun, T., Wollstein, G., Ishikawa, H., Gabriele, M., Dilworth, W., Kagemann, L., & Schuman, J. (2005). Advanced scanning methods with tracking optical coherence tomography. *Optics Express, 13*(20), 7937–7947.
38. Farsiu, S., Bower, B. A., Izatt, J. A., & Toth, C. A. (2008). Image fusion based resolution enhancement of retinal spectral domain optical coherence tomography images. *Investigative Ophthalmology & Visual Science, 49*(5), E(abstract) 1845.
39. Chavala, S. H., Farsiu, S., Maldonado, R., Wallace, D. K., Freedman, S. F., & Toth, C. A. (2009). Insights into advanced retinopathy of prematurity using handheld spectral domain optical coherence tomography imaging. *Ophthalmology*.
40. Chong, G. T., Farsiu, S., Freedman, S. F., Sarin, N., Koreishi, A. F., Izatt, J. A., & Toth, C. A. (2009). Abnormal foveal morphology in ocular albinism imaged with spectral-domain optical coherence tomography. *Archives of Ophthalmology, 127*(1), 37–44.
41. Scott, A. W., Farsiu, S., Enyedi, L. B., Wallace, D. K., & Toth, C. A. (2009). Imaging the infant retina with a hand-held spectral-domain optical coherence tomography device. *American Journal of Ophthalmology, 147*(2), 364–373.
42. Robinson, M. D., Chiu, S. J., Lo, J., et al. (2010). *New applications of super-resolution in medical imaging*. CRC Press.
43. Greenspan, H. (2009). Super-resolution in medical imaging. *The Computer Journal, 52*, 43–63.
44. Wallach, D., Lamare, F., Kontaxakis, G., et al. (2012). Super-resolution in respiratory synchronized positron emission tomography. *IEEE Transactions on Medical Imaging, 31*, 438–448.
45. Trinh, D.-H., Luong, M., Dibos, F., et al. (2014). Novel example-based method for super-resolution and denoising of medical images. *IEEE Transactions on Image Processing, 23*, 1882–1895.

46. Wang, Y.-H., Qiao, J., Li, J.-B., et al. (2014). Sparse representation-based MRI super-resolution reconstruction. *Measurement, 47*, 946–953.
47. Borman, S., & Stevenson, R. L. (1998). Super-resolution from image sequences: A review. In *mwscas* (p. 374). IEEE.
48. Tsai, R. Y., & Huang, T. S. (1984). Multiframe image restoration and registration. In T. S. Huang (Ed.), *Advances in Computer Vision and Image Processing* (Vol. 1(2), pp. 317–339).
49. Freeman, W. T., Pasztor, E. C., & Carmichael, O. T. (2000). Learning low level vision. *International Journal of Computer Vision, 40*, 25–47.
50. Gajjar, P. P., & Joshi, M. V. (2011). New learning based super-resolution: Use of DWT and IGMRF prior. *IEEE Transactions on Image Processing, 19*, 1201–1213.
51. Gou, S., Liu, S., & Wu, Y. (2016). Jiao L image super-resolution based on the pairwise dictionary selected learning and improved bilateral regularization. *IET Image Processing, 10* (2), 101–112.
52. Hertzmann, A., Jacobs, C. E., Oliver, N., Curless, B., & Salesin, D. H. (2001). Image analogies. In *Proceedings of the SIGGRAPH* (pp. 327–340). Los Angeles.
53. Kim, K., & Kwon, Y. (2008). *Example-based learning for single image super-resolution and jpeg artifact removal,*". Technical Report 173. Max Planck Institute.
54. Tang, Y., & Shao, L. (2017). Pairwise operator learning for patch-based single-image super-resolution. *IEEE Transactions on Image Processing, 26*(2), 994–1003.
55. Tang, Y., & Yuan, Y. (2014). Learning from errors in super-resolution. *IEEE Transactions on Cybernetics, 44*(11), 2143–2154.
56. Tian, F., Zhou, W., Shang, Y. X., & Liao, Q. (2016). Anchored neighborhood regression based single image super resolution from self-examples. In *2016 IEEE International Conference on Image Processing (ICIP)* (pp. 2827–2831). Phoenix.
57. Dai, S., Han, M., Xu, W., Wu, Y., & Gong, Y. (2007). Soft edge smoothness prior for alpha channel superresolution. In *Proceedings of the IEEE Conference on Computer Vision and Pattern Recognition* (pp. 1–8).
58. EladM, H.-O. Y. (2001). A fast super-resolution reconstruction algorithm for pure translational motion and common space-invariant blur. *IEEE Transactions on Image Processing, 10*(8), 1187–1193.
59. Fattal, R. (2007). Image up sampling via imposed edge statistics. *ACM Trans Graph, 26*(3), 95:1–95:8.
60. Jia, K., & Gong, S. (2005). Multi-model tensor face for simultaneous super-resolution and recognition. *Proceedings of IEEE International Conference on Computer Vision, 2*, 1683–1690.
61. Lee, S.-J., Kang, M.-C., Uhm, K.-H., & Ko, S.-J. (2016). An edge-guided image interpolation method using Taylor series approximation. *IEEE Transactions on Consumer Electronics, 62*(2), 159–165.
62. Liu, Z.-S., Siu, W.-C., & Huang, J.-J. (2015). Image super-resolution via hybrid NEDI and wavelet-based scheme. In *2015 Asia-Pacific Signal and Information Processing Association Annual Summit and Conference (APSIPA)* (pp. 1131–1136). IEEE Conference Publications. https://doi.org/10.1109/APSIPA.2015.7415447.
63. Sun, J., Xu, Z., & Shum, H.-Y. (2008). Image super-resolution using gradient profile prior. *Proceedings of the IEEE Conference on Computer Vision and Pattern Recognition*, 1–8.
64. Sun, J., Xu, Z., & Shum, H.-Y. (2011). Gradient profile prior and its applications in image super-resolution and enhancement. *IEEE Transactions on Image Processing, 20*(6), 1529–1542.
65. Tai, Y.-W., Liu, S., Brown, M. S., & Lin, S. (2010). Super-resolution using edge prior and single image detail synthesis. *Proceedings of the IEEE Conference on Computer Vision and Pattern Recognition*, 2400–2407.
66. Wang, L., Xiang, S., Meng, G., Wu, H., & Pan, C. (2013). Edge-directed single-image super-resolution via adaptive gradient magnitude self-interpolation. *IEEE Transactions on Circuits and Systems for Video Technology, 23*(8), 1289–1299.

67. Zhang, K., Gao, X., Tao, D., & Li, X. (2013). Single image super-resolution with multiscale similarity learning. *IEEE Transactions on Neural Networks and Learning Systems, 24*(10), 1648–1659.
68. Dirk Robinson, M., Chiu, S. J., Toth, C. A., & Farsiu, S. (2010). New applications of super-resolution in mdical imaging. In *Book chapter in super resolution imaging*. CRC Press.
69. Ahirwar, R., & Choubey, A. (2011). *A novel wavelet-based denoising method of SAR image using interscale dependency*. International Conference on Computational Intelligence and Communication Networks.
70. Bhargava, S., & Somkuwar, A. (2016). Estimation of noise removal techniques in medical imaging data—A review. *Journal of Medical Imaging and Health Informatics, 6*(4), 875–885.
71. Singh, A., & Singh, J. (2019). Survey on single image based super-resolution — Implementation challenges and solutions. *Multimedia Tools and Applications*.
72. Choubey, S. B., & Rao, S. P. V. S. (2018). Implementation of hybrid filter technique for noise removal from medical images. *International Journal of Engineering & Technology, 7*(1.1), 25–29.
73. Ting, H., & Hang, H. (1997). Edge preserving interpolation of digital images using fuzzy inference. *Journal of Visual Communication and Image Representation, 8*(4), 338–355.
74. Chen, J., Chang, J., & Shieh, K. (2000). 2-D discrete signal interpolation and its image resampling application using fuzzy rule-based inference. *Fuzzy Sets and Systems, 114*(2), 225–238.
75. Chen, H.-C., & Wang, W.-J. (2010). Locally edge-adapted distance for image interpolation based on genetic fuzzy system. *Expert Systems with Applications, 37*(1), 288–297.
76. Nejiya, A. K., & Wilscy, M. (2013). Example based super-resolution using fuzzy clustering and sparse neighbour embedding. In *2013 IEEE Recent Advances in Intelligent Computational Systems (RAICS)* (pp. 251–256). IEEE Conference Publications.
77. Purkait, H. P., Pal, N. R., & Chanda, B. (2014). A fuzzy-rule-based approach for single frame super-resolution. *IEEE Transactions on Image Processing, 23*(5).
78. Bhagya Raju, V., Jaya Sankar, K., & Naidu, C. D. (2015). Fuzzy based super-resolution multispectral image compression with improved SPIHT. In *2015 International Conference on Communications and Signal Processing (ICCSP)* (pp. 0263–0266).
79. Li, X., & Fu, W. (2019). Regularized super-resolution restoration algorithm for single medical image based on fuzzy similarity fusion. *EURASIP Journal on Image and Video Processing, 2019*, 83.
80. Greeshma, M. S., & Bindu, V. R. (2020). Super-resolution quality criterion (SRQC): A super-resolution image quality assessment metric. *Multimedia Tools and Applications, 79*(7), 1–22.
81. Razmjooy, N., Ramezani, M., & Ghadimi, N. (2017). Imperialist competitive algorithm-based optimization of neuro-fuzzy system parameters for automatic red-eye removal. *International Journal of Fuzzy Systems, 19*(4), 1144–1156.
82. Mir, M., et al. (2020). Employing a Gaussian Particle Swarm Optimization method for tuning Multi Input Multi Output-fuzzy system as an integrated controller of a micro-grid with stability analysis. *Computational Intelligence, 36*(1), 225–258.
83. Carolina, A., Monteirol, B., Padilha, R., & Estrela, V. V. (2019). Health 4.0: Applications management technologies and review. *Journal of Medicine Technology, 2*(2), 262–276.
84. Estrela, V., et al. (2019). Why software-defined radio (SDR) matters in healthcare? *Medical Technologies Journal, 3*(3), 421–429.

Super-Resolution with Deep Learning Techniques: A Review

Aarti and Amit Kumar

1 Introduction

Importantly super-resolution (SR) technique is used to produce a higher quality picture from lower quality pictures. High pixel thickness is offered by a high-quality picture and accordingly more insights regarding the real scene. The requirement for high quality is essential in computer vision applications for better execution in pattern recognition and investigation of pictures. High-quality pictures have significance in medical imaging for diagnosis. Numerous applications require zooming of a particular area of interest for the picture wherein high quality becomes necessary, for example, forensic, surveillance, and satellite imaging applications. Notwithstanding, high-quality pictures are not constantly accessible. This is because the adjustment for high-quality imaging is costly, and it may not be feasible generally because of the trait constraints of the sensor and optics producing innovation. With the usage of image processing algorithms, these issues can be overcome, which are moderately economical, offering to ascend to the idea of super-resolution. It gives an advantage as it might cost less, and the current low-resolution imaging frameworks can now be used. Vision is generally considered as exceptional of each of the five senses. So it does not surprise that pictures play the absolute most significant job in human perception. High-resolution pictures are required, and constantly attractive, above all else significant applications for military and nonmilitary personnel [1]. Ongoing advances in video and image sensing have heightened client desires on the visual nature of captured information. Because of confinements like power, camera, memory size, cost, and constrained data transfer capacity, it is not constantly conceivable to get a high-quality picture. Super-resolution is basically what you find

Aarti (✉)
Lovely Professional University, Phagwara, India

A. Kumar
Dr B R Ambedkar National Institute of Technology, Jalandhar, India

A. Deshpande et al. (eds.), *Computational Intelligence Methods for Super-Resolution in Image Processing Applications*, https://doi.org/10.1007/978-3-030-67921-7_3

in movies and arrangements like CSI, where somebody zooms into a picture, and it improves in quality, and the details simply show up [2]. Image super-resolution delivers a high-resolution picture utilizing at least one low-quality picture [3]. The subject has become an extremely well-known research zone because of the reality that high-quality pictures contained more information that does not legitimately exist in the LR pictures. Resolution is identified by the pixel thickness [1].

While resizing the images in OpenCV or Scipy, the conventional technique (e.g.,, interpolation) is utilized to approximate the estimations of new elements dependent on close by element values which fail to impress anyone as far as visual quality, as the details (e.g., sharp edges) are frequently not protected [4].

It is commencing with an introduction of a quantitative quality estimation technique to assess and contrast the approaches. A metric computed for the execution of each model that is normally used to quantify the nature of the reproduction of lossy compression codecs called peak signal-to-noise ratio (PSNR). This measurement is a true standard utilized in the research of super-resolution. It quantifies how much the distorted image (potentially of lower quality) deviates from the real high-quality picture. In this arrangement, PSNR is the proportion of most extreme conceivable pixel estimation of the photograph (signal strength) to maximum mean squared error (MSE) among the real image and its evaluated version (noise strength), presented in logarithmic scale. The bigger the PSNR values, the better the remaking, and thus, PSNR maximized value normally prompts to limit MSE as the goal work. This methodology was followed in two of the three models. Image SR is a significant class of image processing technique used to improve the resolution of pictures and recordings in PC vision. Advancement of picture super-resolution using deep learning methods has been seen in recent years [5]. A super-resolution strategy remakes a HR picture or series from the observed LR pictures. As SR has been developed for more than three decades, both multi-edge and single outline SR have noteworthy applications in our daily life [6].

Deep learning is a procedure of data mining which uses designs of a deep neural network, which are explicit sorts of artificial intelligence and machine learning algorithms that have become extremely critical in the last few years. Deep learning permits to teach machines how to finish complicated tasks without explicitly programming them to do so [7].

2 Related Work

Krupali et al. [8] have discussed various super-resolution techniques with their advantages and disadvantages and also presented challenging issues and future research directions for super-resolution. Yungang Zhang has given a short review of the latest in-depth learning based strategies for SISR, such as network type, network structure, and training strategies. The advantages and disadvantages of these strategies are analyzed as well [9]. Zhihao Wang intends to give a far-reaching overview of ongoing advances in image super-resolution using deep learning approaches [5]. Christian et al. [10] showed the utilization of GANs, explicitly SRGAN framework, to produce yield pictures having enriched pixel quality and sometimes even more. Gulrajani et al. [11] projected an enhanced approach for training the discriminator – termed critic by Arjovsky et al. [12] – which behaves stably, even with deep ResNet architectures. GANs have mostly been investigated on pictures, showing significant success with tasks such as image generation [13–15], image super-resolution [10], and many others. A lot of literature is there on SR surveys from different sources, such as forum posts, blogs, journal articles, and conference proceedings. This chapter intends to give an audit of SR from the viewpoint of deep learning methods particularly the primary contributions in recent years, while the more significant part of the previous works [16–19] focuses on looking over conventional SR methods and a few investigations, for the most part, focus on giving quantitative assessments dependent on full reference measurements or human visual observation [20, 21].

2.1 *Single Image Super-Resolution (SISR)*

SISR includes expanding the range of a small picture, while downplaying the orderly degradation in quality. It has various applications, which include satellite and elevated imaging examination, medical image processing, compressed picture/video improvement, and so forth [22]. The goal is to consider a low-quality picture and generate an approximation of a related high-quality photograph. This issue is poorly presented – various high-quality pictures can be created from a similar low-quality picture.

Image super-resolution is the innovation that permits to build the resolution of pictures using deep learning to zoom into pictures [4]. It is a software strategy that allows enhancing the picture spatial resolution with the existing hardware. Low-resolution (LR) picture allows scarcely any details based on small pixel thickness within an image, whereas a high-resolution (HR) image enables a lot of details based on large pixel density within an image. A procedure that is used to remake a high-quality image from one or many low-quality photographs by reestablishing the high frequency details called super-resolution [4]. The goal of SR is to get back a high-quality picture from a low-quality input [23].

Super-resolution relies on the possibility that a mix of a low-quality grouping of pictures of a scene is utilized to produce a high-quality photograph or picture sequence. Accordingly, it endeavors to remake the standard scene picture with high quality given a lot of observed pictures at low quality. The common methodology takes the low-quality pictures because of the resampling of a high-quality photograph. The objective is to recuperate the high-quality picture which when resampled relies on the intake pictures and the imaging model will deliver the low-quality observed pictures. In this way, the precision of the imaging model is imperative for super-resolution and an off base demonstrating state of movement can corrupt the picture further. The observed pictures could be taken from one or various cameras or could be frames of a video succession. These pictures should be mapped to a typical reference outline. This procedure is known as registration. The SR methodology can be applied to a locale of interest for the adjusted complex picture. The way to effective super-resolution comprises precise arrangement, for example, registration and formulation of a suitable advanced picture model.

2.2 *Interpolation*

The strategy interpolation is considered as widely accepted and regularly used for upscaling a picture. It is easy to execute but the strategy leaves much to be desired as far as visual quality because the details (for example, sharp edges) are not regularly saved [22]. Pictures of lower spatial quality can likewise be scaled by a classic upsampling strategy, for example, bilinear or bicubic interpolation. Progressively advanced strategies exploit internal similarities of a given picture or use datasets of low-quality pictures and high-quality counterparts, viably learn a mapping among them. Among examples based on SR algorithms, the sparse coding based strategy is well known that needs a dictionary to be discovered which permit the outline of low-quality pictures into a transitional, sparse depiction. Likewise, the high-quality dictionary is accomplished and permits to reestablish an estimate of a high-quality photograph. Such a pipeline, as a rule, includes a few stages, and not everyone of them can be streamlined. Ideally to join all these steps in one significant advance with the entirety of its parts being optimizable that impact can be accomplished by a neural system, the engineering of which is inspired by sparse coding.

2.2.1 Image Quality Assessment (IQA)

Image quality alludes to the optical traits of pictures and spotlights on the affective appraisals of observers [5]. The nature of pictures is pertinent in building compression and picture upgradation calculations. IQA plays a significant job in many image processing undertakings to quantitatively measure picture quality. Although a few IQA techniques have been produced for quite a long time, every strategy has its trait and there are a couple of researches concentrating on the improvement of value appraisal in a particular kind of picture, for example, medical pictures [24]. By and large, IQA techniques incorporate subjective strategies that rely on human' approach which shows the sensibility of the picture and objective [5] techniques rely on computational models that attempt to determine the picture quality. Although subjective metrics are frequently more perceptually exact, some of them are cumbersome, tedious, or costly to evaluate [25]. The earlier one is more following our need yet regularly tedious and costly; in this manner, the later one is as of now the standard. But these techniques are not reliable between one another, because objective strategies are unable to catch the human visual observation precisely, which may prompt huge distinction in IQA results [10, 26]. Moreover, the objective IQA techniques are further partitioned into three principle areas like full reference IQA techniques performing appraisal utilizing reference pictures, reduced reference IQA strategies rely on comparisons of extricated highlights, and no reference IQA strategies (i.e., blind IQA) with no reference pictures. Reference-based techniques also depend on high-quality photographs to assess the contrast between the two pictures. Unlike reference-based techniques, no reference IQA do not require a base picture for evaluating the quality of a photograph [27], but the distorted picture is accessible [28]. Visual information fidelity (VIF) [29] is a full reference strategy that uses hypothetical information standard for image fidelity estimation. Sharpness degree [30] is one of the instances of a no reference IQA strategy. It is used to show the degree of sharpness of the picture. Blur metric is one of the other tools for estimating blurs and try to get the advancement of the edges. A few most regularly used IQA strategies covering both subjective and objective methods.

Peak Signal-to-Noise Ratio (PSNR)

PSNR is the generally acclaimed reestablished quality estimation of lossy change, for instance, picture inpainting. It is characterized using the most extreme pixel value and the mean squared error (MSE) among pictures [5] for image SR. It quantifies the alteration among the created high-quality photograph to the standard picture (real HR picture) that can be characterized as the proportion among maximum pixel value of the input picture to the MSE among the pixel estimations of the reestablished photograph and the standard photograph represented on a logarithmic scale.

$$\text{PSNR} = 10 \times \log\left(\text{max value}^2/\text{MSE}\right) \tag{1}$$

where *max value denotes* the maximum element value, which is present in the intake picture. Computing the trade-off among them is the motive behind this condition.

Structural Similarity Index (SSIM)

It is one of the examples of a reference-based technique. It is also a subjective metric. Taking into consideration that the human visual system (HVS) is profoundly adjusted to remove picture arrangement [31], the SSIM [26] is recommended for estimating the structural similarity among pictures, rely on autonomous correlations regarding contrast, luminance, and structures. It is viewed as one of the most referenced quality measurements and is identified with the single-scale estimation that understands its best execution when applied at a proper scale. The highest SSIM demonstrates the best denoising method [28].

It is given by:

$$\mathrm{SSIM}\left(I,\hat{I}\right)=\left[C_l\left(I,\hat{I}\right)\right]^{\alpha}\left[C_c\left(I,\hat{I}\right)\right]^{\beta}\left[C_S\left(I,\hat{I}\right)\right]^{\gamma} \tag{2}$$

where α, β, and γ are the adjusted control parameters denoting luminance, contrast, and structure comparison functions, respectively. The SSIM assesses the reestablishment quality from the viewpoint of HVS, as it is good to confirm the necessities of affective evaluation [32, 33], and is likewise generally utilized.

Mean Opinion Score (MOS)

MOS testing is usually used as a subjective image quality assessment strategy, where human raters are approached to allow affective quality points to approved pictures. Typically, the points range from 1 (*bad*) to 5 (*excellent*). Also, the last one is determined as the arithmetic mean of all evaluations. Even though the MOS testing appears to be a dependable IQA strategy which has some intrinsic deformities, for example, inclinations, nonlinearly perceived scales, and fluctuation of appraisal models. As a general rule, some SR approaches operating ineffectively in standard IQA measurements (e.g., PSNR), however, far surpassing others as far as perceptual quality, in that case, the MOS testing is the most dependable IQA technique for precisely estimating the affective quality [10, 34–39].

Other IQA Strategies

Notwithstanding the given IQA techniques, various less well-known SR measurements are also there. The multiscale structural similarity (MS-SSIM) [40] affords more adaptability than a single scale (SSIM) in combining the varieties of review situations. The natural image quality evaluator (NIQE) [41] uses quantifiable deviations from factual regularities noticed in general pictures, without an introduction to altered pictures. The feature similarity (FSIM) [42] removes feature points of human intrigue dependent on stage congruency and picture gradient magnitude to assess picture resolution. Currently Blau et al. [43] demonstrate numerically that distortion such as PSNR, SSIM, and perceptual quality like MOS is at chances with one another, and demonstrate that as the distortion reduces, the perceptual quality must be more terrible.

2.2.2 Loss Function

In the super-resolution discipline, they are utilized to quantify reproduction mistakes and advise the approach streamlining. In early occasions, specialists typically utilize the elementwise L2 loss, yet later on, find that it cannot quantify reproduction quality precisely. In this manner, an assortment of loss functions, such as content loss [44], adversarial loss [10], is embraced for estimating the better reproduction mistake while creating more sensible and high-quality outcomes. They are also used to determine the distinction between the created HR picture and the ground truth HR picture. This distinction (blunder) is then used to upgrade the supervised learning approach. A few categories of loss functions exist where each one is fined an alternate part of the created picture. Currently these loss functions have been playing a significant job and are frequently utilized by weighting and summarizing the blunders got from each separately. It empowers the approach to concentrate on perspectives provided by numerous functions together [25]. Some mainstream functions are used for preparing models like pixel loss, content loss, texture loss, total variation loss, adversarial loss, and perceptual loss.

Pixel Loss

It is the least difficult class of these functions where every element in the produced picture is straightforwardly contrasted with every element in the ground truth picture. Mainstream loss functions, for example, the loss L1 or L2 or advanced variations, the loss of smooth L1 are utilized. The PSNR measurement is profoundly connected with the element distinction thus limiting the element loss straightforwardly boosts the PSNR measurement value. Notwithstanding, pixel loss does not consider the picture quality and the model regularly yields perceptually uninspiring outcomes which frequently is inadequate with regard to high recurrence details [25].

Content Loss

It assesses the picture quality depending on its affective quality. The higher the PSNR, the better the quality of the reproduced picture as it attempts to limit the MSE among the pictures as for the most extreme pixel estimation of the input picture. But increasing PSNR alone does not function as the pictures created could be excessively smooth, which does not look perceptually genuine [4]. An intriguing method is contrasting the significant highlights of the produced picture and ground truth picture. In this way, deep neural systems are utilized as a feature extractor and consider the distinction in the feature map among the produced picture and ground truth picture as a loss [4]. The significant level highlights can be acquired by moving both pictures through a pre-prepared image classification system (e.g., a VGG-Net or a ResNet). This is called content loss.

Content Loss [5]

$$L_{\text{content}}\left(\widehat{I}, I; \varphi, l\right) = \frac{1}{h_l w_l c_l}\sqrt{\sum_{i,j,k}\left(\phi_{i,j,k}^{(l)}\left(\widehat{I}\right) - \phi_{i,j}^{(l)}(I)\right)^2} \quad (3)$$

The above condition evaluates the content loss among a ground truth picture and created a picture, given a pre-prepared system (*Φ*) and a layer (*l*) of the pre-prepared system. It urges a created picture to be perceptually like the ground truth picture [25].

Texture Loss

Texture loss is utilized to empower the produced pictures that have a similar style (color, texture, contrast, etc.) as the ground truth picture. The style recreation of a photograph portrayed by Gatys et al. [45] is characterized as the relationship among various aspect channels. They are typically acquired from feature maps removed using a pre-prepared image classification system (*Φ*) [25].

Gram matrix [5]

$$G_{ij}^{(l)}(I) = \text{vec}\left(\varphi_i^{(l)}(I)\right).\text{vec}\left(\varphi_j^{(l)}(I)\right) \quad (4)$$

Gram matrix (*G*) is the internal item among the vectorized feature maps *i* and *j* on layer *l* that depicts the connection among them. When it determines two pictures then the estimation of the style recreation is direct, as given below [5]:

$$L_{\text{texture}}\left(\widehat{I}, I; \varphi, l\right) = \frac{1}{c_l^2}\sqrt{\sum_{i,j}\left(G_{i,j}^{(l)}\left(\widehat{I}\right) - G_{i,j}^{(l)}(I)\right)^2} \quad (5)$$

By utilizing texture loss, the approach is spurred to make reasonable textures and outwardly fulfilling more outcomes.

Total Variation Loss

It is utilized to restrain the noise in produced pictures which considers the total of the absolute contrasts between nearby pixels and find out the quantity of noise present in the picture [25]. It is calculated for a produced picture and is demonstrated as below [5]:

$$L_{TV}\left(\widehat{I}\frac{1}{hwc}\sum_{i,j,k}\sqrt{\left(\widehat{I}_{i,j+1,k}-\widehat{I}_{i,j,k}\right)^{2}+\left(\widehat{I}_{i+1,j,k}-\widehat{I}_{i,j,k}\right)^{2}}\right. \tag{6}$$

where i denotes height, j denotes width, and k represent channels, respectively.

Adversarial Loss

As discussed earlier that PSNR is not an ideal measurement to foresee whether the picture is genuine or not, it is considered better in such an event to pass the judgment on the quality of the produced picture and ignore the fake ones. This is the reason behind to utilize other systems which foresee the genuineness of the produced picture. This type of system is called a discriminator network which attempts to foresee whether the pictures are sensible or not created by the generator. In this way, a GAN can be successfully utilized, which consists of two systems: the generator and the discriminator. A low-quality picture is considered as input in the generator which constructs a high-quality picture as yield. The discriminator chooses the affective reality of the produced picture and also adds an input to the generator known as adversarial loss that foresees the likelihood of whether the created picture is reasonable. Then, the generator utilizes it as a prompt for improvement. First and foremost, the generator creates awful quality pictures; however, as the preparation proceeds, it begins to deliver reasonable pictures. When the preparation is finished, the discriminator cannot distinguish between the standard high-quality picture and the other made by the generator [4]. Normally, models have a better perceptual quality that is prepared with an adversarial loss even though they may miss out on PSNR contrasted with those prepared on pixel loss. The preparation procedure of GANs is somewhat troublesome and changeable is considered as a minor drawback [25].

Perceptual Loss

When the created picture is perceptually like the ground-truth picture that is known as perceptual loss. It is utilized in contrasting two distinct pictures that seem to be comparative, like a similar picture but moved by one pixel. It is utilized to analyze high-level contrasts like content and style disparities between pictures. A perceptual loss function is fundamentally the same as the per-pixel loss function, as both are utilized for preparing feed-forward neural systems for photo transformation undertakings. The perceptual loss function is a more usually utilized part as it frequently gives more precise outcomes in regard to style transfer [46].

3 SISR with Deep Learning Techniques

The goal is to improve the low-resolution (LR) picture as acceptable (or better) than the objective, known as the ground truth, which in this circumstance is the first picture to be downscaled into the low-quality picture [2]. To achieve this mathematical function takes a low-quality picture that lacks details and daydreams the details and highlights onto it. In doing so, the function discovers detail conceivably never recorded by the original camera. This mathematical function is known as the model, and the upscaled picture is the model's prediction. It is already known that convolution activity consistently decreases the size of the information. Deep learning convolutional neural networks can be utilized to prepare an approach for the super-resolution, which requires overcoming the fact related to convolution activity that diminishes the size of intakes. So, it is needed to utilize a deconvolution or a comparative layer with a goal that the yield picture size is multiple times the information capacity by a specific cause to accomplish the desired outcome.

For the preparation of such a super-resolution approach, high-quality pictures are needed to download from the web and escalate it down to a low-quality picture. Then these low-quality intakes are given to the system and prepared to take pictures of a high quality that contrasts with the standard high-resolution variants [47]. The accomplishment of the system can be quantified by the way it diminishes the mean squared error (MSE) among the yield and the standard pixels. Training data is simple for this: gather an enormous number of high-resolution (HR) pictures of the web, at that point downscale it with 4 that will be low-resolution (LR) pictures. Then give these LR images (30×30) to the system and prepare it to produce the HR images (90×90). The system's goal is to diminish the MSE among the pixels of the created picture and ground truth picture [4].

The MSE equation is:

$$MSE = \sum_{i=1}^{M} \sum_{j=1}^{N} \| f(i,j) - g(i,j) \|^2 / M * N \tag{7}$$

where f denotes the standard image matrix

g denotes the matrix of the rebuild high-quality picture

i denotes the index of the Tuple
j denotes the index of the attribute

M denotes the count of rows of pixels of the picture
N denotes the count of attributes of elements of the picture

Theoretically, getting an error to 0 means that the system is currently ready to produce high-quality pictures. In this way, a measurement for quality is characterized.

3.1 Super-Resolution Convolutional Neural Network (SRCNN)

Convolutional neural network (CNN) is a sort of deep network architecture intended for explicit undertakings like image classification. They were motivated through the association of nerve cells in the optical cortex of the animal mind which provides some alluring features that are valuable for managing a particular type of data such as pictures, sound, and videos [7]. It is made of an input layer, however, for primitive image processing, this intake is normally a 2D array of nerve cells which contrast with the pixels of a picture. It likewise consists of an output layer usually called a 1D setup of output nerve cells. It also utilizes a mix of sparsely associated convolution layers that execute image processing on data sources and also contains a down sampling layer known as the pooling layer to lessen the count of nerve cells essential in subsequent layers of the system. At last, CNN ordinarily contains one or more linked layers to connect the pooling layer with the output layer. Convolution is a method that permits to extricate of optical features from a picture in little lumps. In the convolution layer, every nerve cell is answerable for a little bunch of neurons in the last layer. It consists of kernel or filters that decide the bunch of nerve cells. Filters scientifically accommodate the input of convolution to assist it in identifying particular sort of highlights in the photograph that can restore the unmodified photograph, blur the photograph, sharpen the photograph, and identify the edges. It is accomplished by duplicating the standard photograph values through a convolution network. Pooling alternatively called subsampling or downsampling diminishes the count of nerve cells in the previous convolution layer that also holds the most important information. Different kinds of pooling that can be performed, for instance, and is considering the sum of neurons, average, or most extreme value. On turning around this design to make what is called a deconvolutional neural network, this system executes the opposite of a convolutional network, that is, instead of taking a picture and changing over forecast value, these systems take an input value and try to deliver a picture. CNN's function well for an assortment of assignments, including image processing, image recognition, video analysis, image segmentation, and natural language processing.

SRCNN is a deep learning approach that rebuilds the HR picture from the LR picture.

It consists of the following operations:

1. Preprocessing: Upscaling of LR picture to a required high-quality picture.
2. Feature extraction: A set of feature maps are extracted from the upscaled LR picture.
3. Nonlinear mapping: Mapping of the feature maps representing LR to high-quality patches.
4. Reconstruction: Generation of high-quality pictures from high-quality patches [23].

It was considered the first deep learning technique to surpass conventional ones. This is a CNN comprising three convolutional layers that are patch extraction and representation, nonlinear mapping, and reconstruction [22]. A picture should be upsampled using bicubic interpolation before given to the system. Then it should be changed over to YCbCr shading space, though only luminance channel is utilized through the system. The system's yield is at that point converged with interpolated C_BC_R channels to create a concluding color picture. It is discovered that to train SRCNN is extremely hard.

3.2 Super-Resolution with Generative Adversarial Network (SRGAN)

Generative adversarial networks (GANs) have been progressively utilized for a few pictures based functions which include SR. It is the most frequently utilized deep learning design to prepare deep learning-based super-resolution approaches, whereas the design relies on unsupervised learning. It is a mix of two deep learning neural networks dwelling in a solitary structure and contending in a zero-sum situation [47]. It comprises a generator network (GN) and a discriminator network (DN). The GN assembles arranged data learns to produce SR images from LR images as close as possible to HR. It also tries to create manufactured data endless from real data. The DN tries to recognize whether the data are real or manufactured, which is seen and also turns out to be dynamically improved at distinguishing counterfeit data [7]. During the process of training, a high-quality picture is always changed over into a low-quality picture by downsampling. The generator of the GAN is answerable for changing over the low-quality picture to high-quality picture, and the discriminator is liable for arranging the produced pictures [48]. These networks are adversaries as both challenging to beat each other. The generator exploits ResNet and subpixel convolution for upsampling. It additionally joins perceptual loss with a generative or adversarial loss for the calculation of its loss [49]. GANs can create various artificial artifacts, for example, sound, videos, and pictures which impersonate human-made partners. The objective of this procedure is to take basic input and utilize it to produce a complicated yield with a significant balance of precision [47]. One of the constraints of GANs is that they are adequately and apathetic methodology as their loss function, the critic, is prepared as a major aspect of the procedure and not explicitly built for this reason. This could be one reason numerous models are just acceptable at super-resolution and not image repair. GANs are used to prepare many of the deep learning-based SR approaches [2]. SRGAN utilizes the GAN to deliver high-quality pictures from low-quality pictures [50]. SRGANs can be utilized to deliver high-quality pictures by practicing a blend of an adversarial network and a deep neural network (Fig. 1).

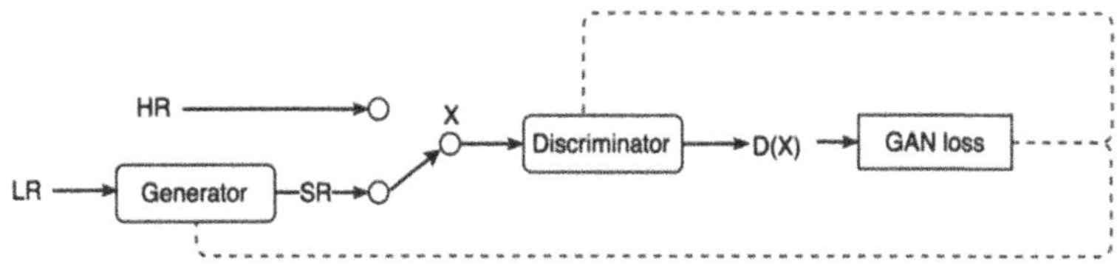

Fig. 1 SRGAN [51]

Following are the steps to prepare an SRGAN

- Consider a lot of high-quality pictures and downscale to low quality.
- Add pictures into the generator to come out with a yield of super-resolution pictures.
- Make the use of discriminator to recognize the standard high-quality and SR pictures, and afterward adopt back propagation to prepare both.
- Repeat the steps until they arrive at adequate outcomes [47].

GAN models that are running regularly include serious and complex assignments, for example, preparing an SRGAN approach. Furthermore, preparation and working of these models need strong equipment to work proficiently, typically requiring numerous days or hours. These models can work on frameworks involving different graphics processing units (GPUs) that are frequently utilized to assist the work of DL approaches much effectively. In any case, this arrangement is profoundly costly. Another arrangement is to make the working of these approaches in cloud-based frameworks. Then it is needed to continually monitor examinations to abstain from sitting around idly and cash. To run GAN models with high effectiveness and maintaining a strategic distance from pointless consumptions, a deep learning platform like MissingLink can be utilized that allows an answer for deep learning techniques and GAN can also work over various machines. Super-resolution is a field in which GAN depicts a very remarkable outcome with commercial chances [52]. The utilization of GAN for SR tackles the inadequacies of the ordinary strategies, which includes the DL techniques, with the absence of high recurrence data [53].

4 SR Residual Networks (ResNet)

Residual networks is a convolutional neural network (CNN) design comprising an arrangement of residual blocks (ResBlocks) depicted underneath with skip associations separating ResNets from different CNNs [2]. At the point when previously concocted ResNet won that year's ImageNet competition by a noteworthy edge as it tended to the vanishing gradient issue, whereas more layers are included preparing eases back and precision does not improve or even deteriorates. It is the systems skip associations that achieve this accomplishment. It is developed to expand the PSNR execution and reduce the MSE which accomplishes the best results on standard benchmarks.

Each ResBlock has two associations from its input, one experiencing a progression of convolutions, batch normalization, and linear functions and the other association skipping over that arrangement of functions and convolutions. These are known as identity, cross, or skip associations [54, 55]. The tensor yields of the two associations are included. Where a ResBlock gives a yield that is a tensor expansion, this can be changed to be a tensor link. With each cross–skip association, the system turns out to be increasingly thick. The ResBlock at that point turns into a DenseBlock, and the system turns into a DenseNet. This permits the calculation to skip over bigger and bigger parts of the design.

5 Conclusion

In this chapter, super-resolution from the viewpoint of deep learning approaches and especially the main improvements in recent years is covered. This chapter started with an introduction which provides the knowledge of image super-resolution with deep learning approaches, SISR, and then interpolation is discussed. After this, various performance evaluation metrics like PSNR, SSIM, MOS, and many more are discussed which can be used for monitoring the performance. Moreover, loss functions like pixel loss, content loss, and many more are also discussed in detail. The super-resolution with deep learning techniques like SRCNN and SRGAN are then discussed. In addition, the super-resolution residual network (ResNet) is also discussed. Finally, it is concluded that image SR is a significant innovation field in image analysis. This method has been broadly utilized in numerous computer vision applications. No doubt, the achievement of deep learning strategies in super-resolution has captivated an ever-increasing number of specialists.

References

1. Zhu, X. (2014). Computational intelligence techniques and applications. *Computational Intelligence Techniques in Earth and Environmental Sciences*, 3–26. https://doi.org/10.1007/978-94-017-8642-3_1.
2. 302 Found. (2019). Retrieved from https://towardsdatascience.com/deep-learning-based-super-resolution-without-using-a-gan-11c9bb5b6cd5
3. Morera-Delfín, L., Pinto-Elías, R., & Ochoa-Domínguez, H.-J. (2018). Overview of super-resolution techniques. *Advanced Topics on Computer Vision, Control and Robotics in Mechatronics*, 101–127. https://doi.org/10.1007/978-3-319-77770-2_5.
4. Deep Learning based image Super-Resolution to enhance photos. (2018, July 25). Retrieved from https://cv-tricks.com/deep-learning-2/image-super-resolution-to-enhance-photos/
5. Wang, Z., Chen, J., & Hoi, S. C. H. (2020). Deep learning for image super-resolution: A survey. *IEEE Transactions on Pattern Analysis and Machine Intelligence, 1*. https://doi.org/10.1109/tpami.2020.2982166.

6. Yue, L., Shen, H., Li, J., Yuan, Q., Zhang, H., & Zhang, L. (2016). Image super-resolution: The techniques, applications, and future. *Signal Processing, 128*, 389–408. https://doi.org/10.1016/j.sigpro.2016.05.002.
7. Goyal, R. (2018, May 9). *Five important techniques that you should know about deep learning* [Blog post]. Retrieved from https://www.zeolearn.com/magazine/five-important-techniques-that-you-should-know-about-deep-learning
8. Ramavat, K., Joshi, M., & Swadas, P. B. (2016). A survey of super-resolution techniques. *International Research Journal of Engineering and Technology, 3*(12), 1035–1039. Retrieved from https://www.irjet.net/archives/V3/i12/IRJET-V3I12238.pdf.
9. Zhang, Y., & Xiang, Y. (2018). Recent advances in deep learning for single image super-resolution. *Advances in Brain Inspired Cognitive Systems*, 85–95. https://doi.org/10.1007/978-3-030-00563-4_9.
10. Ledig, C., Theis, L., Huszar, F., Caballero, J., Cunningham, A., Acosta, A., et al. (2017). Photo-realistic single image super-resolution using a generative adversarial network. *IEEE Conference on Computer Vision and Pattern Recognition (CVPR), 2017*, 105–114. https://doi.org/10.1109/cvpr.2017.19.
11. Gulrajani, I., Ahmed, F., Arjovsky, M., Dumoulin, V., & Courville, A. C. (2017). Improved training of Wasserstein GANs. *ArXiv, abs/1704.00028.*
12. Arjovsky, M., Chintala, S., & Bottou, L. (2017). Wasserstein gan. arXiv preprint *arXiv:1701.07875.*
13. Karras, T., Aila, T., Laine, S., & Lehtinen, J. (2018). Progressive growing of GANs for improved quality, stability, and variation. *ArXiv, abs/1710.10196.*
14. Denton, E. L., Chintala, S., Szlam, A., & Fergus, R. (2015). Deep generative image models using a Laplacian pyramid of adversarial networks. *ArXiv, abs/1506.05751.*
15. Im, D.J., Kim, C.D., Jiang, H., & Memisevic, R. (2016). Generating images with recurrent adversarial networks. *ArXiv, abs/1602.05110.*
16. Park, S. C., Park, M. K., & Kang, M. G. (2003). Super-resolution image reconstruction: A technical overview. *IEEE Signal Processing Magazine, 20*(3), 21–36. https://doi.org/10.1109/msp.2003.1203207.
17. Nasrollahi, K., & Moeslund, T. B. (2014). Super-resolution: A comprehensive survey. *Machine Vision and Applications, 25*(6), 1423–1468. https://doi.org/10.1007/s00138-014-0623-4.
18. Tian, J., & Ma, K.-K. (2011). A survey on super-resolution imaging. *Signal, Image and Video Processing, 5*(3), 329–342. https://doi.org/10.1007/s11760-010-0204-6.
19. Van Ouwerkerk, J. D. (2006). Image super-resolution survey. *Image and Vision Computing, 24* (10), 1039–1052. https://doi.org/10.1016/j.imavis.2006.02.026.
20. Yang, C.-Y., Ma, C., & Yang, M.-H. (2014). Single-image super-resolution: A benchmark. *Computer Vision – ECCV, 2014*, 372–386. https://doi.org/10.1007/978-3-319-10593-2_25.
21. Thapa, D., Raahemifar, K., Bobier, W. R., & Lakshminarayanan, V. (2016). A performance comparison among different super-resolution techniques. *Computers & Electrical Engineering, 54*, 313–329. https://doi.org/10.1016/j.compeleceng.2015.09.011.
22. Kańska, K. (2019, April 25). *Cookie and Privacy Settings* [Blog post]. Retrieved from https://deepsense.ai/using-deep-learning-for-single-image-super-resolution/
23. Salaria, S. (2019, August 22). *Using the super-resolution convolutional neural network for image restoration* [Blog post]. Retrieved from https://medium.com/datadriveninvestor/using-the-super-resolution-convolutional-neural-network-for-image-restoration-ff1e8420d846
24. Razmjooy, N., Estrela, V. V., & Loschi, H. J. (2020). Entropy-based breast cancer detection in digital mammograms using world cup optimization algorithm. *International Journal of Swarm Intelligence Research (IJSIR), 11*(3), 1–18.
25. Raj, B. (2019, July 1). *An Introduction to Super-Resolution using Deep Learning* [Blog post]. Retrieved from https://medium.com/beyondminds/an-introduction-to-super-resolution-using-deep-learning-f60aff9a499d

26. Wang, Z., Bovik, A. C., Sheikh, H. R., & Simoncelli, E. P. (2004). Image quality assessment: From error visibility to structural similarity. *IEEE Transactions on Image Processing, 13*(4), 600–612. https://doi.org/10.1109/tip.2003.819861.
27. 302 Found. (2019, June 1). Retrieved from https://heartbeat.fritz.ai/research-guide-image-quality-assessment-c4fdf247bf89
28. Mahmoudpour, S., & Kim, M. (2015). A study on the relationship between depth map quality and stereoscopic image quality using upsampled depth maps. *Emerging Trends in Image Processing, Computer Vision and Pattern Recognition*, 149–160. https://doi.org/10.1016/b978-0-12-802045-6.00010-7.
29. Sheikh, H. R., & Bovik, A. C. (2006). Image information and visual quality. *IEEE Transactions on Image Processing, 15*(2), 430–444. https://doi.org/10.1109/tip.2005.859378.
30. Tsai, C., Liu, H., & Tasi, M. (2011). Design of a scan converter using the cubic convolution interpolation with canny edge detection. In *Proceedings of the international conference on electric information and control engineering* (pp. 5813–5816).
31. de Jesus, et al. (2020, April). Using Transmedia Approaches in STEM. In *2020 IEEE Global Engineering Education Conference (EDUCON)* (pp. 1013–1016). IEEE.
32. Sheikh, H. R., Sabir, M. F., & Bovik, A. C. (2006). A statistical evaluation of recent full reference image quality assessment algorithms. *IEEE Transactions on Image Processing, 15* (11), 3440–3451. https://doi.org/10.1109/tip.2006.881959.
33. Zhou, W., & Bovik, A. C. (2009). Mean squared error: Love it or leave it? A new look at signal fidelity measures. *IEEE Signal Processing Magazine, 26*(1), 98–117. https://doi.org/10.1109/msp.2008.930649.
34. Wang, X., Yu, K., Dong, C., & Change Loy, C. (2018). Recovering realistic texture in image super-resolution by deep spatial feature transform. *IEEE/CVF Conference on Computer Vision and Pattern Recognition, 2018*, 606–615. https://doi.org/10.1109/cvpr.2018.00070.
35. Wang, Z., Liu, D., Yang, J., Han, W., & Huang, T. (2015). Deep networks for image super-resolution with sparse prior. In *2015 IEEE International Conference on Computer Vision (ICCV)* (pp. 370–378). https://doi.org/10.1109/iccv.2015.50.
36. Xu, X., Sun, D., Pan, J., Zhang, Y., Pfister, H., & Yang, M.-H. (2017). Learning to super-resolve blurry face and text images. In *2017 IEEE International Conference on Computer Vision (ICCV)* (pp. 251–260). https://doi.org/10.1109/iccv.2017.36.
37. Dahl, R., Norouzi, M., & Shlens, J. (2017). Pixel recursive super resolution. In *2017 IEEE International Conference on Computer Vision (ICCV)* (pp. 5449–5458). https://doi.org/10.1109/iccv.2017.581.
38. Lai, W.-S., Huang, J.-B., Ahuja, N., & Yang, M.-H. (2019). Fast and accurate image super-resolution with deep laplacian pyramid networks. *IEEE Transactions on Pattern Analysis and Machine Intelligence, 41*(11), 2599–2613. https://doi.org/10.1109/tpami.2018.2865304.
39. Sajjadi, M. S. M., Scholkopf, B., & Hirsch, M. (2017). EnhanceNet: Single image super-resolution through automated texture synthesis. In *2017 IEEE International Conference on Computer Vision (ICCV)* (pp. 4501–4510). https://doi.org/10.1109/iccv.2017.481.
40. Wang, Z., Simoncelli, E. P., & Bovik, A. C. (2003). Multiscale structural similarity for image quality assessment. *The Thirty-Seventh Asilomar Conference on Signals, Systems & Computers, 2003*, 1398–1402. https://doi.org/10.1109/acssc.2003.1292216.
41. Mittal, A., Soundararajan, R., & Bovik, A. C. (2013). Making a "completely blind" image quality analyzer. *IEEE Signal Processing Letters, 20*(3), 209–212. https://doi.org/10.1109/lsp.2012.2227726.
42. Lin, Z., Zhang, L., Mou, X., & Zhang, D. (2011). FSIM: A feature similarity index for image quality assessment. *IEEE Transactions on Image Processing, 20*(8), 2378–2386. https://doi.org/10.1109/tip.2011.2109730.
43. Blau, Y., & Michaeli, T. (2018). The perception-distortion Tradeoff. *IEEE/CVF Conference on Computer Vision and Pattern Recognition, 2018*, 6228–6237. https://doi.org/10.1109/cvpr.2018.00652.

44. Johnson, J., Alahi, A., & Fei-Fei, L. (2016). Perceptual losses for real-time style transfer and super-resolution. *Computer Vision – ECCV, 2016*, 694–711. https://doi.org/10.1007/978-3-319-46475-6_43.
45. Gatys, L. A., Ecker, A. S., & Bethge, M. (2016). Image style transfer using convolutional neural networks. *IEEE Conference on Computer Vision and Pattern Recognition (CVPR), 2016*, 2414–2423. https://doi.org/10.1109/cvpr.2016.265.
46. Perceptual Loss Functions. (2019, May 17). Retrieved from https://deepai.org/machine-learning-glossary-and-terms/perceptual-loss-function
47. Super-Resolution Deep Learning: Making the Future Clearer. [Blog Post]. Retrieved from https://missinglink.ai/guides/computer-vision/super-resolution-deep-learning-making-future-clearer/
48. Yashwanth, N., Navya, P., Rukhiya, M., Prasad, K. S., & Deepthi, K. S. (2019). Survey on generative adversarial networks. *International Journal of Advance Research, Ideas and Innovations in Technology, 5*, 239–244.
49. An Evolution in Single Image Super-Resolution using Deep Learning. (2019, December 3). Retrieved from https://towardsdatascience.com/an-evolution-in-single-image-super-resolution-using-deep-learning-66f0adfb2d6b
50. Sinha, V. (2019, December 17). *Super Resolution GAN (SRGAN)* [Blog post]. Retrieved from https://medium.com/analytics-vidhya/super-resolution-gan-srgan-5e10438aec0c
51. Hui, J. (2018, July 2). *GAN — Super Resolution GAN (SRGAN)* [Blog post]. Retrieved from https://medium.com/@jonathan_hui/gan-super-resolution-gan-srgan-b471da7270ec
52. Shaikh, F. (2020, May 11). *Top 5 Interesting Applications of GANs for Every Machine Learning Enthusiast!* Retrieved from https://www.analyticsvidhya.com/blog/2019/04/top-5-interesting-applications-gans-deep-learning/
53. Gonog, L., & Zhou, Y. (2019). A review: Generative adversarial networks. *2019 14th IEEE Conference on Industrial Electronics and Applications (ICIEA)*, 505–510.
54. Monteiro, et al. (2018). Health 4.0: applications, management, technologies and review. *Personalized Medicine, 2*(4), 262–276.
55. Razmjooy, N., et al. (2020). Computer-aided diagnosis of skin cancer: A review. *Current Medical Imaging, 16*(7), 781–793.

A Comprehensive Review of CAD Systems in Ultrasound and Elastography for Breast Cancer Diagnosis

Rajeshwari Rengarajan, Geetha Devasena M S, and Gopu Govindasamy

1 Introduction

Cancer is a set of many diseases characterized by the uncontrolled growth and spread of cells in the human body. It can start in any organ or tissue of the body when cells become abnormal and divide uncontrollably, go beyond their boundaries to invade adjoining parts of the body, and spread to other regions. World Health Organization (WHO) and the American Cancer Society (ACS) reported that 9.6 million people died globally from cancer in 2018. Lung, breast, liver, prostate, cervical, thyroid, and stomach cancer are the most common types of cancer in humans. The estimated number of new cases each year is expected to rise from 9.6 million in 2018 to 15 million by 2030 [6].

Breast cancer (BC) is a kind of cancer in which breast cells grow, spread, and divide out of control. These abnormal cells proliferate and form an extensive tissue or lump, named as tumors. A tumor can be benign and malignant according to its features. Benign tumors do not spread to other parts of the body, and the cells do not affect other tissues. Benign tumors are non-cancerous and can be removed. Malignant (cancerous) tumors can invade and damage nearby tissues or organs. Breast tumors form either in the ducts or lobules of the breast. It can occur in both men and women, but it is rare in men. BC is one of the leading causes of death in women after lung cancer. 1 in 8 women will have BC in their lifetime, and 1 in 39 women will die from BC every year [14]. The WHO and ACS reported that 3.8 million US women

R. Rengarajan (✉)
KPR Institute of Engineering & Technology, Coimbatore, Tamil Nadu, India
e-mail: rajeshwari.r@kpriet.ac.in

G. Devasena M S · G. Govindasamy
Sri Ramakrishna Engineering College, Coimbatore, Tamil Nadu, India
e-mail: msgeetha@srec.ac.in

A. Deshpande et al. (eds.), *Computational Intelligence Methods for Super-Resolution in Image Processing Applications*, https://doi.org/10.1007/978-3-030-67921-7_4

live with BC, and approximately 41,760 women are expected to die from BC in 2019 [6].

The risk factors of BC are diet, smoking, alcohol, birth control, family history, menstrual periods, drinking, and lifestyle. Hormonal changes can promote BC. The most common BC symptoms are skin irritation, nipple pain, nipple discharge other than breast milk, prolonged pain in one or both breasts, nipple retraction, change of breast skin, inflammation of the whole breast, redness, peeling, presence of lumps of any size, shape, and texture in one or both of breasts [41]. BC is a curable disease once diagnosed early. Since BC remains unknown, early diagnosis is the key to increasing the survival rate and reducing mortality. Therefore, it is of prime importance to have suitable imaging techniques to detect BC early. An accurate BC diagnosis needs an efficient imaging modality to acquire the image with high resolution. The primary focus of this chapter is to assess CAD systems for BC detection from sonogram and elastography images. It is essential to design a new sophisticated CAD system to reduce manual interpretation and enhance classification accuracy by providing super-resolution.

Early BC diagnosis with an appropriate screening method and super-resolution images could reduce the mortality rate significantly. Various medical imaging techniques such as mammography [19, 7, 49], magnetic resonance imaging (MRI) [48], ultrasound (US) imaging or sonogram [10], and elastography[13] are widely used to diagnose BC. A mammography is an efficient tool for BC detecting. However, mammography has some limitations, including less sensitivity to small tumors. Ionizing radiation can increase the health risk. Mammography is not suitable for identifying BC in adolescent women with dense breasts. A sonogram is an alternative tool to mammography because it is inexpensive, accessible, safer, faster, and tolerated by women.

Additionally, US does not utilize any ionizing radiation. Therefore, US is an alternative tool to mammography that can increase the sensitivity and BC detection rate. Sonogram has high sensitivity in the detection of tumors but has low specificity as solid lesions are benign. Some studies have shown that sonography is an effective tool for diagnosing BC at an earlier stage than mammography [15, 18, 28, 51]. Sample of benign and malignant tumors of US images appears as in Fig. 1. However, a US has some limitations: several tumor attributes must be estimated based on breast imaging, reporting, and data systems (B1-RADS) criteria to obtain an accurate result. In addition to this, the sonogram analysis heavily depends on the physician. Even an experienced person may have different effects or higher inter observation variation rate.

Stiffness-based imaging (elastography) alleviated the issues mentioned above and obtained more accurate results. US elastography integrates sonogram with the principle of elastography. Breast US elastography is a non-invasive technique that detects the tumors based on tissue stiffness. Figure 2 shows the sample elastography image. Different US elastography techniques, including shear wave elastography (SWE) and strain elastography (SE), have been developed for differentiating benign from malignant tumors. SE measures the stiffness of tissue when an external force is applied, whereas SWE evaluates the speed of shear waves propagated transversely in

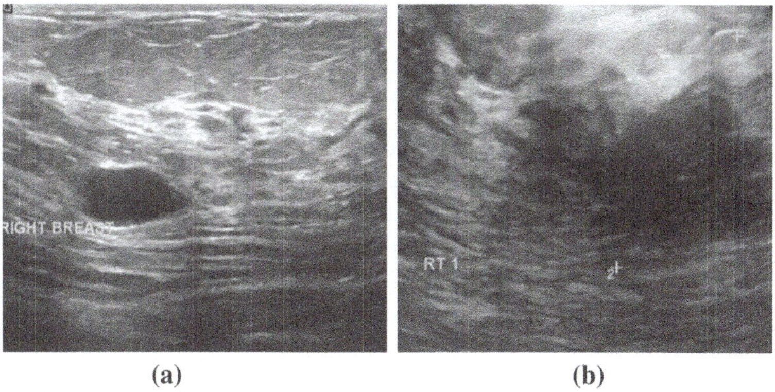

Fig. 1 Ultrasound images: (**a**) benign and (**b**) malignant

Fig. 2 Breast elastography

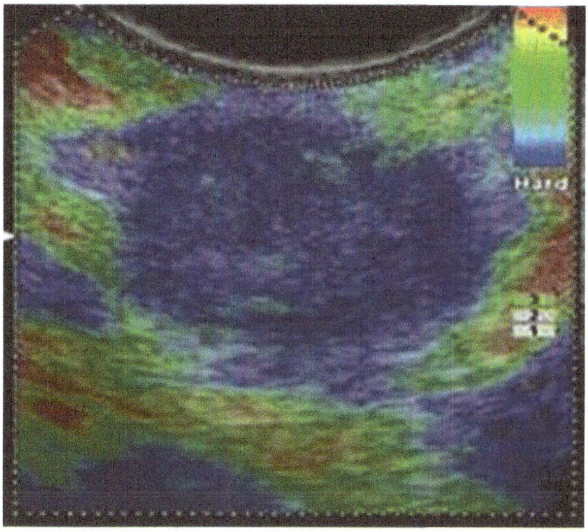

the tissue. US and elastography are efficient BC detection techniques, which have drawn more attention as an alternative tool to mammography. However, both the imaging modalities are relying on the operator than mammography. Analyzing US and elastography images requires an experienced physician. Therefore, a computer-aided diagnosis (CAD)-based classification system is necessary to aid the physician in the BC diagnosis. Standard methods like logistic or Cox regression are used. Recently, linear discriminant analysis (LDA), principal component analysis (PCA), support vector machine (SVM), and ANN have been successfully applied for BC detection. Amidst, ANN methods provide outstanding performance than statistical methods in the BC diagnosis.

In specific, many papers related to BC detection have been surveyed, categorized, and outcomes appear in Tables. The remainder of the manuscript is structured as follows. Section 2 presents the motivation for this text. Section 3 explains about CAD system for BC diagnosis. Section 4 reviews recent works on BC diagnosis using US and elastography images and condenses reviewed papers in Tables. Section 5 explains the idea and suggested method to have super-resolution in BC detection, followed by relevant references.

2 Motivation

Several CAD systems have appeared; no specific guidelines or exact rules for differentiating benign from malignant tumors have been established. Standard methods like logistic methods have been used for analyzing BC. More sophisticated algorithms based on machine learning methods have been adopted in recent years to obtain more accurate results. Among many machine-learning methods, artificial neural network (ANN) has been applied in BC detection due to its flexibility characteristics. It can capture the hidden information via the training process. However, there is limited research conducted to enhance classification accuracy. Specificity, sensitivity, and receiver operating characteristic (ROC) are the relevant indices to be considered for the BC diagnosis. The aim of this chapter is to explore the previous CAD systems for BC diagnosis along with their merits and demerits so that the researchers can choose the required model for their work and solve limitations of the earlier methods' limitations.

3 CAD System

Many CAD systems have been developed for US to aid the physician in image analysis and spot the suspicious parts that require additional examination and increase specificity and sensitivity. Further to this, the CAD system can help the physician find the exact location of lesions that cannot be found naked eye. Generally, a CAD system possesses five significant processes, as depicted in Fig. 3.

3.1 Preprocessing

The main disadvantage of sonogram images is low image quality and interference with noise. Preprocessing is an essential step in any image analysis. The goal of preprocessing is to improve the input image quality and reduce redundant data or noise without losing important information. Histogram equalization can enhance the brightness of the input image. Filters can reduce noise [10, 34]. Mean filter, median

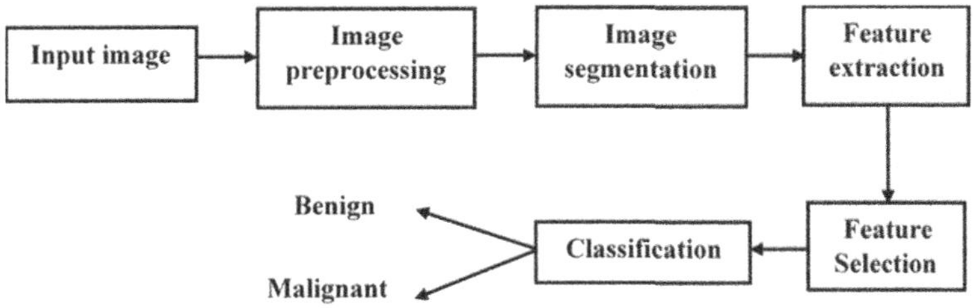

Fig. 3 Computer-aided diagnosis system

filter, wavelet methods, and compounding methods are widely utilized to reduce the noise. Filtering methods are easy to implement. These methods are sensitive to the size of the kernel. Wavelet methods convert image to wavelet domain and eliminate noise at different scales. However, these methods increase computational cost. Compounding methods need hardware support, which increases time complexity during image segmentation.

3.2 Segmentation

Segmentation is the process of separating tumor regions from the background tissue in the sonogram. It is an important and one of the difficult processes in BC detection. These are four major approaches in sonogram segmentation: (i) boundary-based methods (edge detection, thresholding), (ii) active contour model, (iii) Markov random field (MRF), and (iv) ANN [26, 29].

3.3 Feature Extraction

Feature extraction techniques are employed to detect breast tumors' most critical features that can differentiate benign and malignant. The four essential features of breast images are (i) texture features, (ii) model-based features, (iii) descriptor features, and (iv) morphological features. The most popular feature extraction methods are PCA [5, 30], LDA [24, 52–54], and independent component analysis (ICA) [25, 43, 45].

3.4 Feature Selection

Feature selection is a crucial step. For soft computing models, including ANN, the feature vector dimension affects the performance of the classifier. It also increases the training time. Therefore, it is necessary to extract useful features for CAD systems. Either filter method or wrapper method [14, 30, 50, 52] is employed for selecting valuable features.

3.5 Classification

After feature extraction and section, the obtained features are taken as feature vectors and used as an input to a classifier to differentiate the images into benign or malignant. Linear and nonlinear classifiers are commonly used. A linear classifier is not suitable for complex data. Nonlinear classifiers learn the relationship between input and output data through the training process. ANN [2, 19, 28, 36, 39], SVM [1, 54–58], and deep learning [2, 1, 21] have been widely used for BC detection and classification.

4 CAD System for Breast Cancer

4.1 Ultrasound Images

Research studies have shown that SVM is a useful tool for the classification task. In 2009, Akay [4] used SVM for categorizing breast lesions to benign and malignant. The F-score method helped to reduce the dimension feature vector. Simulations were performed on the Wisconsin Breast Cancer Dataset (WBCD), and the effectiveness of the designed system is measured using some measures including accuracy, specificity, sensitivity, positive predictive value (PPV) and negative predictive value (NPV), and confusion matrix. Results showed that SVM could be useful to the radiologists for making the final decision.

Computational intelligence (CI) techniques such as ANN, fuzzy logic, and bio-inspired algorithms, including particle swarm optimization (PSO), genetic algorithm (GA), and genetic programming (GP) have been commonly employed to solve many problems, pattern recognition, classification, and control [37]. Among many CI techniques, ANNs have been applied for the BC diagnosis in past decades. Though ANN is a promising tool for BC diagnosis, it has some shortcomings like local minima problem, slow convergence rate and long training time. Bio-inspired algorithms tune the parameters of ANN to solve these problems. He et al. [28] presented an algorithm for detecting and classifying breast tumors based on GSO and ANN. Results showed that the GSO algorithm is sufficiently good enough for solving BC problems.

Chiao et al. [17] developed an automated CAD system for differentiating benign from malignant tumors using a sonogram. The developed model combines mask regions with convolutional neural networks to obtain better classification results. Mask R-Convolutional Neural Network (CNN) has the merits of automatic image segmentation and drawing a tumor boundary before classification. The developed model is validated using US images. This system obtained an accuracy of 85%.

BC is a harmful cause of death in women worldwide, and treatment has become a severe problem for researchers and experts. Cedeno et al. [38] introduced an ANN-based classification algorithm named artificial metaplasticity multilayer perceptron (AMMLP). In this approach, a multilayer perceptron (MLP) is designed with an input neuron representing BC data attributes, a hidden layer with 8 hidden neurons, and an output layer with one neuron. The proposed algorithm was validated with the WBCD. Outcomes showed excellent performance, obtaining the following results: specificity 97.89%, sensitivity 100%, and 99.26% accuracy.

In 2019, Agarap [3] compared six methods: linear regression, MLP, nearest neighbor, softmax regression, SVM, and Gated Recurrent Unit-Support Vector Machine (GRU-SVM) to distinguish benign from malignant breast tumors utilizing the preprocessed WBCD to avoid errors. The result showed that all the machine learning models performed well on classifying breast tumor tasks. MLP achieved the highest classification accuracy of 99.03%. MLP outperforms earlier methods in terms of classification accuracy.

Kim et al. [32] evaluated the CAD system's performance in BC diagnosis using sonogram images. Sobel operators are used to detect and extract boundaries. Otu's thresholding converted a grayscale image to a binary image. Several morphological operations such as dilation, filling, and region growing achieved accurate segmentation. After the segmentation process, statistical features are obtained and applied as input to the classifier. SVM classifier is employed to classify data. Accuracy and sensitivity achieved by this CAD system were 80% and 82.87%, respectively.

Though ANN is a promising tool for categorizing breast lesions, it has some shortcomings, such as local minima, slow convergence rate, and prolonged training time. Deep learning (DL)-based CAD systems have been proposed by Choi et al. [16] to prevent the ANN drawbacks. The proposed CAD system is validated on a dataset that includes 173 non-cancer and 90 cancer breast lesions. The sensitivity of 85%, the specificity of 95.4%, and the accuracy of 92.1% resulted from this method. Results showed that adopting a deep learning-based CAD system to US image analysis significantly improves classification accuracy and specificity.

Zaher and Eldeib [2] presented a CAD system to diagnose BC early. In this system, MLP has trained with the Levenberg Marquart (LM) algorithm, and weights are optimized by using DBN. The performance of the proposed CAD system is analyzed by a varying number of training and testing samples. Results demonstrated that the developed system's efficacy is superior to earlier models in terms of sensitivity, specificity, and classification accuracy.

Paulin and Santhakumaran [37, 44] developed a BC classification method grounded on feed forward neural network (FFNN). The developed method was tested on the WBCD. A set of features is obtained and used as input to the FFNN. The network was designed with nine input neurons representing a number of input feature vectors, a hidden layer with six hidden neurons. An output layer with one output neuron where 0 represents benign, and 1 represents malignant. FFNN is trained with different training algorithms, including batch gradient descent (BGD), batch gradient descent with momentum (BGDM), quasi-Newton (QN), Resilient Back Propagation (RBP), conjugate gradient (CG), and LM, to find suitable for BC diagnosis. Among six training algorithms, LM provided the best result with a classification accuracy of 99.28%.

Mart et al. [40] investigated the effect of feature reduction technique, ICA, on BC diagnosis. The authors examined 569 samples from the WBCD, including 357 benign and 212 malignant lesions. Attributes have been calculated from a digitized image of the fine needle aspirate (FNA) of breast lesions. Ten real features were obtained from each image, resulting in 30 features. The dimension of the feature vector may affect the efficacy of ANN. To address such an issue, they adopted ICA to reduce the dimension of the feature vector. Reduced feature vector (1 independent component) fed as input to the classifier. Four classifiers, ANN, K-nearest neighbor (K-NN), radial basis function neural network (RBFNN), and SVM, are used to categorize benign and malignant tumors. All the classifiers' effectiveness is evaluated by measuring specificity, sensitivity, accuracy, F-score, Youden's index, discriminant power, and receiver operating characteristic (ROC). Results showed that the classifiers with 30 features provide the worst performance than the reduced features. It also demonstrated that the RBFNN classifier outperforms the other classifier when one-dimensional feature vectors are used as an input.

Nahato et al. [42] developed a classification system based on ANN. The rough set (RS) indiscernibility relation method with a backpropagation neural network (BPNN) inspired the classifier. Initially, the data set is smoothed and an indiscernibility method is employed for extracting attributes from the data. Finally, BPNN assisted in classifying the data. BPNN comprised an input layer with nine units corresponding to input vectors, a hidden layer with 25 hidden units, and an output layer with one unit representing the output class. Tangent sigmoidal is applied at the hidden layer, and the linear activation function is applied at the output layer. The proposed classifier is verified on WBCD. The sensitivity, specificity, and accuracy obtained from the proposed classifier are 98.76%, 98.57%, and 98.60%, respectively.

4.2 *Elastography Images*

Numerous investigations in BC diagnosis and developing the most suitable and efficient methods for increasing classification accuracy have been conducted. Elastography is a non-invasive technique in which stiffness images help determine abnormal parts based on the elasticity patterns[7, 18, 21, 9, 25, 27, 60]. Recently, elastography is considered an adjunct diagnostic tool for the US. Liu et al. [35] classified breast lesions into benign or malignant based on patients' elastography images' characteristics. All patients were examined with both US and elastography. A threshold of 4.15 was chosen for differentiating benign from malignant breast tumors. Empirical findings demonstrated that US, combined with elastography, provides high accuracy for BC diagnosis.

Evans et al. [23] proposed a technique to discriminate breast lesions. SWE is a technique of acquiring measurable tissue elasticity during US examination. Elastography images are acquired of each lesion, and then the mean elasticity value was computed. After many experiments, an average elasticity cutoff value of 50 Kpa was chosen for BC classification. Empirical findings demonstrated that SWE's efficacy is superior to grayscale BI-RADS in terms of accuracy, PPV, NPV.

Klotz et al. [33] examined the strength of SWE in detecting benign and malignant breast tumors and characterized the factors influencing elasticity values. They examined 167 (65 – benign, 102 – malignant) breast tumors. Breast lesions are classified into cancer and non-cancer lesions based on maximum elasticity, mean elasticity, and ratio. Results confirmed that the integration of US and SWE could enhance the classification accuracy in BC diagnosis.

Botticelli et al. [8] compared the diagnostic characteristic of three screening methods: US, color Doppler (CD), and elastography. A total of 212 patients' data were used in the analysis. US and CD images were classified according to the Breast Imaging, Reporting, and Data systems (BIRADS) found by ACS. ES images were classified based on elasticity score. CD images were categorized as positive (malignant) or negative. Results showed that US with ES could provide more information on the breast lesion, particularly in small size. ES achieved the highest specificity of 99% compared to other models: US and US-ES.

Chang et al. [11] compared US and SWE's efficacy in discriminating benign from malignant. The authors used average elasticity (E_{mean}) for SWE analysis and BI-RADS criteria for US image diagnosis. SWE analysis proved that (E_{mean}) of a malignant tumor is higher than (E_{mean}) of benign tumors. Experimental results revealed that SWE provides more information for BC diagnosis when compared to US images.

Evans et al. [22] assessed the effectiveness of SWE integrated with BIRADS categorization of US images for classifying breast lesions into benign and malignant lesions. E_{mean} was calculated from SWE. For E_{mean} versus BI-RADS classification performance: Sensitivity: 95% vs 95% and specificity: 77% vs 69%. The accuracy of 86%, the specificity of 61%, and the sensitivity of 100% were obtained by combining US with SWE.

Chang et al. [12] performed a comparative study between the SWE and SE approaches for discriminating benign and malignant breast lesions. In their study, US and SWE were performed in 150 breast lesions. For the analysis, E_{mean} on SWE and elasticity score on SE of all breast lesions were calculated. Diagnostic performance was compared using the ROC curve. The area under the curve (AUC) for the SWE was similar to that of the SE. The combined use of US and ES or SWE enhances diagnostic accuracy compared to US alone. A threshold of 80 kPa was selected for SWE and a score between 3 and 4 for ES. Results proved that the efficacy of SE and SWE was the same. SE or SWE can enhance BC detection accuracy when combined with US.

Youk et al. [59] assessed and compared 3D SWE's effectiveness with 2D SWE in BC classification. Diagnostic ability of both SWE and SE was gauged by means of the ROC curve. 2D SWE showed high performance than 3D SWE. The AUC of 2D

SWE was higher when compared to 3D SWE in almost all features. The analysis illustrated that the classification accuracy of 3D SWE is lower compared to 2D SWE ever after integrating with US.

Supersonic SWE has been emerged as a new technique to gauge the elasticity of tissue and become a good candidate for BC diagnosis. Xiao et al. [57] presented a CAD system for discriminating between benign and malignant BC based on supersonic SWE images. For each lesion, a set of ten features were calculated and analyzed. SVM classifier carried out the classification task. The accuracy, sensitivity, and specificity of BC classification were 95.2%, 90.9%, and 97.5% for CAD and 79.2%, 90.9%, and 72.8% for BIRADS assessment. Results indicated that the designed CAD method could enhance BC classification accuracy.

SWE provides both quantitative and qualitative information on the elastic properties of tissue in the real-time assessment. Sobczak et al. [20] evaluated the relevance of SWE in the BC diagnosis for the US. Mean elasticity of lesions (E_{avl}), and tissue (E_{avf}), maximum ($E_{max.adj}$), and minimum ($E_{av.adj}$) elasticity of lesions and adjacent tissues were calculated. Based on the calculated values, breast lesions are categorized as benign and malignant lesions. Cutoff values for $E_{av.adj}$, $E_{max.adj}$, and E_{avl} are computed, and images are classified based on the obtained cutoff values. Results proved that $E_{av.adj}$ displayed lower specificity and sensitivity than the US. Moreover, $E_{max.adj}$ enhanced the specificity of breasts for US with a loss of sensitivity.

Zhang et al. [59] studied the elastic property of breast lesions in SWE and gauged SWE's strength in differentiating benign from malignant lesions. Five texture features such as mean (T_{mean}), maximum (T_{max}), median (T_{med}), third quartile (T_{qt}), and standard (T_{sd}) were computed from the directional sub-bands of the SWE image. The classification performance of texture features is compared with conventional features. T_{mean}, T_{max}, T_{med}, T_{qt}, and T_{sd} yielded an accuracy of 92.5%, 82%, 88.2%, 90.7%, and 90.7%, respectively, while conventional features obtain an accuracy of 86.3%. Among the contourlet-based texture features, T_{mean} reached the highest accuracy of 92.5% with a sensitivity of 89.1% and specificity of 94.3%. Numerical results showed that the texture features enhanced the diagnostic performance contrasted with the conventional feature set.

Youk et al. [58] provided a brief review of SWE and its BC diagnosis applications. The authors discussed the merits and demerits of SWE. Moreover, not only SWE has been proven as a powerful technique, but it has also been demonstrated to obtain more important details that can be utilized for further investigations.

Kim et al. [31] compared the diagnostic performance of the SE, the SWE, and US approaches in the classification of breast lesions. Cutoff points for SE and SWE were computed and evaluated. SWE images are classified into benign or malignant based on E_{ratio}, whereas the SE image relies on elasticity score. Diagnostic performance of SE and SWE was similar when discriminating benign from malignant tumors. Results revealed that the diagnostic performance improved by combining either SE or SWE with the US.

4.3 Summary of Reviewed Papers

In this work, the summary of several CAD systems of BC on US and elastography is done on the following aspects:

- Number of features utilized
- Type of classifier used to classify the breast lesions
- Parameters that are employed to gauge the effectiveness of CAD system

Over the past years, numerous CAD systems have been proposed for BC detection via US and elastography to support physicians in image analysis and highlight affected areas. The CAD system can assist the physician or radiologist in diagnosing BC accurately. With the advent of new technology, many researchers and scientists have attempted to develop an efficient technique that can be utilized in US and elastography to enhance sensitivity and accuracy[45, 47, 24, 48, 46]. The summary of the reviewed papers emerges in Table 1. Based on the reviewed articles, it is noticed that the CAD system with ANN is shown to be a better classifier in US and elastography for classifying breast tumors into benign and malignant tumors. Numerous ANN methods have been proposed to improve the classification accuracy. Most of the research works that use ANN have provided better results. Fusion methods, which integrate two or more ways, have been developed recently for the BC diagnosis. Fusion methods are found to be a promising approach rather than using a single process. It is also noticed that the higher number of input features will affect the performance of ANN. Many ANN methods for US and elastography applications provide better results in terms of sensitivity and accuracy. The main advantage of using ANN in the BC detection is training it using different training algorithms to obtain a better outcome. However, ANN has some limitations like local minima, slow convergence, overfitting, and extended training time. To solve these problems, some authors have used hybrid methods and deep learning neural networks. The trend now is emerging toward fusion methods like ANN with bio-inspired algorithms. The integration of bio-inspired algorithms such as PSO is being used together with ANN too. The fusion method needs lesser time and hence is useful in BC diagnosis. Besides, researchers use deep learning neural networks to obtain high classification accuracy. Deep learning neural networks do not need any image preprocessing most of the time (although at the expense of a much higher computational load) and can recognize patterns with extreme variability and some geometric transformations such as scaling, rotation, translation, and noise.

The efficacy of BC diagnosis methods is evaluated based on specific performance indices. Figure 4 illustrates the statistical indicators that assess any CAD system's ability to classify breast lesions as benign and malignant lesions. This chapter surveyed a recent investigation in the BC diagnosis and noticed the evaluation parameters utilized to gauge the BC detection system, listed in Table 2. Any CAD system tries to enhance the following measures: classification accuracy, sensitivity, and specificity. Some authors try to reduce computational cost. Mert et al. [40], differently from others, used the F-score and Youden's index.

Table 1 A summary of recent works

Researcher	Year	Image	Input features	Data set details			Data division		Classifier
				Lesion	Benign	Malignant	Train (%)	Test (%)	
Akay et al. [4]	2009	US	F score	683	444	239	80	20	SVM
He et al. [28]	2009	US	9	699	458	241	75	25	GSO-ANN
Evans et al. [23]	2010	SWE	E_{mean}	53	23	30	–	–	Statistical analysis
Cedeno et al. [22]	2011	US	9	683	444	239	60	40	ANN
Paulin and Santhakumaran [44]	2011	US	9	683	444	239	80	20	FFNN
Chang et al. [11]	2011	SWE	E_{mean}	182	93	89	–	–	–
Evans et al. [22]	2012	US and SWE	E_{mean}	175	–	–	–	–	–
Chang et al. [12]	2013	SE and SWE	E_{mean} *a* E_{score}	150	79	71	–	–	–
Youk et al. [59]	2013	2D and 3D SWE	E_{max}	163	115	48	–	–	
Liu et al. [35]	2014	US with elastography	Stiffness	431	155	276	–	–	Statistical analysis
Klotz et al. [33]	2014	US + SWE	E_{max}, E_{min}, E_{ratio}	167	65	102	–	–	
Kim et al. [32]	2014	US	18	89	42	27	80	20	SVM
Xiao et al. [57]	2014	Supersonic SWE	–	125	81	44	–	–	–
Botticelli et al. [8]	2015	US and Elastography	E_{score}	395	197	198	–	–	–
Mert et al. [40]	2015	US	Independent component	569	357	212	80	20	FFNN, RBFNN, SVM
Nahato et al. [42]	2015	US	9	699	458	241	80	20	RS-BPNN
Sobczak et al. [20]	2015	SWE and US	E_{max}, E_{min}, $E_{max.adj}$	84	43	41	–	–	
Youk et al. [58]	2015	SWE	E_{max}	161	106	55	–	–	
Zaher & Eldeib [2]	2016	US	Image	683	444	239	–	–	DBN
Kim et al. [31]	2018	SE, SWE	E_{ratio} and score	108	64	44	–	–	

Chiao et al [15]	2019	US	Image	307	178	129	80	20	R-CNN
Agarap [3]	2019	US	10	569	357	212	70	30	MLP, GRU-SVM
Choi et al. [16]	2019	US	Image	253	173	90	–	–	Deep learning

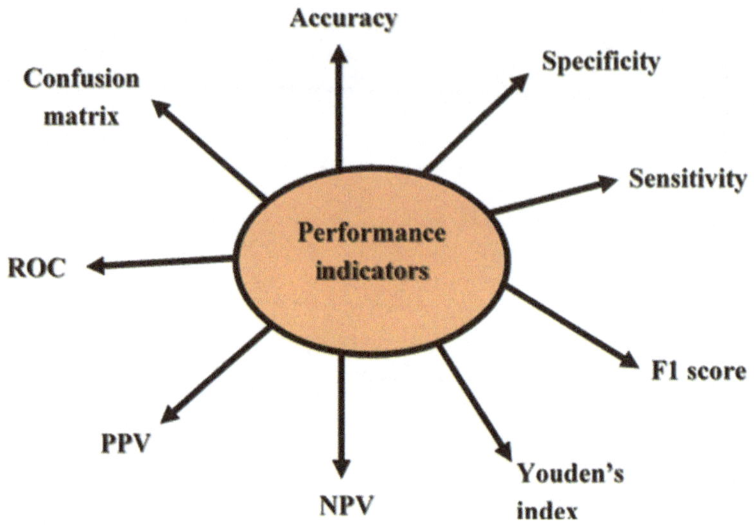

Fig. 4 Performance indicators

5 Conclusion and the Future with Super-Resolution for Breast Cancer Detection

In this investigation, a detailed review of different approaches used to detect BC early has been presented. This chapter focused on input image type, classification methods, and statistical metrics employed for assessing performance. It is noticed that each researcher has utilized various performance measures for their developed methods. From the review, it is observed that there are still many issues that need to be addressed to improve classification accuracy. Many research problems related to BC detection are given below. It has been inferred that there is no standard method to solve all the difficulties involved in BC detection and classification. Researchers have employed simulation tools to validate and measure the efficiency of their developed practices. These methods must be validated in real-time to show an accurate assessment. Further to this, most of the researchers compared their outcome with standard procedures and did not correspond with advanced techniques, which are far better compared to standard methods. Therefore, it is essential to design a new sophisticated CAD system to reduce manual interpretation and enhance classification accuracy.

Super-resolution in BC imaging is necessary to find even the minor abnormalities with the given dose of inputs, thus improving the decision support systems. Consider combining the images taken with two different modalities and fusing them to improve the decision's quality. Mixing certain techniques is useful for enriching resolution. Instead of a single feature selection method, considers texture and morphological features from breast US image. Texture features result from the gray-level co-occurrence matrix. Pulse-coupled neural network tuned by differential

Table 2 Evaluation parameters

Researcher	Year	Performance indices		
		Sensitivity (%)	Specificity (%)	Accuracy (%)
Akay et al. [4]	2009	100	97.91	99.5
He et al. [28]	2009	97	83	91
Evans et al. [23]	2010	97	83	91
Cedeno et al. [22]	2011	100	97.89	99.26
Paulin and Santhakumaran [44]	2011	99.28	–	–
Chang et al. [11]	2011	88.8	84.9	86.8
Evans et al. [22]	2012	100	61	86
Chang et al. [12]	2013	95.8	84.8	–
Youk et al. [59]	2013	93	90.4	96.6
Liu et al. [35]	2014	92.60	86.40	90
Klotz et al. [33]	2014	93.1	87.7	93
Kim et al. [32]	2014	82.87	–	80
Xiao et al. [57]	2014	90.9	97.5	95.2
Botticelli et al. [8]	2015	80	99	–
Mert et al. [40]	2015	100	93.39	97.53
Nahato et al. [42]	2015	98.76	98.57	98.6
Sobczak et al. [20]	2015	86.05	87.8	86.1
Youk et al. [58]	2015	89.1	94.3	92.5
Zaher and Eldeib [2]	2016	100	99.47	99.68
Kim et al. [31]	2018	93.2	70.3	–
Chiao et al. [15]	2019	–	–	85
Agarap [3]	2019	–	–	99.03(MLP) 93.75(GRU-SVM)
Choi et al. [16]	2019	85	95.4	92.1

evolution can be implemented to isolate the lesion area with better edge coverage. SVM can be desinged and trained to learn benign and malignant lesion feature vectors. The contrast-limited adaptive histogram equalization (CLAHE) employs adaptive histogram equalization (AHE). CLAHE can do preprocessing rather than a standard histogram method to improve clarity. Differential evolution-tuned pulse-coupled neural network (DE-tuned PCNN) can support the decision on the segmentation process, and differential evolution is engaged for optimizing the PCNN parameters.

References

1. Abdelwahed, N. M., & Eltoukhy, W. M. (2015). Computer aided system for breast cancer diagnosis in ultrasound images. *Journal of Ecology of Health & Environment, 3*(37), 71–76.
2. Abdel-Zaher, A. M., & Eldeib, A. M. (2016). Breast cancer classification using deep belief networks. *Expert Systems with Applications, 46*(38), 139–144.

3. Agarap, A. F. M. (2018). On breast cancer detection: An application of machine learning algorithms on the Wisconsin diagnostic dataset. In *Proceedings of the 2nd International Conference on Machine Learning and Soft Computing* (Vol. 24, pp. 5–9).
4. Alexe, G., Dalgin, G. S., Ganesan, S., Delisi, C., & Bhanot, G. (2007). Analysis of breast cancer progression using principal component analysis and clustering. *Journal of Biosciences, 32*(1), 1027–1039.
5. American Cancer Society. (2019). *Breast Cancer Facts & Figures 2019–2020*. Atlanta: American Cancer Society, Inc. https://www.cancer.org/content/dam/cancer-org/research/cancer-facts-and-statistics/breast-cancer-facts-and-figures/breast-cancer-facts-and-figures-2019-2020.pdf.
6. American Cancer Society. (2019). *Cancer Facts & Figures (2019)*. Atlanta: American Cancer Society. https://www.cancer.org/content/dam/cancer-org/research/cancer-facts-and-statistics/annual-cancer-facts-and-figures/2019/cancer-facts-and-figures-2019.pdf.
7. Barr, R. G. (2019). Future of breast elastography. *Ultrasonography, 38*(2), 93. (16).
8. Botticelli, A., Mazzotti, E., Di Stefano, D., Petrocelli, V., Mazzuca, F., La Torre, M., & Bonifacino, A. (2015). Positive impact of elastography in breast cancer diagnosis: An institutional experience. *Journal of Ultrasound, 18*(4), 321–327. (21).
9. Bowles, D., & Quinton, A. (2016). The use of ultrasound in breast cancer screening of asymptomatic women with dense breast tissue: A narrative review. *Journal of Medical Imaging and Radiation Sciences, 47*(3), S21–S28). (12).
10. Cal6ope, P. B., Medeiros, F. N., Marques, R. C., & Costa, R. C. (2004). A comparison of filters for ultrasound images. In *International Conference on Telecommunications* (pp. 1035–1040). Berlin, Heidelberg: Springer. (28).
11. Chang, J. M., Moon, W. K., Cho, N., Yi, A., Koo, H. R., Han, W., et al. (2011). Clinical application of shear wave elastography (SWE) in the diagnosis of benign and malignant breast diseases. *Breast Cancer Research and Treatment, 129*(1), 89–97. (42).
12. Chang, J. M., Won, J. K., Lee, K. B., Park, I. A., Yi, A., & Moon, W. K. (2013). Comparison of shear-wave and strain ultrasound elastography in the differentiation of benign and malignant breast lesions. *American Journal of Roentgenology, 201*(2), W347–W356.
13. Chen, Y. L., Gao, Y., Chang, C., Wang, F., Zeng, W., & Chen, J. J. (2018). Ultrasound shear wave elastography of breast lesions: Correlation of anisotropy with clinical and histopathological findings. *Cancer Imaging, 18*(1), 11.
14. Cheng, H. D., Shan, J., Ju, W., Guo, Y., & Zhang, L. (2010). Automated breast cancer detection and classification using ultrasound images: A survey. *Pattern Recognition, 43*(1), 299–317.
15. Chiao, J. Y., Chen, K. Y., Liao, K. Y. K., Hsieh, P. H., Zhang, G., & Huang, T. C. (2019). Detection and classification the breast tumors using mask R-CNN on sonograms. *Medicine, 98* (19), e15200.
16. Choi, J. S., Han, B. K., Ko, E. S., Bae, J. M., Ko, E. Y., Song, S. H., et al. (2019). Effect of a deep learning framework-based computer-aided diagnosis system on the diagnostic performance of radiologists in differentiating between malignant and benign masses on breast ultrasonography. *Korean Journal of Radiology, 20*(5), 749–758.
17. Christensen-Jeffries, K., Brown, J., Harput, S., Zhang, G., Zhu, J., Tang, M., Dunsby, C., & Eckersley, R. E. (2019). Poisson statistical model of ultrasound super-resolution imaging acquisition time. *IEEE Transactions on Ultrasonics, Ferroelectrics, and Frequency Control, 66*, 1246–1254.
18. Christensen-Jeffries, K., Harput, S., Brown, J., Wells, P. N., Aljabar, P., Dunsby, C., et al. (2017). Microbubble axial localization errors in ultrasound super-resolution imaging. *IEEE Transactions on Ultrasonics, Ferroelectrics, and Frequency Control, 64*(11), 1644–1654.
19. Dheeba, J., Singh, N. A., & Selvi, S. T. (2014). Computer-aided detection of breast cancer on mammograms: A swarm intelligence optimized wavelet neural network approach. *Journal of Biomedical Informatics, 49*, 45–52.
20. Dobruch-Sobczak, K., & Nowicki, A. (2015). Role of shear wave sonoelastography in differentiation between focal breast lesions. *Ultrasound in Medicine & Biology, 41*(2), 366–374.

21. Kanoulas, E., Butler, M., Rowley, C., Voulgaridou, V., Diamantis, K., Duncan, W. C., Mcneilly, A. S., Averkiou, M., Wijkstra, H., Mischi, M., Wilson, R. S., Lu, W., & Sboros, V. (2019). Super-resolution contrast-enhanced ultrasound methodology for the identification of in vivo vascular dynamics in 2D. *Investigative Radiology, 54*, 500.
22. Evans, A., Whelehan, P., Thomson, K., Brauer, K., Jordan, L., Purdie, C., et al. (2012). Differentiating benign from malignant solid breast masses: Value of shear wave elastography according to lesion stiffness combined with greyscale ultrasound according to BI-RADS classification. *British Journal of Cancer, 107*(2), 224–229.
23. Evans, A., Whelehan, P., Thomson, K., McLean, D., Brauer, K., Purdie, C., et al. (2010). Quantitative shear wave ultrasound elastography: Initial experience in solid breast masses. *Breast Cancer Research, 12*(6), R104.
24. Gallardo-Caballero, R., García-Orellana, C. J., García-Manso, A., González-Velasco, H. M., & Macías-Macías, M. (2012). Independent component analysis to detect clustered microcalcification breast cancers. *The Scientific World Journal, 2012*, 1.
25. Goddi, A., Bonardi, M., & Alessi, S. (2012). Breast elastography: a literature review. *Journal of Ultrasound, 15*(3), 192–198. (13).
26. Gonzalez, R. C., & RE, W. (2002). *Digital Image Processing, 2*, 550–570.
27. Harput, S., Tortoli, P., Eckersley, R. J., Dunsby, C., Tang, M., Christensen-Jeffries, K., Ramalli, A., Brown, J., Zhu, J., Zhang, G., Leow, C. H., Toulemonde, M., & Boni, E. (2019). 3-D super-resolution ultrasound imaging with a 2-D sparse array. *IEEE Transactions on Ultrasonics, Ferroelectrics, and Frequency Control, 67*, 269–277.
28. He, S., Wu, Q. H., & Saunders, R. J. (2009). Breast cancer diagnosis using an artificial neural network trained by global search optimizer. *Transactions of the Institute of Measurement and Control*, 1–15.
29. Horsch, K., Giger, M. L., Venta, L. A., & Vyborny, C. J. (2001). Automatic segmentation of breast lesions on ultrasound. *Medical Physics, 28*(8), 1652–1659.
30. Jensen, J. A., Ommen, M. L., Øygard, S. H., Schou, M., Sams, T., Stuart, M. B., et al. (2019). Three-dimensional super-resolution imaging using a row–column array. *IEEE Transactions on Ultrasonics, Ferroelectrics, and Frequency Control, 67*(3), 538–546.
31. Kim, H. J., Kim, S. M., Kim, B., La Yun, B., Jang, M., Ko, Y., & Cho, N. (2018). Comparison of strain and shear wave elastography for qualitative and quantitative assessment of breast masses in the same population. *Scientific Reports, 8*(1), 1–11.
32. Kim, J. H., Cha, J. H., Kim, N., Chang, Y., Ko, M. S., Choi, Y. W., & Kim, H. H. (2014). Computer-aided detection system for masses in automated whole breast ultrasonography: Development and evaluation of the effectiveness. *Ultrasonography, 33*(2), 105.
33. Klotz, T., Boussion, V., Kwiatkowski, F., Fraissinette, V. D., et al. (2014). Shear wave elastography contribution in ultrasound diagnosis management of breast lesions. *Diagnostic and Interventional Imaging, 95*, 813–824.
34. Liu, B., Cheng, H. D., Huang, J., Tian, J., Liu, J., & Tang, X. (2009). Automated segmentation of ultrasonic breast lesions using statistical texture classification and active contour based on probability distance. *Ultrasound in Medicine & Biology, 35*(8), 1309–1324.
35. Liu, X. J., Zhu, Y., Liu, P. F., & Xu, Y. L. (2014). Elastography for breast cancer diagnosis: A useful tool for small and BI-RADS 4 lesions. *Asian Pacific Journal of Cancer Prevention, 15* (24), 10739–10743.
36. Luke, G. P., Hannah, A. S., & Emelianov, S. Y. (2016). Super-resolution ultrasound imaging in vivo with transient laser-activated nanodroplets. *Nano Letters, 16*(4), 2556–2559.
37. Madjar, H. (2010). Role of breast ultrasound for the detection and differentiation of breast lesions. *Breast Care, 5*(2), 109–114.
38. Marcano-Cedeño, A., Quintanilla-Domínguez, J., & Andina, D. (2011). WBCD breast cancer database classification applying artificial metaplasticity neural network. *Expert Systems with Applications, 38*(8), 9573–9579.

39. Mehdy, M. M., Ng, P. Y., Shair, E. F., Saleh, N. I., & Gomes, C. (2017). Artificial neural networks in image processing for early detection of breast cancer. *Computational and Mathematical Methods in Medicine., 2017*, 1.
40. Mert, A., Kılıç, N., Bilgili, E., & Akan, A. (2015). Breast cancer detection with reduced feature set. *Computational and Mathematical Methods in Medicine, 2015*, 1.
41. Mitsuk, A. (2016). *Breast cancer information for young women*, Ph.D Thesis, a project for Terveysnetti.
42. Nahato, K. B., Harichandran, K. N., & Arputharaj, K. (2015). Knowledge mining from clinical datasets using rough sets and backpropagation neural network. *Computational and Mathematical Methods in Medicine, 2015*, 1.
43. Pan, H. B. (2016). The role of breast ultrasound in early cancer detection. *Journal of Medical Ultrasound, 24*(4), 138–141.
44. Paulin, F., & Santhakumaran, A. (2011). Classification of breast cancer by comparing back propagation training algorithms. *International Journal on Computer Science and Engineering, 3*(1), 327–332.
45. Rajaguru, H., & Prabhakar, S. K. (2017, October). Bayesian linear discriminant analysis for breast cancer classification. In *2017 2nd International Conference on Communication and Electronics Systems (ICCES)* (pp. 266–269). IEEE.
46. Ramya, S., & Nanda, S. (2017). Breast cancer detection and classification using ultrasound and ultrasound Elastography images. *IRJET, 4*, 596–601.
47. Rasmussen, E. B., Lawyer, S. R., & Reilly, W. (2010). Percent body fat is related to delay and probability discounting for food in humans. *Behavioural Processes, 83*(1), 23–30.
48. Roganovic, D., Djilas, D., Vujnovic, S., Pavic, D., & Stojanov, D. (2015). Breast MRI, digital mammography and breast tomosynthesis: Comparison of three methods for early detection of breast cancer. *Bosnian Journal of Basic Medical Sciences, 15*(4), 64.
49. Rouhi, R., Jafari, M., Kasaei, S., & Keshavarzian, P. (2015). Benign and malignant breast tumors classification based on region growing and CNN segmentation. *Expert Systems with Applications, 42*(3), 990–1002.
50. Sahiner, B., Chan, H. P., Roubidoux, M. A., Hadjiiski, L. M., Helvie, M. A., Paramagul, C., & Blane, C. (2007). Malignant and benign breast masses on 3D US volumetric images: Effect of computer-aided diagnosis on radiologist accuracy. *Radiology, 242*(3), 716–724.
51. Sloun, R. V., Solomon, O., Bruce, M., Khaing, Z. Z., Eldar, Y. C., & Mischi, M. M. (2019). Deep learning for super-resolution vascular ultra sound imaging. In *ICASSP 2019–2019 IEEE International Conference on Acoustics, Speech and Signal Processing (ICASSP)* (pp. 1055–1059).
52. Uncu, Ö., & Türkşen, I. B. (2007). A novel feature selection approach: Combining feature wrappers and filters. *Information Sciences, 177*(2), 449–466.
53. Van Sloun, R. J., Solomon, O., Bruce, M., Khaing, Z. Z., Wijkstra, H., Eldar, Y. C., & Mischi, M. (2018). Super-resolution ultrasound localization microscopy through deep learning. *arXiv preprint arXiv, 1804*, 07661.
54. Veloz, A., Orellana, A., Vielma, J., Salas, R., & Chabert, S. (2011). Brain tumors: How can images and segmentation techniques help? *Diagnostic Techniques and Surgical Management of Brain Tumors, 67.*
55. Viessmann, O. M., Eckersley, R. J., Christensen-Jeffries, K., Tang, M. X., & Dunsby, C. (2013). Acoustic super-resolution with ultrasound and microbubbles. *Physics in Medicine & Biology, 58*(18), 6447.
56. Weigert, J., & Steenbergen, S. (2012). The Connecticut experiment: The role of ultrasound in the screening of women with dense breasts. *The Breast Journal, 18*(6), 517–522.
57. Xiao, Y., Zeng, J., Niu, L., Zeng, Q., Wu, T., Wang, C., et al. (2014). Computer-aided diagnosis based on quantitative elastographic features with supersonic shear wave imaging. *Ultrasound in Medicine & Biology, 40*(2), 275–286.
58. Youk, J. H., Gweon, H. M., & Son, E. J. (2017). Shear-wave elastography in breast ultrasonography: The state of the art. *Ultrasonography, 36*(4), 300.

59. Youk, J. H., Gweon, H. M., Son, E. J., Chung, J., Kim, J. A., & Kim, E. K. (2013). Three-dimensional shear-wave elastography for differentiating benign and malignant breast lesions: Comparison with two- dimensional shear- wave elastography. *European Radiology, 23*(6), 1519–1527.
60. Zahran, M. H., El-Shafei, M. M., Emara, D. M., & Eshiba, S. M. (2018). Ultrasound elastography: How can it help in differentiating breast lesions? *The Egyptian Journal of Radiology and Nuclear Medicine, 49*(1), 249–258.

Part II
State-of-the-Art Computational Intelligence in Super-Resolution Imaging

Pictorial Image Synthesis from Text and Its Super-Resolution Using Generative Adversarial Networks

Khushboo Patel and Parth Shah

1 Introduction

The problem of text-to-image synthesis (TIS) can be defined as the process of deciphering thc structure of sentences to associate them with picturing pixels. Photo editing, drawing bot, animated films based on a screenplay, computer games, and teaching assistant can be an application of the said problem. Here, we are interested in translating the given text into appealing image pixels. To cite an example, "this red bird has pointed beak, and black belly," or "this flower has long yellow petals with red dots." or "this is a flower with small white petals, and yellow pistils and anther filaments." can be represented in the form of picture as shown in Fig. 1.

Word embedding maps characterize the words as vectors in a multidimensional space. Text embedding (TE) improves that concept by mapping phrases, paragraphs, and manuscripts to vectors, a fundamental issue in natural language processing (NLP). This quantification process allows one to transform a set of sentences into a numerical and measurable vector representation (called an embedding). Such a depiction encompasses the writing's semantic content. If a pair of words or documents contain a similar TE, they are semantically analogous. For example, the TEs associated with "wings" and "bird" are close, while the ones for the words "wings"

K. Patel (✉)
U and P U. Patel Department of Computer Engineering, Chandubhai S. Patel Institute of Technology (CSPIT), Faculty of Technology and Engineering (FTE), Charotar University of Science & Technology (CHARUSAT), Changa, Gujarat, India
e-mail: khushboopatel.ce@charusat.ac.in

P. Shah
Smt. Kundanben Dinsha Patel Department of Information Technology, Chandubhai S. Patel Institute of Technology (CSPIT), Faculty of Technology and Engineering (FTE), Charotar University of Science & Technology (CHARUSAT), Changa, Gujarat, India
e-mail: parthshah.ce@charusat.ac.in

A. Deshpande et al. (eds.), *Computational Intelligence Methods for Super-Resolution in Image Processing Applications*, https://doi.org/10.1007/978-3-030-67921-7_5

Fig. 1 The problem of performing text-to-image synthesis

and "dog" are not. Likewise, one concept expressed in different languages, for example "finestra" and "window," corresponds to close embedding.

A frequent NLP challenge is the fact the same meaning can give rise to many written expressions. TE techniques aid in text analysis tasks such as assessing the semantic similarity among one or more texts written in different languages. The knowledge about the TEs of two manuscripts allows one to appraise their similarity vis-à-vis meaning or content. Hence, TE ameliorates name matching, content examinations, sorting, information filtering, and so forth.

NLP's conversion into a plausible image means that these sentences are first converted to TEs using different techniques. Some popular TE approaches are Char-CNN-RNN [1], Char-CNN-GRU [5], deep attentional multimodal similarity model (DAMSM) [13], and semantic text embedding module (STEM) [13], among others. The generated embedding is then provided as a condition of the generator along with some noise. The discriminator stage compares both images, that is, the generated fake image and the original one. Here, the primary responsibility goes to the discriminator. It is mainly responsible for judging the difference between the synthetic image and the original image, as based on the loss value generator is going to generate a new image.

The literature has shown that training more discriminator per generator epoch may increase accuracy [13]. The generator generates images of different sizes, specifically 64 × 64, 128 × 128, and 256 × 256 based on the model specifications. The problem in terms of image sharpness and accuracy becomes more sensible as the image size increases [2].

Generating realistic pictures from content would be fascinating and helpful, yet current artificial intelligence (AI) frameworks are still a long way from this objective [1]. Failure of convergence, Nash equilibrium, mode collapse, and vanishing gradient problems are some of the inherent training issues from the generative adversarial network (GAN) [2]. The results of the literature show that generated images are either blurred or not aligned to the given text script. To further improve the sharpness of the blurred image, one can apply image super-resolution (SR) techniques.

A super-resolution generative adversarial network (SRGAN) is a GAN that incorporates SR. Its generator module tries to produce a high-resolution (HR) image from low-resolution (LR) images produced via a text-to-image generation technique. Its discriminator stage differentiates between real and fake images [9].

Here, the main challenging problem can be further divided into three sub-problems. First, learning text features into significant visual subtleties. Second, the use of generated text features for creating plausible images, and third, applying image super-resolution techniques to get the HR image. Fortunately, current deep learning resources have made tremendous advances in all of these problems.

This chapter is divided into sections as follows. Section 2 introduces various text embedding techniques used in the existing TIS models. It also cast a light on different image generation models and why GAN is best to choose. In addition to that, it elaborates on various types of super-resolution techniques and the importance of an SRGAN for problem undergoing study. Section 3 illustrates and depicts a GAN along with loss function and SRGAN. Section 4 comes out with state-of-the-art text-to-image generation techniques along with its findings. Section 5 puts forward the method to generate an HR image from the given text.

Furthermore, this work addresses several image quality synthesis metrics such as the inception score (IS) [15], Frechet inception distance (FID) [16], and R-precision [13]. In Sect. 6, we discuss possible improvements to the proposed technique.

2 Literature Survey

2.1 Text Embedding

There have been noteworthy advancements in this field to generate text features from given information. Tomas Mikolov et al. (2013) at Google developed the Word2Vec [17], an unpretentious strategy for efficiently learning a single word embedding from the content. Word2Vec tries to predict a single word based on the input text. It preserves the linguistic meaning of the said text. However, it does not maintain the sequence of long text sentences in embedding. Pennington et al. at Stanford represented data utilizing matrix factorization technique, latent semantic analysis (LSA) of global text statistics and Word2Vec for local context-based learning [18].

Visual-text pair embedding generates a good joint representation of sentence and image as joint embedding. Different researchers have done great work by

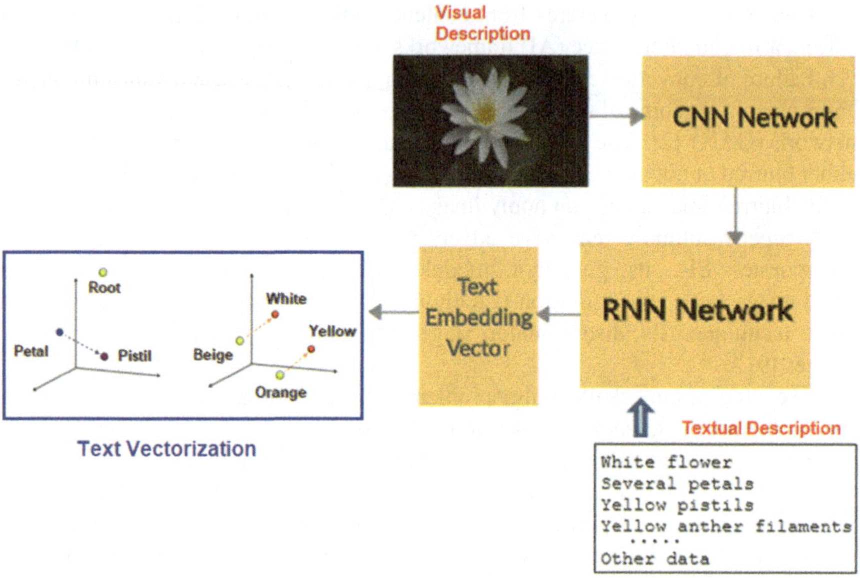

Fig. 2 Char-CNN-RNN model for text embedding

determining the goal of the multi-model depiction of text and image pairs. They achieved this by mingling the convolutional neural networks (CNN) and recurrent neural network (RNN) methods along with their variations. Reed et al. (2016) have developed a deep symmetric structured joint embedding technique, that is, the char-CNN-RNN model as one of the fundamental and paramount TE methods [19]. Here, CNN and RNN stand for convolutional neural network and recurrent neural network, respectively. The model considers text and image both as joint embedding to generate text embedding, as shown in Fig. 2 [19]. It trains fine-grained and class-specific pictures based on natural language text. The model outperforms existing techniques for zero-shot image retrieval for specific text. First, the char-CNN-RNN strategy takes an image as input for the CNN model and text as input for the RNN model. The RNN model finally outputs a text-embedding vector. This technique is simple for generating fixed-size text embedding vectors. It also preserves the text sequence order. This set of characteristics uses a deep convolutional recurrent text encoder on the structured joint embedding of text captions [1–4]. In spite of this, the model does not focus on the specific important word. It gives equal importance to all the words.

Scott Reed et al. (2016) used the char-CNN-Gated Recurrent Unit (GRU) [5] and made an average of specific category sentences as a text embedding. GRU's benefit over RNN is that GRU uses forget and update gates to transfer the information so that it can judge whether the data should be passed or not to the next gate. Hao Dong et al. (2017) used the long-term short memory (LSTM) technique opposed to the GRU used in the previous one [6]. LSTM networks are more suitable for more extended natural language text sequences. It decodes a fixed-length vector sequence to the required sentence. Alex Fu et al. (2016) used two LSTM networks along with CNN to process forward direction and backward direction with forget gate. This model helps the generator to focus on furnishing the creatable pieces of information. It also preserves the spatial relationship between the different objects, ultimately essential for the target problem definition [7]. The state-of-the-art TIS model used deep attentional multimodal similarity model (DAMSM) proposed by Tao Xu et al. (2017). The technique uses an attention mechanism so that it develops word-level embedding in addition to sentence-level embedding. Both types of embedding are added into the hidden layer of the network, which ultimately helps to generate fine-grained visual details in the output image [13].

2.2 Generative Image Models

Generating an appealing image by using generative models has a considerable interest among the researchers. There has been astonishing progress in this direction to engender a sharp and high-quality image. The model variational autoencoders (VAEs) figures the issue with probabilistic graphical models to expand the lower bound of information probability. It effectively inferences the latent variable but produces poor image quality. The generated images are blurred and hence this technique cannot fit this work's problem, as it seeks sharp images [8]. Autoregressive models such as the pixel RNN learn explicit distribution among pixels imposed by model structure [24]. Even though it has a simple and straightforward training process, and its likelihood is tractable, the image sequence generation process is slow.

Moreover, the generated picture is not so good. Currently, GANs have established a new horizon in terms of sharp synthetic image creation. It can create a compelling sensible image of different objects such as bedrooms, birds, flowers, faces, and many more. Compared to variational autoencoders (VAE) and auto-regressing models, the GAN architecture primes for its sharp image generation. Though GANs are hard to train, plenty of works have been proposed to balance out the improvement of the picture characteristics [2]. Mode collapse, vanishing gradient, and failure to converge are some of the training issues, to count a few. To solve the above issues, Martin Arjovsky et al. (2017) proposed Wasserstein GAN (WGAN), which uses gradient clipping so that the gradient will not have a significant update at the time of backpropagation. Here, the training is more stable compared to Vanilla GAN [14]. The authors trained the discriminator stage five times more than the generator phase. Additionally to this, Ishaan Gulrajani et al. (2017) have trained GAN employing Wasserstein distance and gradient penalty (WGAN-GP) to remove GAN training issues [20]. Somehow, we found that WGAN-GP is too slow to train, as it uses gradients for weights also.

2.3 Image Super-Resolution

Image super-resolution is a highly complex task that generates a high-resolution image from a given low-resolution image. It has numerous real-world applications such as medical, remote sensing, microscopy, and video enhancement. Despite great effort, producing a high-resolution image is challenging because, as the image's size increases, it loses the granular image information. Deep learning has shown promising results for this problem.

The primary strategy for single image SR (SISR) includes prediction-based methods such as bicubic, which is fast and yields pixel arrangements with excessively smooth textures [9]. Chao Dong et al. (2016) developed a CNN-based image super-resolution that first fed the image to bicubic interpolation, where it will be up-sampled. The up-sampled image will now be input for CNN, which generates a high-resolution image [21]. To recreate reasonable texture information while maintaining edge issue, Tai et al. (2010) mixed edge-directed SR technique and gradient profile [23]. Considering the success of deep learning-based image resolution techniques, Christian Ledig et al. (2017) have developed a residual block-based image SR technique that has shown benchmark results in the non-GAN-based methods addressing the above problems. The same paper also discussed a GAN-based image super-resolution approach (SRGAN), which first downsamples the same image to deliver the high-resolution image. The generator model has incorporated ResNet and sub-pixel convolution, which calculates perceptual loss to generate state-of-the-art high-resolution images [9].

3 Background

3.1 Generative Adversarial Network

Ian J. Goodfellow et al. developed a semi-supervised framework for assessing generative models through adversarial training. They simultaneously trained two structures: The generator (G) that generates a fake image and the discriminator (D) that judges either the generated image is real or fake by competing once against another. G and D can be any type of deep neural networks such as simple vanilla feed-forward network, convolutional network, or any other network type. Generator G tries to produce data distribution (fake image) from random noise $\mathbf{z}$, which should look like the original. The discriminator minimizes the probability of generating unreal imageries. The discriminator tries to maximize the original image's probability of being the true match. This trade-off is also known as a min-max game or a zero-sum game. The generator and discriminator are both trained simultaneously. In the initial stages of learning, the discriminator can confidently reject the generated image. After a certain number of iterations, generator becomes powerful enough to fill the discriminator, the discriminator can hardly identify the fake image. This stage is known as a Nash equilibrium, where the generator mostly produces meaningful images. Overall, the training process has the following value function *V(D(.), G(.))* or *V(D, G)* for short [12] as follows:

$$\min_{G} \max_{D} V(D, G) = \mathrm{E}_{x \sim p_{\text{data}}(x)}[\log D(x)] + \mathrm{E}_{z \sim \mathrm{p}_{z(z)}}[\log (1 - D(G(z)))] \quad (1)$$

Here, x represents a real image of any size (32 × 32, 64 × 64, 128 × 128, 256 × 256 etc.) from the true data distribution p_{data}, and z is random noise (generally a 100-dimensional vector) taken from a Gaussian or other arbitrary data distribution. Here, $\mathrm{E}_{x\tilde{p}_{\text{data}}}$ denotes the expectation that the image is coming from the original data and $\mathrm{E}_{x\tilde{p}_z}$ means the image is coming from noise. To overcome the vanishing gradient problem, practically, generator G is updated to maximize log (D ($G(z)$)) rather than minimizing log (1 − D ($G(z)$) [9–11].

Mehdi Mirza et al. (2014) developed conditional GAN where the generator network and discriminator network both are provided some additional, conditional information c, forming G (z, c) and $D(x, c)$ [12]. In short, generator G uses this condition c to generate attractive images.

3.2 Super-Resolution Generative Adversarial Network (SRGAN)

The super-resolution GAN (SRGAN) generates appealing HR images by using an adversarial network. This framework yields more convincing photorealistic images from LR images compared to other techniques without GAN. Here, the training process first includes the down sampling of the HR image to the LR image. Then, the GAN generator up-samples the LR frame to generate the SR loss l^{SR}. Here, the discriminator differentiates HR-LR and back propagates the adversarial loss for training the generator and discriminator. Both networks generally use convolution, batch normalization, and a parameterized leaky rectified linear unity (ReLU). Similar to the ResNet, skip connections are applied to the generator.

SRGAN can be related to the visual geometry group (VGG) architecture through the content loss function known as VGG loss that applies to generate and real images. In an SRGAN, the perceptual loss function l^{SR} (obtained from VGG_{19} content loss) [9] can be defined as

$$l^{\text{SR}} = \underbrace{\underbrace{l_x^{\text{SR}}}_{\text{content loss}} + \underbrace{10^{-3} l_{\text{Gen}}^{\text{SR}}}_{\text{adversarial loss}}}_{\text{perceptual loss(for VGG based content loss)}} . \quad (2)$$

Where x is an image generated from simple text-to-image synthesis module, Gen is generated super-resolution image. The content loss is defined as in [9].

$$l^{SR}_{VGG/\ i,j} = \frac{1}{W_{i,j}H_{i,j}} \sum_{x=1}^{W_{i,j}} \sum_{y=1}^{H_{i,j}} \left(\varnothing_{i,j}(I^{HR})_{x,y} - \varnothing_{i,j}(G_{\theta_G}(I^{LR}))_{(x,y)} \right)^2 \quad (3)$$

which is the Euclidean distance between the feature representations of a reconstructed image G_G, I^{LR} is the LR image, and the query (SR) image is I^{HR}. $\varnothing_{i,\,j}$ is the feature map whose dimensions are $W_{i,\,j}$ and $H_{i,\,j}$ resulting from the *j-th* convolution (subsequently to the activation) and before the *i-th* max-pooling layer of the VGG_{19} framework. There are certain possible values of $\varnothing_{ij}$ for VGG_{19} network such as $\varnothing_{1,2}$, $\varnothing_{2,2}$, $\varnothing_{3,4}$, $\varnothing_{4,4}$ and $\varnothing_{5,\,4}$, among which features generated by $\varnothing_{5,\,4}$ the most meaningful features to be considered, as deeper layer features contain more information [9]. The features are extracted from HR and SR image by $\varnothing_{5,\,4}$ layer, which is known as $\varnothing_{i,\,j}(I^{HR})$ and $\varnothing_{i,j}(G_{\theta_G}(I^{LR})$,where value of i and j is 5 and 4, respectively.

The pixel-wise content loss is calculated as the mean squared error (MSE) between the HR and LR images. The MSE [10] can be defined as the difference between the actual sought-after image (which in the case of this work is the HR image) and predicted (super-resolved image for the matter of this work) as

$$\text{MSE} = \frac{1}{N}(\mathbf{y} - \widehat{\mathbf{y}})^2, \quad (4)$$

where N is the size of the image, with $\mathbf{y}$ and $\widehat{\mathbf{y}}$ signifying the actual and predicted images, respectively. Now, the adversarial loss becomes [9].

$$l^{SR}_{Gen} = \sum_{n=1}^{N} - \text{loglog}\, D_{\theta_D}\left(G_{\theta_G}(I^{LR})\right). \quad (5)$$

Here, $D_{\theta_D}(G_{\theta_G}(I^{LR}))$ denotes the probability that the reconstructed image $G_{\theta_G}(I^{LR})$ is the original high-resolution image.

4 Techniques for Text-to-Image Synthesis

4.1 Generative Adversarial Text-to-Image Synthesis

Scott Reed et al. (2016) have tried to generate the images from text caption using GAN very first time. First, the text is encoded with the char-CNN-RNN model and submerged the noise (*z*). Noise dimension can be any. The combination of embedding and noise goes into the generator to render a plausible image based on a given text caption or written condition. Discriminator takes embedding, generated image (fake), and the original (real) image as input and differentiates between real and unreal. They used a convolutional network in both generator and discriminator. This process is further extended with the matching aware discriminator, also known as GAN with conditional latent space (GAN-CLS) and earning with manifold interpolation (GAN-INT).

Fig. 3 Text-to-image generation using the GAN-CLS method for CUB dataset

Here, the generated image size is 64 × 64. While experimenting, it has been observed that the model does not reflect the fine-grained text details in the generated image. The model is also sensitive to the mode collapse issue. Authors have suggested that among all the variants of the model, the GAN with manifold interpolation (GAN-INT) and the GAN-INT with a matching aware discriminator (GAN-INT-CLS) yield plausible results [1].

4.1.1 Experiments with the CUB Dataset

Figure 3 represents experimental results for the vanilla TIS model for the CUB dataset.

"This bird has a green leg and blue body."

4.2 StackGAN and StackGAN++: Realistic Image Synthesis with Stacked Generative Adversarial Networks

Although GANs have presented an outstanding performance in various problems, they still have a significant issue generating attractive images [3]. The StackGAN is a two-stage process for constructing photorealistic pictures from the text. The stage-I GAN draws an image with basic colors and primary shapes from a given text, bringing out LR images. Stage-II GAN takes generated images from stage I and text captions as input and creates plausible pictures with realistic details. The significant contribution of the Stack GAN is the sketch refinement process where the defects of stage I can be autocorrected by stage II, resulting in a sharp and meaningful image of 256 × 256. Prior techniques for generating text captions create static captions to feed the generator network. Still, Stack GAN uses conditional augmentation that first converts the text into embedding, then makes the Gaussian distribution to add the conditioning variable at the time of training. This fact encourages a small variation in the training stage, aiming at a more robust output image. Further, in more advanced StackGAN++, multiple generators and multiple discriminators can be arranged in a tree-like structure for creating the same scene with different dimensions from different branches of the tree [3]. This type of architectural change can transfer background from stage I to stage II.

4.3 Learning What and Where to Draw

Techniques for generating images from the text do not control the pose and location of the object in the picture. Adversarial what-where network (GAWWN) tries to incorporate "what to draw" and "where to draw" information in the generated image. The composed images are conditioned on text description as well as on object location. Areas can be precisely constrained by either a bounding box or some key points such as a beak and a tail. The generated image size is 128 × 128, which counts as a top achievement [5].

4.4 Attentional Generative Adversarial Network (AttnGAN)

The attentional generative adversarial network (AttnGAN) by Xu et al. (2017) is similar to StackGAN++ in case of a structure with some additional elements. As with other techniques for generating text captions, the text encoder receives a text. In addition to that, the text encoder also generates separate text encodings condition on a particular single word that can be achieved using the bidirectional LSTM network. Stage I generates an image by using static text captions and random noise. The output of stage I embedding, along with word-level captions, is provided to the attention model, which generates a word–context matrix to map the word and object in the image. This is, then, delivered to the next stage network along with the information generated by stage I. Each of the stages works the same but gradually increases the images' resolution compared to the previous step [13].

4.5 Hierarchically Nested Adversarial Network (HDGAN)

A hierarchically nested adversarial network (HDGAN) mainly focuses on the text's persisting semantic meaning in the image. The generator favors a vanilla GAN, which does not require any multi-stage training or multiple text conditions. The authors have developed a hierarchically nested discriminator at intermediate layers to encourage adversarial games and generate a real training data distribution for the generator module.

For the training purpose, a novel convolution network has been proposed in the generator. It can be seen that defining a discriminator at a different level has positive feedback for image generation.

5 Text-to-Image Super-Resolution

As discussed above, techniques for generating images from a given text, the generated images are blurred or do not reflect the fine-grained details. Based on that, the problem of text-to-image generation can be broadly divided into two categories: First, generating sharp and high-definition images based on text and second, converting all the text content in the picture by preserving a semantic meaning [11].

For a successful text-to-image super-resolution method, it is necessary to achieve both. The process is divided into two stages. First, generating an appealing image from source text by using either of the technique. Second, applying the generated image to SRGANs for high-resolution. Nevertheless, the second stage is useful to make the image sharp and plausible. Even though this technique will also help create an image for the latter category since the second category mainly focuses on transferring text details into the image rather than its resolution.

The diagram from Fig. 4 outlines the generation of an HR image from the text. The detailed architecture of a vanilla GAN for TIS is shown in Fig. 5. First, to generate an LR image from the source text, GAN is the best choice as it generates accurate images conditioned on the text. Here, the generator network and discriminator can be any well-known network like a fully connected network (FC) or a CNN. The generator uses random uniformly distributed noise as input and text as a condition. The vector size of noise can be anything. The natural language is first converted into embedding by using the char-CNN-RNN technique as shown in Fig. 2. The vector with noise and embedding are deeply concatenated to feed as an

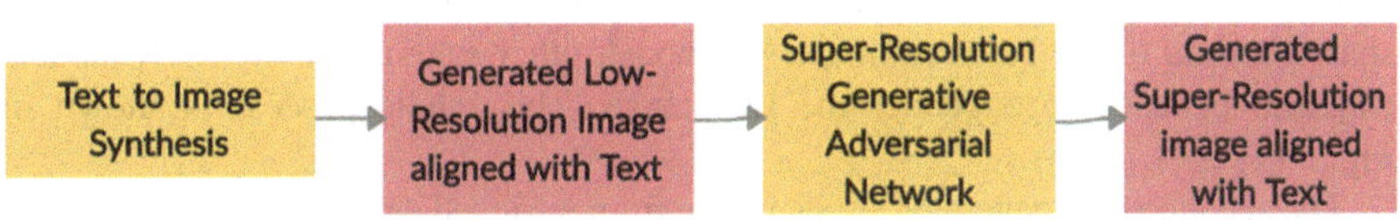

Fig. 4 Text-to-image super-resolution process

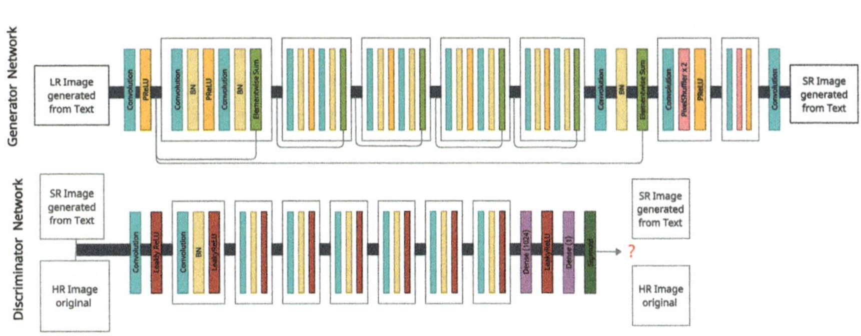

Fig. 5 The basic structure of the text-to-image synthesis technique

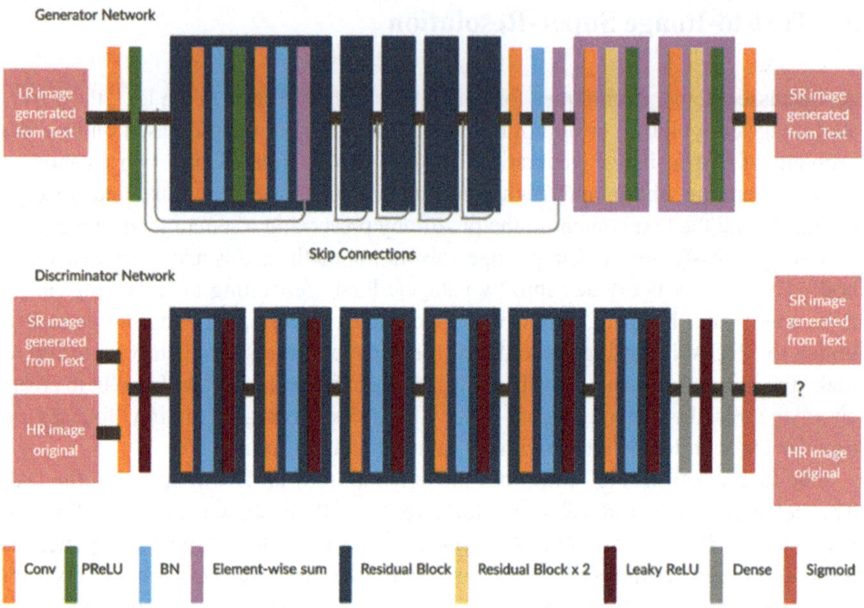

Fig. 6 Detail structure of SRGAN

input for the generator. The discriminator takes the real image and fake image both as input and generates a prediction score. The generator and discriminator are trained based on their loss function shown in Eq. (1). After several epochs, the LR image conditioned on text starts generating. This LR image will look fuzzy and not so sharp. Instead of the basic vanilla GAN model, other state-of-the-art GAN models can also be used to engender the LR image. This whole process is shown in Fig. 5 [1].

The second stage generates an SR image from retrieved candidate LR pictures via a TIS technique. The GAN-based super-resolution technique takes the LR image and generates HR using a convolution series, ReLU, batch normalization, and element-wise sum operation. The discriminator network takes both the HR and LR images and executes a series of convolution, leaky ReLU, batch normalization layers. It has a dense layer, at last, to generate the prediction score that the image is HR or super-resolution. Figure 6 illustrates the detailed architecture of the SRGAN technique for super-resolution [9].

6 Quantitative Evaluation Measures

TIS is a highly multimodal problem. A good algorithm or method contemplates vivid text details in the generated picture. It should generate a sharp and good visual quality image. So the text-to-image generation, evaluation parameters can be divided into two categories: (1) They evaluate the quality/sharpness of image and (2) assess the quality of the image based on the reflection of the text.

6.1 *Evaluating Quality/Sharpness of the Image*

These matrices generate a score based on the diversity of the picture's generated image and quality [24, 25].

6.1.1 Inception Score

In short, the inception score (IS) is an objective evaluation metric for assessing the nature of explicitly manufactured pictures yield by generative adversarial networks. It considers the pre-trained inception-V3 model as an evaluator [22]. The inception network takes a good quantity of images (i.e., approx. 50000) for the classification purpose to calculate the probability of images corresponding to each class. Many generated images are classified using the model to calculate each image's probability corresponding to the particular category. At last, to get the final score, it takes the exponential of the KL divergence of generated probabilities and calculates its average. It gives a rough approximation of two significant properties. First, image quality and second, the diversity of the image [15]. If the generated images have variety and good quality, then IS will be high. It provides a good approximation for the said criteria. Still, it is inclined to the inception network, meaning that if generated image objects are not present in the pre-trained model, then the inception score will be low, no matter how good the image is. The IS per se does not guarantee the generated image is aligned to text or not.

6.1.2 Frechet Inception Distance

The Frechet inception distance (FID) calculates the distance between the original image features and the generated image. Martin Heusel et al. (2017) claimed that the IS does not consider a crucial statistical calculation comparison with the original image and fake image [16]. Statistics mean the average, mean, and the variance defining any statistical Gaussian distribution. FID considers the second last layer of the pre-trained inception-V3 model (i.e., before the output layer) and calculates the mean and variance of generated images and original images. In short, FID calculates the score based on the extracted feature difference between both real and fake images. The IS drawbacks are also inherent in the FID calculation.

6.2 Evaluating the Quality of the Image Based on the Text Reflection

The R-precision is a common ranking retrieval evaluation metric. First, generated images are queried on their respective text descriptions. First, global text features and global image features are calculated based on a pre-trained deep attentional multimodal similarity model (DAMSM) model, and cosine similarities are calculated between them. Finally, the image–text pairs are ranked in descending order to find the top r relevant descriptions for computing the R-precision [13]. R-precision ensures the presence of text in the synthesized image.

7 Future Work

There are certain chances of improvements in this system by considering many future directions as each sub-problem can be an individual point of improvement. First, for the simple text-to-image synthesis issue, the stacking of multiple generator and discriminator network will improve the image quality. Stacking both the networks works as a sketch refinement process where the initial stages of generators engender rough shapes and colors. The remaining generators are responsible for drawing highly precise and sharp images. If the initial stage images are sharp, then applying super-resolution on top of that will surely enhance the image quality. Second, text embedding techniques can be improved by considering the attention mechanism. An attention mechanism-based text embedding techniques can generate importance for highly prime vocabularies. Hence the generated images can ensure the presence of a particular object in a synthesized image. Third, adding a residual network in the discriminator part can improve the discriminator's classification accuracy as well as it can remove the vanishing gradient problem. The residual network decides whether generated features are important for the final decision or not. If not, it updates weight values at zero and thus speeds up the computation and minimizes the over-fitting problem. Forth, to improve GAN's training process (to overcome vanishing gradient issue), one should also consider training by using WGAN or WGAN-GP. Both of these significantly improve the training process by gradient penalty and compromise with training time (WGAN and WGAN-GP are highly slow but precise). Fifth, we believe that apart from looking for a powerful TIS technique, future research should also look for less intricate GAN-based image super-resolution models.

8 Conclusion

This chapter proposes applying a GAN-based text to image super-resolution technique for the generated image from the given text. GAN is a powerful model to create sharp and plausible pictures compared to other available image engendering models. The authors considered a simple text-to-image generator model and created the low-resolution images by considering both the generator network and the discriminator network as a simple convolution neural network. In addition to that, Char-CNN-RNN text embedding techniques are used to generate text embedding of dimension 1024 for the given text data. The random noise and text embedding combinedly generate LR images that are not so powerful and fail to portray real-world scenarios. A pre-trained SRGAN network is applied on top of that to improve the generated image's resolution, where the generator network is implemented with certain skip connections, whereas the discriminator does not. Content loss and reconstruction both are considered for image super-resolution loss to perfectly differentiate real image and fake image. The output of SRGAN generates a high-resolution image aligned to the given text. According to us, SRGAN is the most robust image super-resolution technique available till now to perform super-resolution. We believe that this is the first attempt in history to generate a super-resolution image from the text description. The suggested architecture can render an extremely crisp shot from the given text description.

References

1. Reed, S., Akata, Z., Yan, X., Logeswaran, L., Schiele, B., & Lee, H. (2016). Generative adversarial text to image synthesis. arXiv preprint arXiv:1605.05396.
2. Zhang, H., Xu, T., Li, H., Zhang, S., Wang, X., Huang, X., & Metaxas, D. N. (2017). Stackgan: Text to photo-realistic image synthesis with stacked generative adversarial networks. In *Proceedings of the IEEE international conference on computer vision* (pp. 5907–5915).
3. Zhang, H., Xu, T., Li, H., Zhang, S., Wang, X., Huang, X., & Metaxas, D. N. (2018). Stackgan++: Realistic image synthesis with stacked generative adversarial networks. *IEEE Transactions on Pattern Analysis and Machine Intelligence, 41*(8), 1947–1962.
4. Cha, M., Gwon, Y., & Kung, H. T. (2017, Sept). Adversarial nets with perceptual losses for text-to-image synthesis. In *2017 IEEE 27th international workshop on machine learning for signal processing (MLSP)* (pp. 1–6). IEEE.
5. Reed, S. E., Akata, Z., Mohan, S., Tenka, S., Schiele, B., & Lee, H. (2016). Learning what and where to draw. In *Advances in neural information processing systems* (pp. 217–225).
6. Dong, H., Zhang, J., McIlwraith, D., & Guo, Y. (2017, Sept). I2t2i: Learning text to image synthesis with textual data augmentation. In *2017 IEEE international conference on image processing (ICIP)* (pp. 2015–2019). IEEE.
7. Fu, A., & Hou, Y. (2016). *Text-to-image generation using multi-instance StackGAN* (p. 26). Stanford, CA: Department of Computer Science Stanford University.
8. Kingma, D. P., & Welling, M. (2013). Auto-encoding variational Bayes. arXiv preprint arXiv:1312.6114.
9. Ledig, C., Theis, L., Huszár, F., Caballero, J., Cunningham, A., Acosta, A., Aitken, A., Tejani, A., Totz, J., Wang, Z., & Shi, W. (2017). Photo-realistic single image super-resolution using a

generative adversarial network. In *Proceedings of the IEEE conference on computer vision and pattern recognition* (pp. 4681–4690).

10. Herrmann, A. E., & Estrela, V. V. (2016). Content-based image retrieval (CBIR) in remote clinical diagnosis and healthcare. In M. M. Cruz-Cunha, I. M. Miranda, R. Martinho, & R. Rijo (Eds.), *Encyclopedia of E-health and telemedicine*. Hershey: IGI Global. https://doi.org/10.4018/978-1-4666-9978-6.ch039.
11. Agnese, J., Herrera, J., Tao, H., & Zhu, X. (2020). *A survey and taxonomy of adversarial neural networks for text-to-image synthesis* (p. e1345). Wiley Interdisciplinary Reviews: Data Mining and Knowledge Discovery.
12. Goodfellow, I., Pouget-Abadie, J., Mirza, M., Xu, B., Warde-Farley, D., Ozair, S., Courville, A., & Bengio, Y. (2014). Generative adversarial nets. In *Advances in neural information processing systems* (pp. 2672–2680).
13. Xu, T., Zhang, P., Huang, Q., Zhang, H., Gan, Z., Huang, X., & He, X. (2018). Attngan: Fine-grained text to image generation with attentional generative adversarial networks. In *Proceedings of the IEEE conference on computer vision and pattern recognition* (pp. 1316–1324).
14. Arjovsky, M., Chintala, S., & Bottou, L. (2017). Wasserstein GAN. arXiv preprint arXiv:1701.07875.
15. Salimans, T., Goodfellow, I., Zaremba, W., Cheung, V., Radford, A., & Chen, X. (2016). Improved techniques for training GANS. In *Advances in neural information processing systems* (pp. 2234–2242).
16. Heusel, M., Ramsauer, H., Unterthiner, T., Nessler, B., & Hochreiter, S. (2017). GANs trained by a two time-scale update rule converge to a local Nash equilibrium. In *Advances in neural information processing systems* (pp. 6626–6637).
17. Mikolov, T., Chen, K., Corrado, G., & Dean, J. (2013). Efficient estimation of word representations in vector space. arXiv preprint arXiv:1301.3781.
18. Pennington, J., Socher, R., & Manning, C. D. (2014). Glove: Global vectors for word representation. In *Proceedings of the 2014 conference on empirical methods in natural language processing (EMNLP)* (pp. 1532–1543).
19. Reed, S., Akata, Z., Lee, H., & Schiele, B. (2016). Learning deep representations of fine-grained visual descriptions. In *Proceedings of the IEEE conference on computer vision and pattern recognition* (pp. 49–58. 25).
20. Gulrajani, I., Ahmed, F., Arjovsky, M., Dumoulin, V., & Courville, A. C. (2017). Improved training of Wasserstein GANS. In *Advances in neural information processing systems* (pp. 5767–5777).
21. Dong, C., Loy, C. C., He, K., & Tang, X. (2015). Image super-resolution using deep convolutional networks. *IEEE Transactions on Pattern Analysis and Machine Intelligence, 38*(2), 295–307.
22. Szegedy, C., Vanhoucke, V., Ioffe, S., Shlens, J., & Wojna, Z. (2016). Rethinking the inception architecture for computer vision. In *Proceedings of the IEEE conference on computer vision and pattern recognition* (pp. 2818–2826).
23. Tai, Y. W., Liu, S., Brown, M. S., & Lin, S. (2010, June). Super resolution using edge prior and single image detail synthesis. In *2010 IEEE computer society conference on computer vision and pattern recognition* (pp. 2400–2407). IEEE.
24. Deshpande, A., Patavardhan, P., Estrela, V. V., & Razmjooy, N. (2020). Deep learning as an alter-native to super-resolution imaging in UAV systems. In V. V. Estrela, J. Hemanth, O. Saotome, G. Nikolakopoulos, & R. Sabatini (Eds.), *Imaging and sensing for unmanned aircraft systems* (Vol. 2, 9, pp. 177–212). London: IET. https://doi.org/10.1049/PBCE120G_ch9.
25. Laghari, A. A., Khan, A., He, H., Estrela, V. V., Razmjooy, N., Hemanth, J., & Loschi, H. J. (2020). Quality of experience (QoE) and quality of service (QoS) in UAV systems. In V. V. Estrela, J. Hemanth, O. Saotome, G. Nikolakopoulos, & R. Sabatini (Eds.), *Imaging and sensing for unmanned aircraft systems* (Vol. 2, 10, pp. 213–242). London: IET. https://doi.org/10.1049/PBCE120G_ch10.

Analysis of Lossy and Lossless Compression Algorithms for Computed Tomography Medical Images Based on Bat and Simulated Annealing Optimization Techniques

S. N. Kumar, Ajay Kumar Haridhas, A. Lenin Fred, and P. Sebastin Varghese

1 Introduction

Every day, a massive amount of medical images are acquired by physicians for the diagnosis of diseases and therapeutic applications. Usually, medical images are stored in lossless formats, which requires lossless compression models. Efficient lossy compression algorithms can also compress medical images. The picture archiving and communication system (PACS) requires an efficient compression algorithm, thereby minimizing the degradation of reconstructed image quality. It is problematic to install PACS in rural area hospitals and scan centers due to the high cost and maintenance. Many telemedicine suites are there in the market; however, the price is not affordable for rural area scan centers and hospitals. In [1], a detailed survey was performed on the lossless compression of medical images.

The super resolution (SR)-based image processing techniques compress low-resolution (LR) images successfully [2–4]. Deep convolution neural networks can aid in rendering super-resolution, minimizing artifacts, performing tests in real-time, and executing benchmark data sets [5]. A novel lossy compression technique was proposed based on the hybrid image resizing technique and yielded better results than the classical JPEG compression technique [6].

S. N. Kumar (✉)
Amal Jyothi College of Engineering, Kanirappally, Kerala, India

A. K. Haridhas · A. Lenin Fred
Mar Ephraem College of Engineering and Technology, Elavuvilai, Kanyakumari, Tamil Nadu, India

P. Sebastin Varghese
Metro Scans and Research Laboratory, Thiruvananthapuram, Kerala, India

A. Deshpande et al. (eds.), *Computational Intelligence Methods for Super-Resolution in Image Processing Applications*, https://doi.org/10.1007/978-3-030-67921-7_6

A comparative analysis of compression techniques comprising Convolutional Autoencoders, Generative Adversarial Networks (GANs), and super-resolution techniques appeared in [7]. A prediction-based lossless compression technique from [8] handled the volumetric medical image compression. The sample-based weighted prediction for Enhancement Layer (EL) coding (SELC) was recommended for the lossless compression of video; efficient results were produced when compared with the JPEG lossless approach [9]. The singular value decomposition yields proficient results for compacting fingerprint images [10, 11]. The concept of holography and signal processing challenges in the holographic system are discussed in [12, 13]. The learned resolution compression model (L3C) from [14] yields a better result than the classical approaches. A broad survey of medical image compression techniques describes medical images, formats, and compression techniques as in [15]. The different kinds of brain image compression are discussed in [16].

The chapter organization is as follows: Section 2 describes the contributions of metaheuristic bat optimization (BO) algorithms (BOA)s. Section 3 highlights the lossy and lossless compression algorithms for medical images. Section 4 describes the results and discussion with compression algorithms validation on real-time datasets with performance metrics. Finally, the conclusion arises in Sect. 6.

2 Bat Optimization Algorithm

The metaheuristic optimization algorithms help to solve the multi-objective problem with high nonlinear constraints and approximate the optimal solution [17]. The metaheuristic optimization algorithms eliminate the computational complexity and efficiently find the local minima than the traditional mathematical optimization techniques. Nature-inspired optimization algorithms imitate the characteristic natural, for example, survival of fittest, and collectively learning from the population of individuals. The bat algorithm is a metaheuristic optimization procedure based on the bats' echolocation behavior [17]. The microbats mostly use a sonar wave (sound pulse) to calculate the distance by the reflection of waves from the obstruction or prey. The microbats' ability to distinguish hunt and obstacles by echolocation is related to the objective function to be optimized to find the best possible solution. The frequency of the pulse relies on the distance of the stumbling block or prey. Hence, the i-th bat has a location $\boldsymbol{x}_i$, a velocity $\boldsymbol{v}_i$, loudness A_i and frequency f_i.

The BOAs [18] is based on the following steps:

1. Echolocation helps to calculate the distance and difference between an obstacle and prey.
2. The bats move randomly with the velocity at the position.

3. The bats emit sonar waves with a frequency ranging from $f_{\max}$ to $f_{\min}$, wavelength λ, and loudness A_0 for hunting the prey.
4. The rate of the pulse r_i is uniformly distributed over [0, 1] for each bat i. It is altered based on the distance of the prey by adjusting the corresponding wavelength.
5. The positive loudness A_i varies from A_0 to $A_{\min}$ for each bat i. A_0 is a given constant.
6. The new solution stems from updating frequencies, velocities, and locations/solutions. At time t, the new solution $\boldsymbol{x}_i^t$ and velocity $\boldsymbol{v}_i^t$ are given by

$$\boldsymbol{x}_i^t = \boldsymbol{x}_i^{t-1} + \boldsymbol{v}_i^t,$$

and

$$\boldsymbol{v}_i^t = \boldsymbol{v}_i^{t-1} + \left(\boldsymbol{x}_i^t - \boldsymbol{x}_*\right) f_i$$

where $\boldsymbol{x}_*$ is the best solution with n bats, a given search space and t iterations.
7. The frequency updating f_i occurs as follows

$$f_i = f_{\min} + (f_{\max} - f_{\min})\beta,$$

where $f_{\min}$, and $f_{\max}$ are, respectively, the minimum and maximum pulse frequencies. $\beta \in [0, 1]$ is a random number drawn from a uniform distribution.
8. The new solution for the local search produced by random walk at time t becomes

$$\boldsymbol{x}_{\mathbf{new}} = \boldsymbol{x}_{\mathbf{old}} + \varepsilon \boldsymbol{A}^t,$$

with

ε is a random number uniformly distributed between [−1,1], and
$\boldsymbol{A}^t$ is the average loudness vector whose entries are A_i.

9. Loudness A_0 of bat decreases with an increase in the rate of the pulse, when bat finds the prey, the best optimal solution is achieved.

$$A_i^{t+1} = \alpha A_i^t,$$

and

$$r_i^{t+1} = r_i^0[1 - \exp(-\gamma t)]$$

where

$0 < \alpha < 1$, $\gamma > 0$, and both are constant.

Pseudocode

```
Objective function f(X), X = (x1,x2,...,xn)^T.
Initialize the bat population xi and velocity vi for i=1,
...,n.
Define the pulse frequency fi at xi.
Initialize the pulse rate ri and the loudness Ai
while (t < Max_number_of_iterations)
    Generate new solutions by updating frequencies, ve-
    locities, and locations/solutions.
    if (rand > ri)
          Select the best solution.
        • Generate a local solution around the selected
               best solution.
    end if
    Generate a new solution by flying bats randomly.
    if ((rand <Ai) & (f(xi) < f(x*))
         Accept new solutions.
         Increase ri and decrease Ai.
    end if
    Rank the bats and find the current best x*.
end.
```

For the border detection of melanoma and other skin diseases, the BOA is used for optimizing the rational Bezier curves. The bioinspired optimization gives better RMSE than the classical polynomial method [19]. The optimized binary bat algorithm (OBBA) uses dimensionality reduction by discriminative features to classify leukocytes. The proposed OBBA is compared with KNN, logistic regression, random forest, and decision tree procedures with outstanding performance. The OBBA algorithm is accurate and fast when compared with the optimized crow search algorithm and optimized cuttlefish algorithm [20]. The bioinspired BOA optimizes the parameters of the pulse-coupled neural network for improving the segmentation of the image [21]. For the detection of the abnormality in the brain MR images, the chaotic bat algorithm fine-tunes the parameters for the following arrangements: (i) Hybrid batch normalized AlexNet and extreme learning machine (ELM), (ii) Schmidt neural network (SNN), (iii) Random vector functional link (RVFL) and (iv) the proposed BN-AlexNet-ELM-CBA. The novel architecture gives an excellent performance as follows: sensitivity of 97.14%, specificity of 95.71%, and overall accuracy of 96.43% [22].

3 Lossy and Lossless Compression Algorithms for Medical Images

In this work, lossy and lossless compression algorithms are used for medical images. Vector quantization (VQ)-based algorithms are proposed in this chapter as a part of the lossy approach. The vector quantization (VQ), contextual vector quantization (CVQ), Bat-VQ optimization with static and dynamic codebook designs, and contextual vector quantization coupled with the simulated annealing (CVQ-SA) algorithm are the lossy compression approaches used in this work. In this work, the prediction scheme employs a lossless procedure relying on least squares (LS) [23–25]. For comparative analysis, JPEG lossy and JPEG lossless schemes are also used in this work [26, 27].

3.1 Vector Quantization (VQ)

The VQ is a lossless compression algorithm used in many applications; many variations of vector quantization algorithms are there. VQ calls for partitioning the image into vector components, and for the mapping of each code vector, code words generated from the codebook are used. The generalized Lloyd algorithm (GLA) – also named as Linde Buzo Gray (LBG) algorithm – is used in VQ for the codebook optimization [28]. Figure 1 illustrates the VQ principle.

Let $\boldsymbol{y} = (y_1, y_2, \ldots, y_n)$ represent the training vector, where y_i is the i-th entry. The LBG algorithm steps are summarized, as stated below.

Step 1: Random generation of the initial codebook ($\mathbf{CB_0}$).

Step 2: The Euclidean distance between the training vector and the codeword in the codebook is expressed by

$$\mathbf{ED}(\boldsymbol{x}, \boldsymbol{c}) = \sqrt{\sum_{t=1}^{k} (\boldsymbol{x}_t - \boldsymbol{c}_t)^2}. \tag{1}$$

In the above equation, $\boldsymbol{x}_t$ represents the training vector and $\boldsymbol{c}_t$ represents the code word in the codebook. The closest codeword $\mathbf{CB}_i$ is determined from the Euclidean distance calculation. The flowchart of the codebook optimization is depicted in Fig. 2.

Step 3: The codebook tracked in the previous step is subdivided into $\boldsymbol{N}$ cells.

Step 4: Estimate the cluster centroid to find the new codebook $\mathbf{CB}_{i+1}$.

Step 5: Estimate the mean misrepresentation for $\mathbf{CB}_{i+1}$.

Step 6: The above-said steps are repeated until the distortion value is minimized, else $i = i + 1$, and go to step 3.

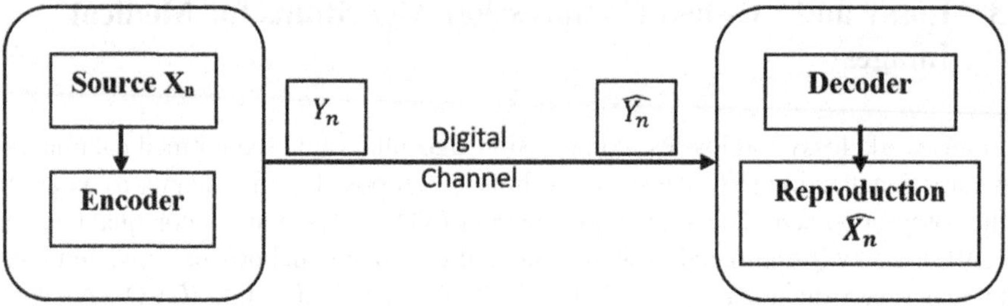

Fig. 1 Principle of vector quantization

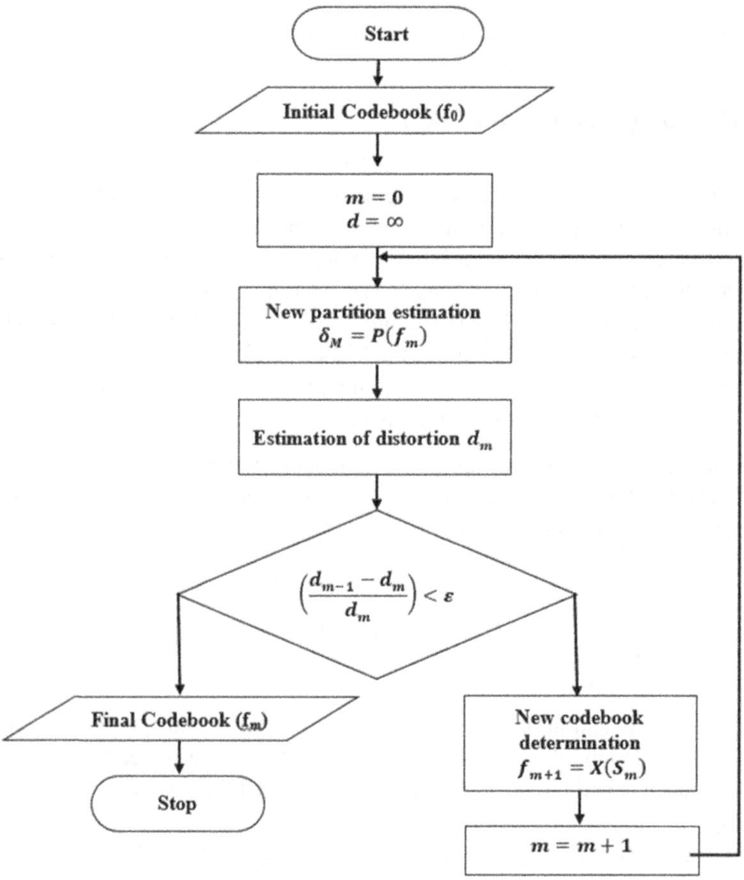

Fig. 2 Flowchart representing the optimization of the codebook

3.2 Bat-VQ Compression Model Based on Static Codebook Design: A Lossy Compression Model

The BOA is utilized in the codebook design of VQ. The parameter tuning is simple in the Bat-VQ algorithm; hence, it is an excellent choice for real-time applications. The steps in the static codebook design are as follows:

Step 1: Parameters are initialized in this step; each bat characterizes a codebook.
Step 2: Calculate the fitness values of all codebooks. The $\boldsymbol{x}_{\mathbf{best}}$ represents the codebook with a good fitness value.
Step 3: The remaining codebooks are tuned toward $\boldsymbol{x}_{\mathbf{best}}$ by the proper selection of parameters.
Step 4: The codebooks are relocated by the $\boldsymbol{x}_{\mathbf{best}}$, when the random number generated is superior to the pulse rate.
Step 5: The random number, when generated, surpasses loudness, and the new codebook fitness is efficient than the previous one, it is accepted as the best solution.
Steps 6: Ranking of bats is done and estimates the current best $\boldsymbol{x}_{\mathbf{best}}$.
Steps 7: Repeat steps (2)–(6) until a termination criterion is reached.

The flowchart depicting the BOA is shown in Fig. 3.

3.3 Bat-VQ Compression Model Based on Dynamic Codebook Design: A Lossy Compression Model

The VQ with static codebook design employs BOA to optimize the codebook, whereas the size of the codebook has to be defined by the user. The performance of the VQ with static codebook design incorporated a BOA for codebook size selection.

The VQ with dynamic codebook design utilizes dynamic optimization for (i) optimum selection of codebook and (ii) tuning of codebook size. The dynamic codebook size selection in the Bat-VQ has superior compression than the classical VQ, JPEG lossy, and Bat-VQ with static codebook size.

The BOA was integrated into the VQ algorithm [29, 30] for the selection of codebook size, and the user defines the lower and upper thresholds. The probability function becomes

$$P(S) = \frac{N - \mathrm{CB_{LL}}}{\mathrm{CB_{UL}} - \mathrm{CB_{LL}}}, \tag{2}$$

where N represents the number of elements in the codebook, and $\mathbf{CB_{LL}}$ and $\mathbf{CB_{UL}}$ represent the lower and upper threshold of the codebook size. The velocity of the bat is reduced when the generated random variable is less than $P(S)$.

Fig. 3 Flowchart depicting the bat optimization algorithm

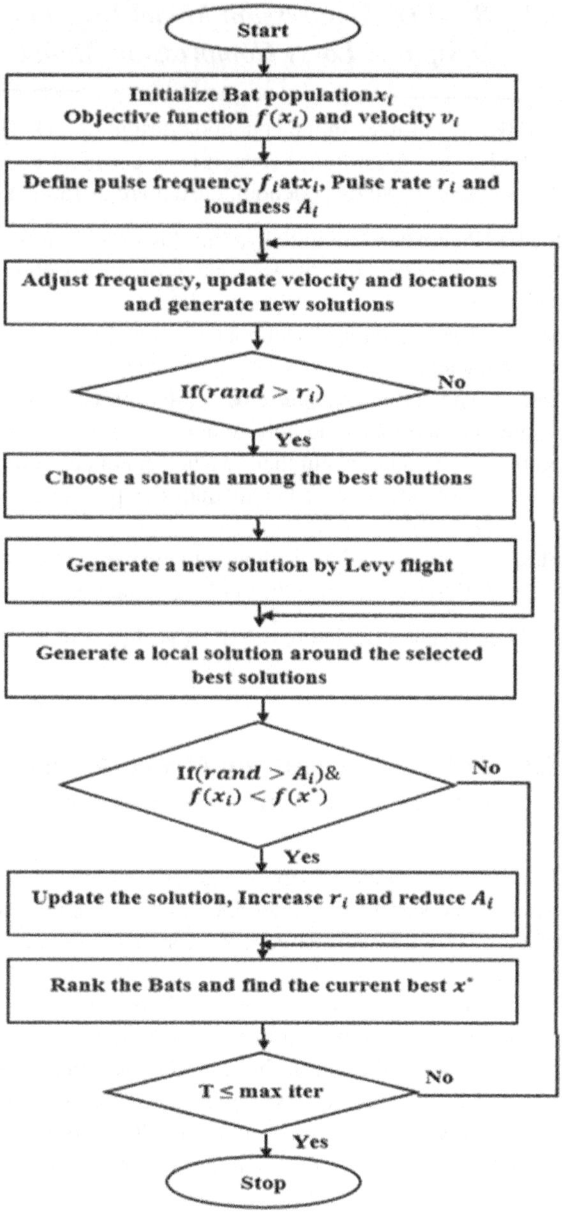

3.4 Contextual Vector Quantization Coupled with Simulated Annealing Optimization: A Lossy Compression Model

The CVQ emphasis on ROI-based compression, the foreground, and background regions are isolated by a segmentation technique [31]. The foreground region was encoded with a high bit rate and low compression ratio. The background region was encoded with a low bit rate and high compression ratio. The SA procedure optimized the merged codebook for the optimum selection of codebook. The performance metrics evaluation reveals that the proposed CVQ-SA outperforms classical compression techniques. The region-growing algorithm was used for the extraction of foreground and background. The foreground and background regions are encoded with a codebook size of 4 and 16; training series sizes of 32 and 128 were used. The CVQ-SA compression model is represented in Fig. 4.

The flowchart depicting simulated annealing optimization appears in Fig. 5. The SA algorithm requires few parameters to be tuned, and the optimized codebook was

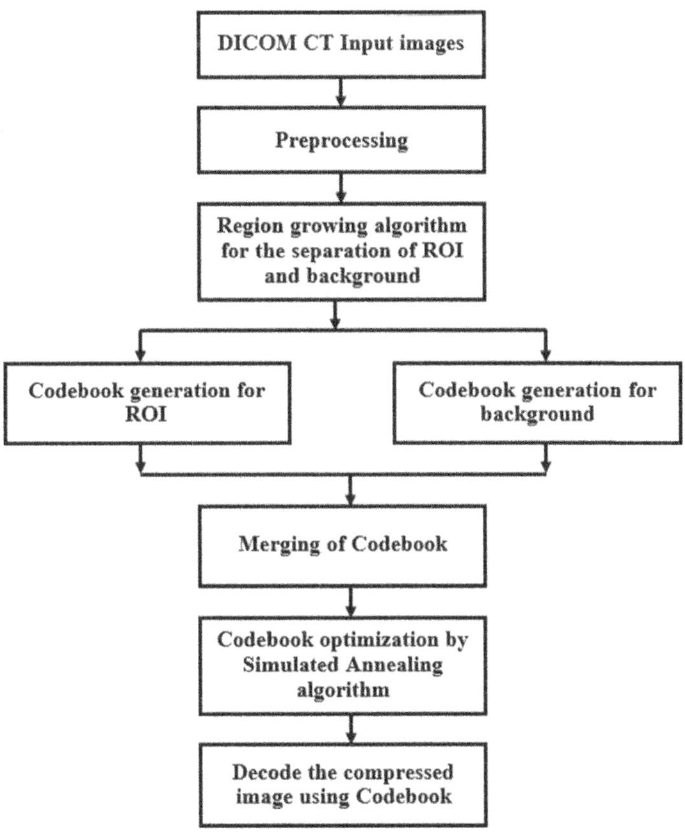

Fig. 4 Flow diagram of the CVQ-SA compression scheme

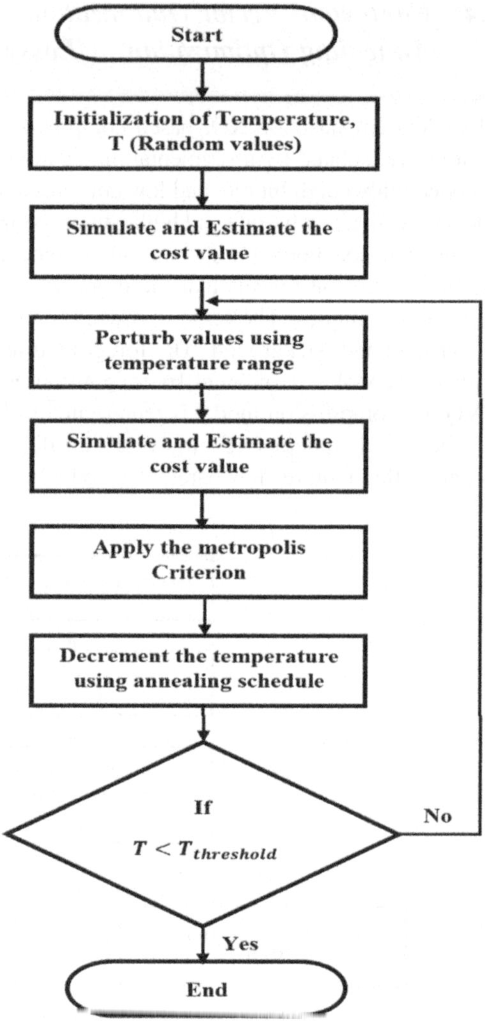

Fig. 5 Flowchart depicting simulated annealing optimization

used for the generation of the compressed image. The SA generates new movement stochastically, and it is a trajectory-based algorithm. It is an extension of the classical metropolis algorithm widely used in many engineering applications. The probability of acceptance of new movement is estimated from the Boltzmann probability distribution function. The SA was found to be effective in the generation of the optimum codebook for vector quantization [32]. The CVQ-SA algorithm generates efficient compression results for abdomen CT images.

The proposed CVQ-SA algorithm outperforms the JPEG lossy, JPEG lossless, VQ, and CVQ approaches. The SA algorithm made the optimum codebook selection from the merged codebook for compression.

3.5 *Least Squares-Based Prediction: A Lossless Compression Model*

The LS-based compression relies on the prediction technique. An appropriate preprocessing scheme can be employed before compression for the filtering of images based on the imaging modality so that there will be an improvement in the quality of the reconstructed image. Different types of predictors are there in the literature, and here in this work, a gradient-adjusted predictor is used. The LS prediction selects the prediction coefficients optimally, and Huffman encodes them. The LS technique groups the pixels' gray values into 7 bins, and it depends on the strength of edges. The flow diagram of the LS-based prediction lossless compression model appears in Fig. 6.

The stages of the LS-based compression technique are prediction, modeling the context of pixels, and entropy encoding. The spatial redundancy is minimized in the prediction approach, and the edge information is traced in the context modeling stage.

The Huffman coding eliminated the statistical redundancy. The gradient adjusted predictor was found to be efficient, since it uses nine pixels for prediction, whereas the median predictor uses 3 pixels for prediction. The median and gradient-adjusted prediction techniques are depicted below in Fig. 7a and b.

The mathematical gradient operation is employed in the gradient-adjusted predictor, and the pixels are categorized into 7 bins based on the calculated slope values.

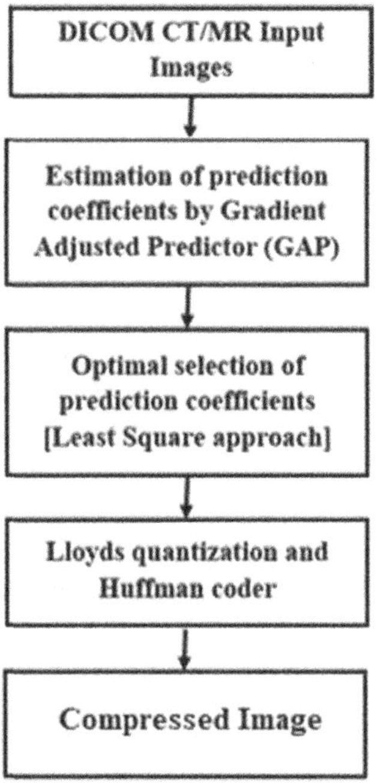

Fig. 6 Flow diagram of the LS-based prediction: a lossless compression model

The first stage is the calculation of gradient in the horizontal and vertical direction of the image. The prediction scheme governs the characteristic features in terms of the slope from the above equations according to

$$S = dv - dh. \tag{3}$$

dv and *dh* represent the horizontal and vertical gradient components of the image. The prediction is performed based on the slope values.

```
if (S > 80)  {Sharp horizontal edge}
    Y(p,q) = 1;   C1 = 1;
else  {1}
    if (S > 32 && S < 80)  {Horizontal edge}
        Y(p,q) = 2;   C2 = 2;
    else {2}
        if (S > 8 && S < 32)     {Weak horizontal edge}
            Y(p,q) = 3;   C3 = 3;
        else {3}
            if (S > −8&& S < 8)  {Soft edge}
                Y(p,q) = 4;    C4 = 4;
            else  {4}
                if (S > −32&& S < −8) {Weak vertical edge}
                    Y(p,q) = 5;    C5 = 5;
                else  {5}
                    if (S>=-80&& S<-32)  {Vertical edge}
                        Y(p,q) = 6;  C6 = 6;
                    else  {6}
                        if (S < −80)
                            Y(p,q) = 7;    C7 = 7;
                        end
                    end  {6}
                end  {5}
            end  {4}
        end  {3}
    end  {2}
end  { 1}
```

C_1, C_2, C_3, C_4, C_5, C_6, and C_7 represent the slope bin values ranging from 1 to 7. The LS method generates a polynomial for the row data in an image as

$$v_i = w_0 + w_1 u_1 + w_2 u_2 + \ldots + w_n u_n. \tag{4}$$

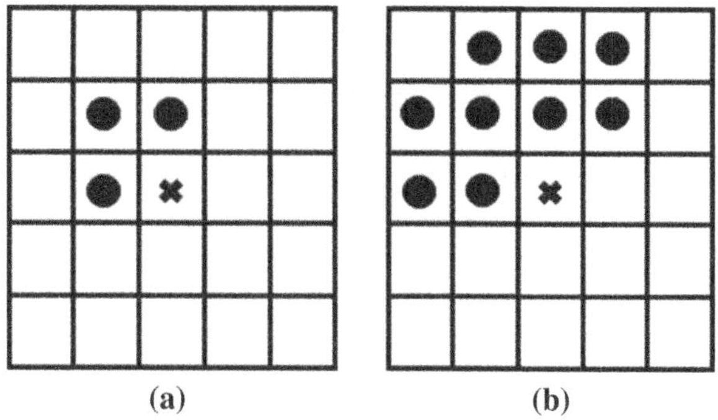

Fig. 7 Prediction schemes (**a**) median adjusted prediction scheme, (**b**) gradient adjusted prediction scheme

The term $\boldsymbol{v}_i$ represents the row vector of an image, and $\boldsymbol{u}$ means the row data. In the above equation, the weights are calculated from the linear equation. The above equation in matrix form is expressed as

$$\boldsymbol{V} = \boldsymbol{U}\boldsymbol{W}, \text{ or} \tag{5}$$

$$\begin{bmatrix} \boldsymbol{v}(0,1) \\ \vdots \\ \boldsymbol{v}(0,p) \end{bmatrix} = \begin{bmatrix} \boldsymbol{u}(1,1) & \dots & \boldsymbol{u}(1,p) \\ \vdots & \vdots & \vdots \\ \boldsymbol{u}(p,1) & \dots & \boldsymbol{u}(p,p) \end{bmatrix} \begin{bmatrix} \boldsymbol{w}(1) \\ \vdots \\ \boldsymbol{w}(p) \end{bmatrix}, \tag{6}$$

where $\boldsymbol{u}(m,n)$ represent the elements of the covariance matrix as

$$\boldsymbol{u}(m,n) = \sum_{m,n\in W} \boldsymbol{x}(n-s)\boldsymbol{x}(n-T). \tag{7}$$

The weight coefficients are estimated from

$$\boldsymbol{W} = \boldsymbol{U}^{-1}\boldsymbol{V}. \tag{8}$$

The computation complexity is $\boldsymbol{O}(n^2)$, and the predicted coefficients are subjected to quantization before the transmission of encoded data. The prediction scheme is depicted below in Fig. 8.

The prediction image is

$$\widehat{Y}(p,q) = \boldsymbol{f}(\varnothing(p,q)), \tag{9}$$

where $\varnothing(p,q)$ represents the casual context of neighboring pixels. The error component of the image becomes

$$\boldsymbol{e}(p,q) = \boldsymbol{Y}(p,q) - \widehat{\boldsymbol{Y}}(p,q). \tag{10}$$

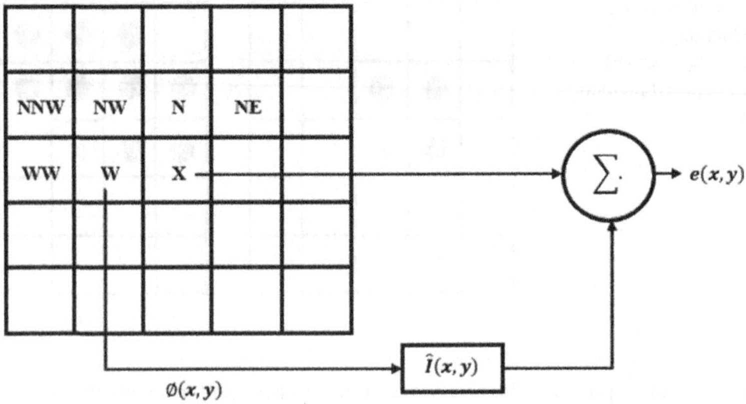

Fig. 8 Estimation of the prediction coefficients

The Huffman coder encodes the predicted coefficients, and the decoder section performs the reconstruction of the image.

4 Results and Discussion

This research work proposes lossy and lossless compression algorithms for the DICOM medical images developed in Matlab 2015a and tested on real-time DICOM CT/MR images. The performance validation of compression results reveals the efficiency of compression techniques. The qualitative analysis is not sufficient for the validation of compression algorithms, and hence, quantitative analysis is needed. The quality of the reconstructed image is vital, and, thus, performance metrics are employed for the validation of the compression model.

For analysis, the following algorithms are taken into account: Bat-VQ based on static codebook design, Bat-VQ based on dynamic codebook design, CVQ, CVQ-SA, JPEG lossy (JPEG), JPEG lossless (JPEG-LL), and LS-based prediction (LS). The real-time abdomen CT images came from Metro Scans and Research Laboratory for analysis of algorithms.

Let $\boldsymbol{I}(i,j)$ and $\boldsymbol{R}(i,j)$ characterize the pixel values of the uncompressed input image and reconstructed image after decompression. Table 1 clarifies the performance metrics.

The performance metrics are determined for compression algorithms [33] and are depicted below in graphs (Figs. 9, 10, 11, 12, 13, 14, 15, 16, 17, and 18).

According to the peak signal-to-noise ratio (PSNR) plot from Fig. 9, the CVQ-SA algorithm excelled (in efficiency) the CVQ algorithm. Out of the lossy compression techniques, Bat-VQ based on dynamic codebook design outshined the CVQ, CVQ-SA, Bat-VQ based on static codebook design, and JPEG lossy algorithms in

Table 1 Performance metrics for compression algorithms

Sl no	Metrics	Formula
1	Peak signal-to-noise ratio (PSNR)	$\mathbf{10\log\left(\frac{255^2}{MSE}\right)}$
2	Mean square error (MSE)	$\frac{1}{mn}\sum_{i=0}^{m-1}\sum_{j=0}^{n-1}[I(i,j)-R(i,j)]^2$
3	Compression ratio.	$\frac{\text{Size of the uncompressed image}}{\text{Size of the compressed image}}$
4	Normalized cross-correlation (NCC)	$\sum_{i=1}^{m}\sum_{j=1}^{n}\frac{I(i,j)\times R(i,j)}{I(i,j)^2}$
5	Normalized absolute error (NAE)	$\frac{\sum_{i=1}^{m}\sum_{j=1}^{n}(I(i,j)-R(i,j))}{\sum_{i=1}^{m}\sum_{j=1}^{n}I(i,j)}$
6	Average difference (AD)	$\frac{1}{mn}\sum_{i=1}^{m}\sum_{j=1}^{n}[I(i,j)-R(i,j)]$
7	Structural content (SC)	$\frac{\sum_{i=1}^{m}\sum_{j=1}^{n}(I(i,j))^2}{\sum_{i=1}^{m}\sum_{j=1}^{n}(R(i,j))^2}$
8	Laplacian mean square error (LMSE)	$\frac{\sum_{i=1}^{m}\sum_{j=1}^{n}[L(I(i,j))-L(R(i,j))]^2}{\sum_{i=1}^{m}\sum_{j=1}^{n}[L(I(i,j))]^2}$
9	Space saving	$1-\frac{\text{Size of compressed image}}{\text{Size of uncompressed image}}$
10	Total matching percentage	$\left(\frac{\text{Matching pixels}}{\text{Total pixels}}\right)*\mathbf{100}$

terms of PSNR. The LS-based compression was found to be proficient when compared with the other algorithms. The higher value of PSNR and the lower value of MSE signify the quality of the compression technique. The PSNR reflects the quality of the reconstructed image: the higher its value, the more proficient the compression algorithm. The compression ratio shows the efficiency of the algorithm in reducing the input file size. Apart from the classical picture performance metrics like PSNR, MSE, and CR, other evaluation metrics like NCC, NAE, AD, LMSE, SC, space savings, and total matching percentage are also used. The value of NCC should be greater than 0.9; the value of SC should be closer to 1; the values of NAE and LMSE should be low for an efficient compression algorithm. The space savings represents the efficiency of the compression algorithm, and the total matching percentage qualifies the quality of the reconstructed image.

The SC plot shows up in Fig. 10. The SC plots reveal that the compression algorithms proposed in this work are all yielding a fair value of structural content. The higher value of SC indicates that the reconstructed image is of poor quality.

Figure 11 displays the NK plot. The ideal value of NK is 1, and it reveals the efficiency of the compression algorithm.

The AD plot from Fig. 12 describes the efficiency of the compression algorithms (the ideal value of AD is 0).

The LMSE plot is depicted in Fig. 13. The lower the value of LMSE, the better is the compression technique.

The NAE plot appears in Fig. 14. The lower the value of NAE, the better is the compression technique.

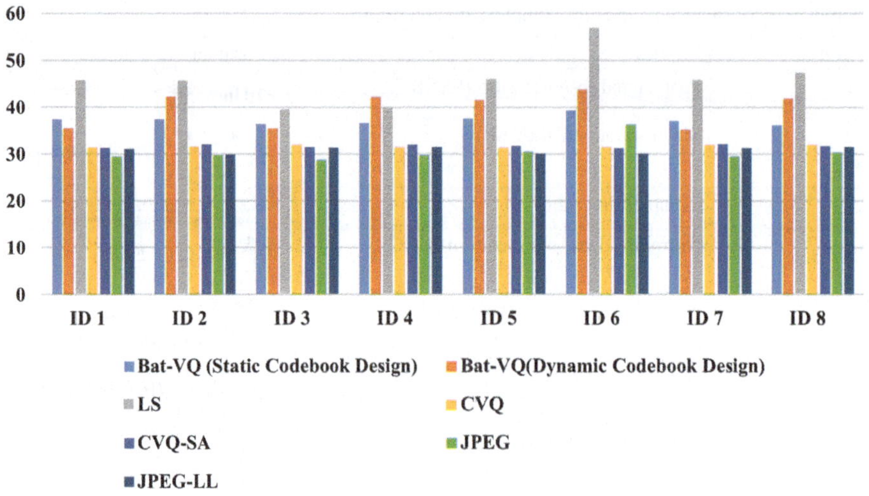

Fig. 9 PSNR plot for the investigated compression techniques

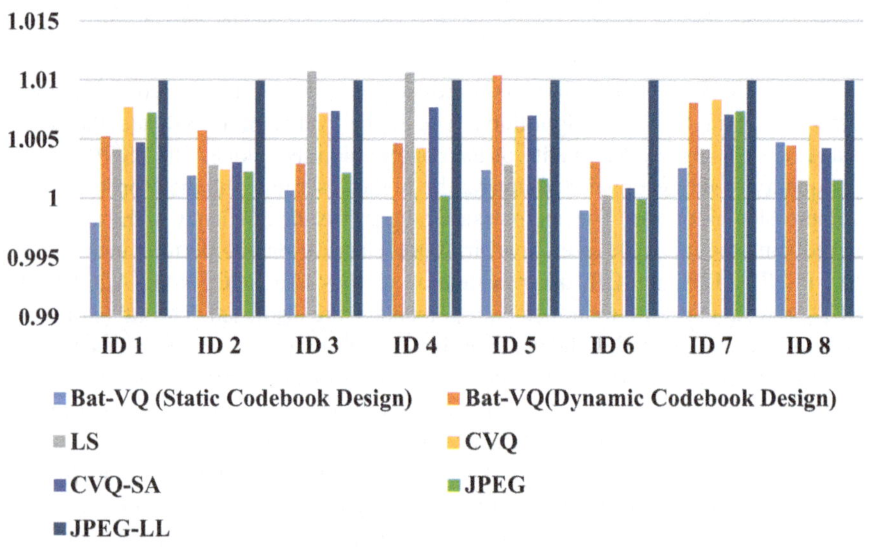

Fig. 10 SC plot of compression techniques

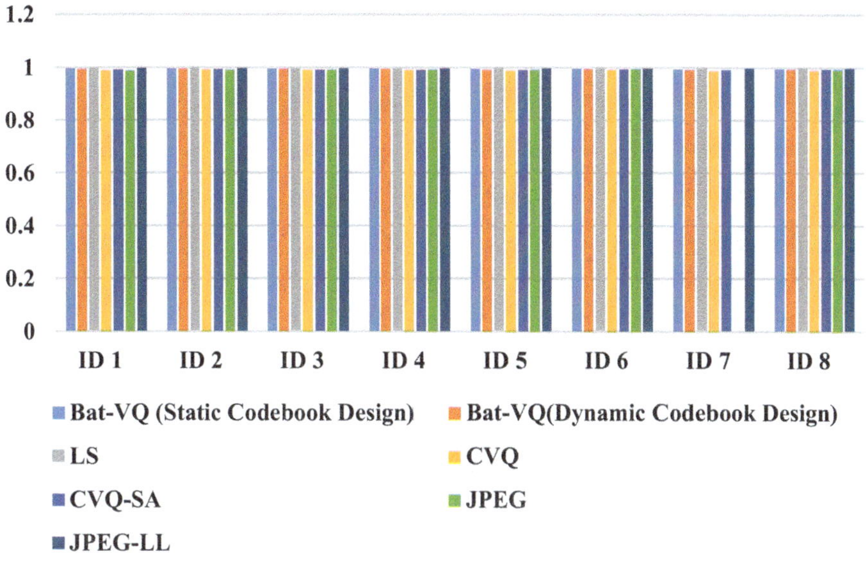

Fig. 11 NK plot of compression techniques

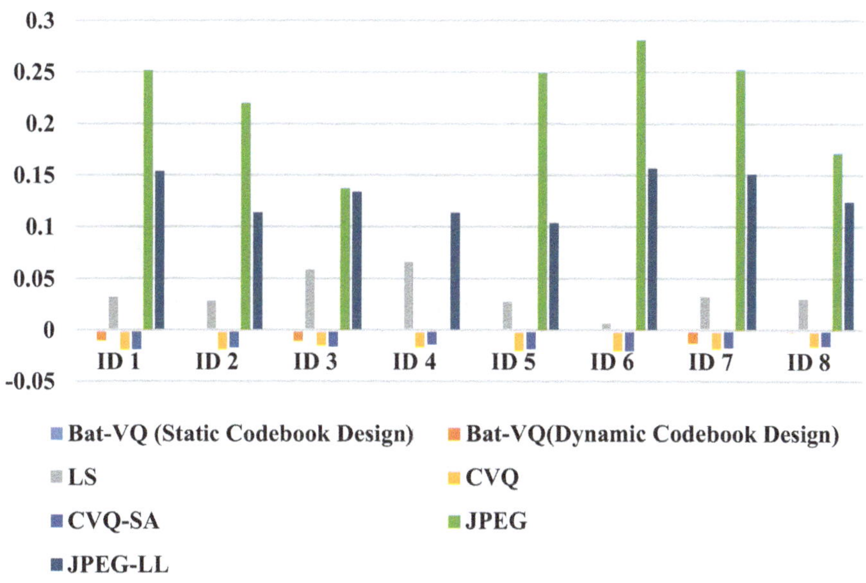

Fig. 12 AD plot for compression techniques

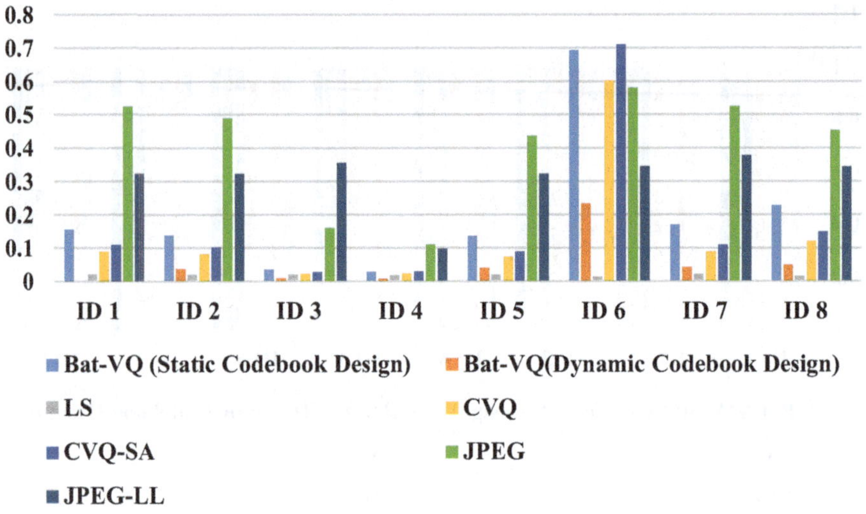

Fig. 13 LMSE plot of compression techniques

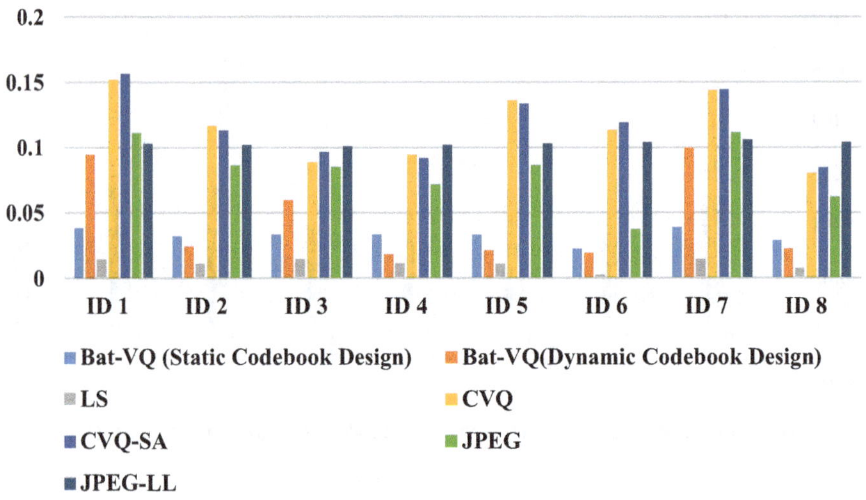

Fig. 14 NAE plot of compression techniques

The space savings plot is shown in Fig. 15.

The nbits/pixels plot is represented in Fig. 16. The bit per pixel is the total number of bits in the compressed image, divided by the number of pixels in the compressed image.

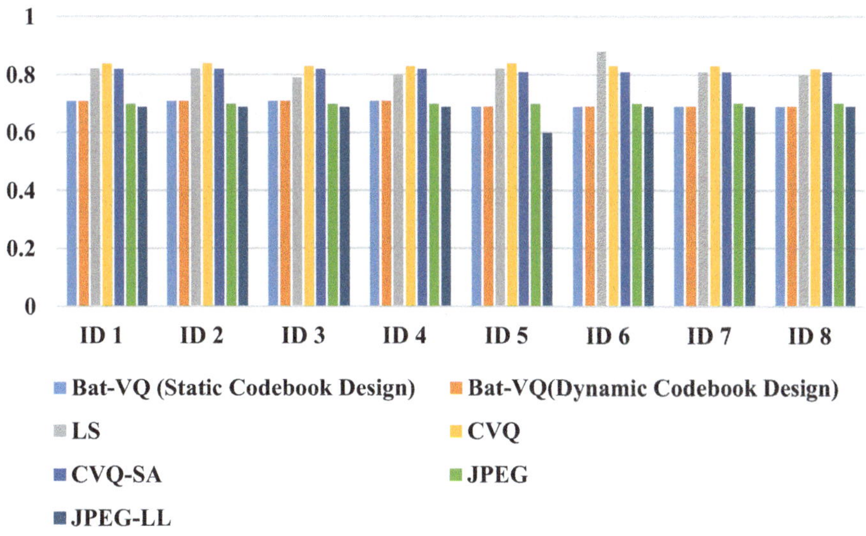

Fig. 15 Space savings plot of compression techniques

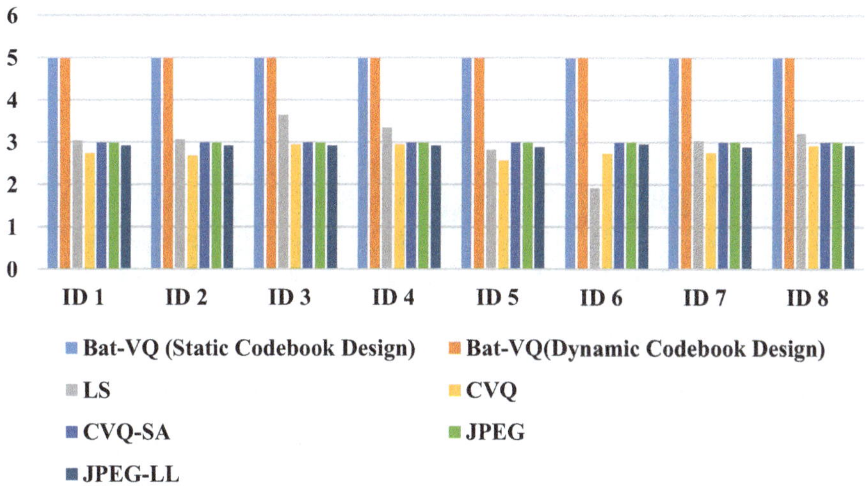

Fig. 16 nbits/pixel of compression techniques

The total matching percentage appears in Fig. 17. The expression for the total matching percentage is shown above. The compression ratio plot looks like in Fig. 18. The input abdomen CT images are depicted in Fig. 19.

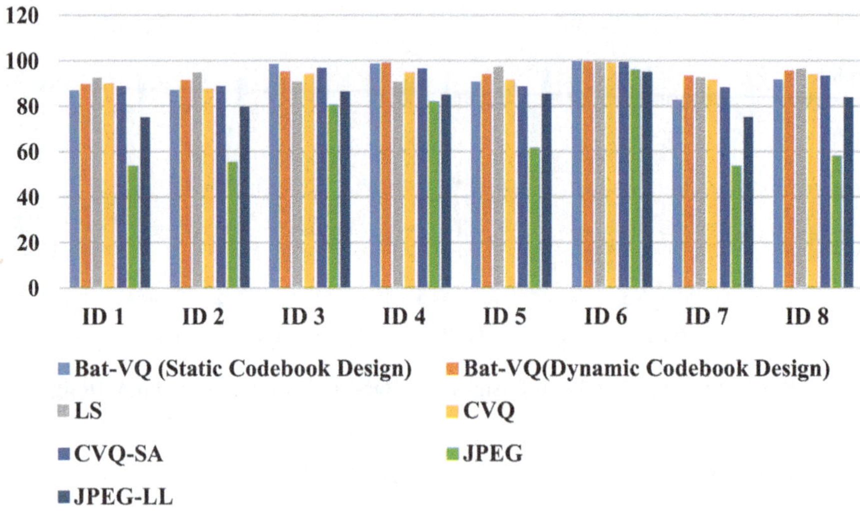

Fig. 17 Percentage of matching for compression techniques

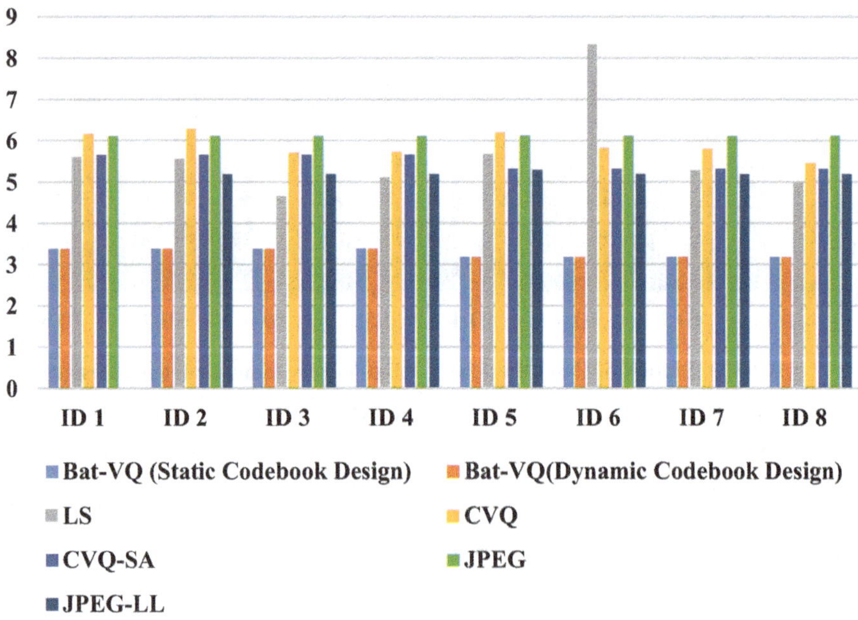

Fig. 18 CR plot of compression techniques

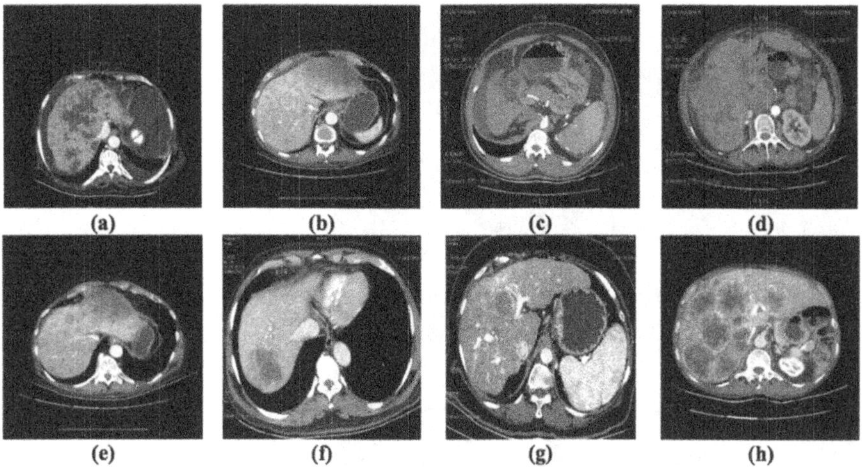

Fig. 19 Input abdomen CT images (a-h) (ID1–ID9)

In Figs. 20 and 21, the first column depicts the compressed image, the second column depicts the reconstructed image, and the third column depicts the bat algorithm optimization plot.

In Figs. 22 and 23, the first column depicts the compressed image, the second column depicts the reconstructed image, and the third column depicts the BOA plot.

In Figs. 24 and 25, the first column depicts the compressed image, the second column represents the prediction image, the third column depicts the error image, and the fourth column represents the reconstructed image.

In Figs. 26 and 27, the first column depicts the normalized outputs, the second column represents the extraction of background, the third column depicts the compressed image, and the fourth column describes the reconstructed image.

In Figs. 28 and 29, the first column depicts the normalized outputs, the second column represents the extraction of background, the third column depicts the compressed image, and the fourth column describes the reconstructed image.

The mean value of performance metrics is depicted in Table 2. The following inferences are made from Table 2. The LS-based prediction lossless compression technique and Bat-VQ based on dynamic codebook design lossy compression technique are efficient in terms of PSNR and were also found to be proficient in terms of other metrics.

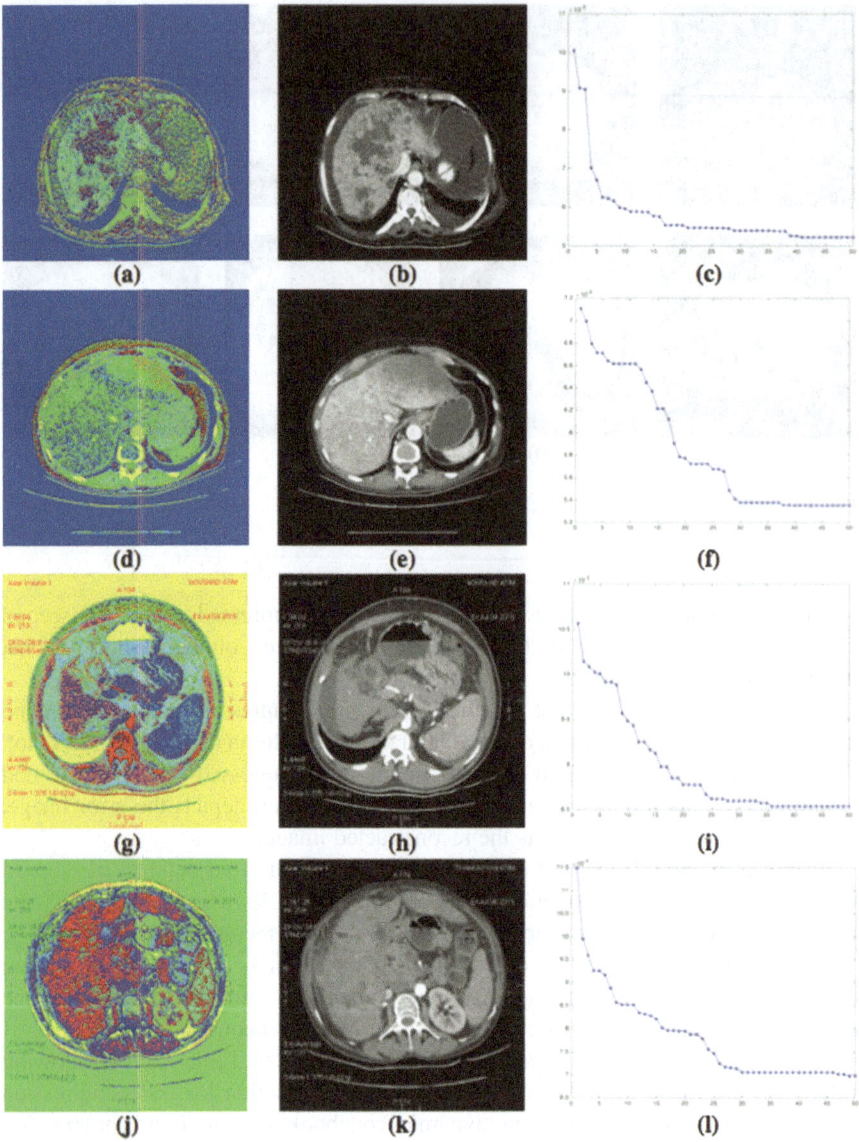

Fig. 20 Bat-VQ based on static codebook design outputs (a-l) (ID1–ID4)

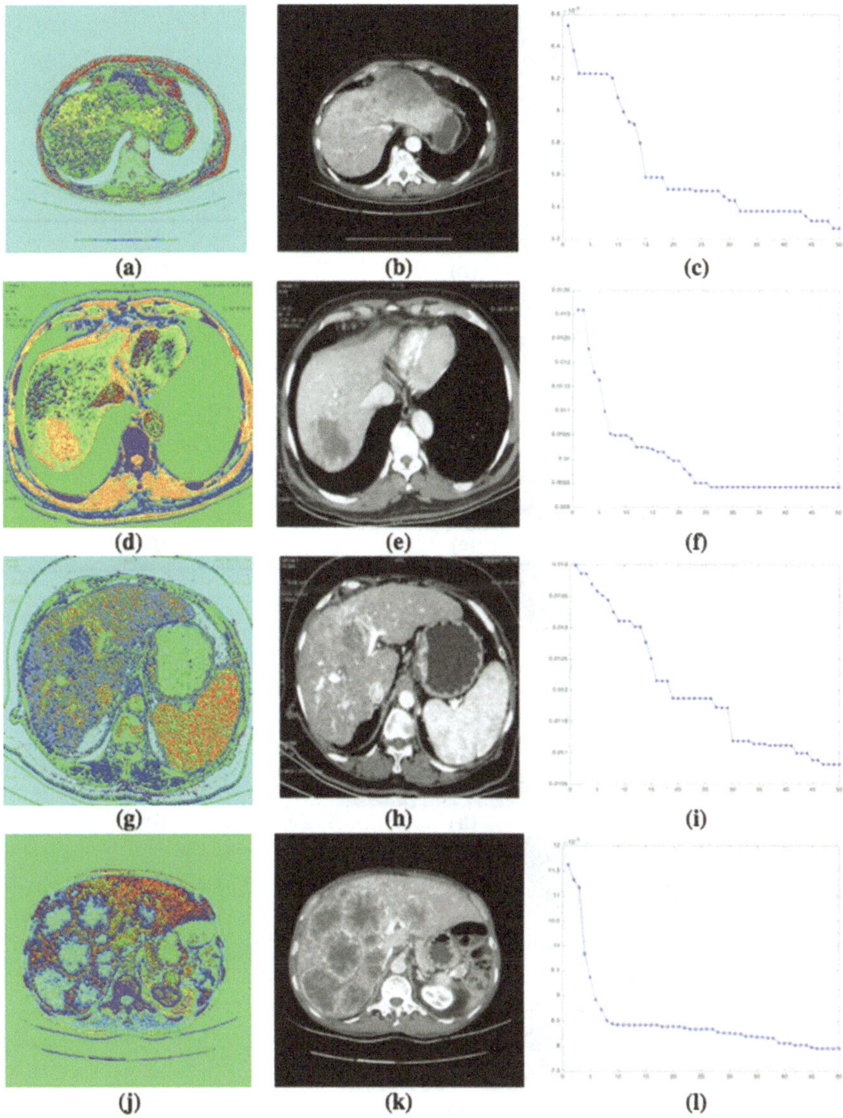

Fig. 21 Bat-VQ based on static codebook design outputs (a-l) (ID5–ID8)

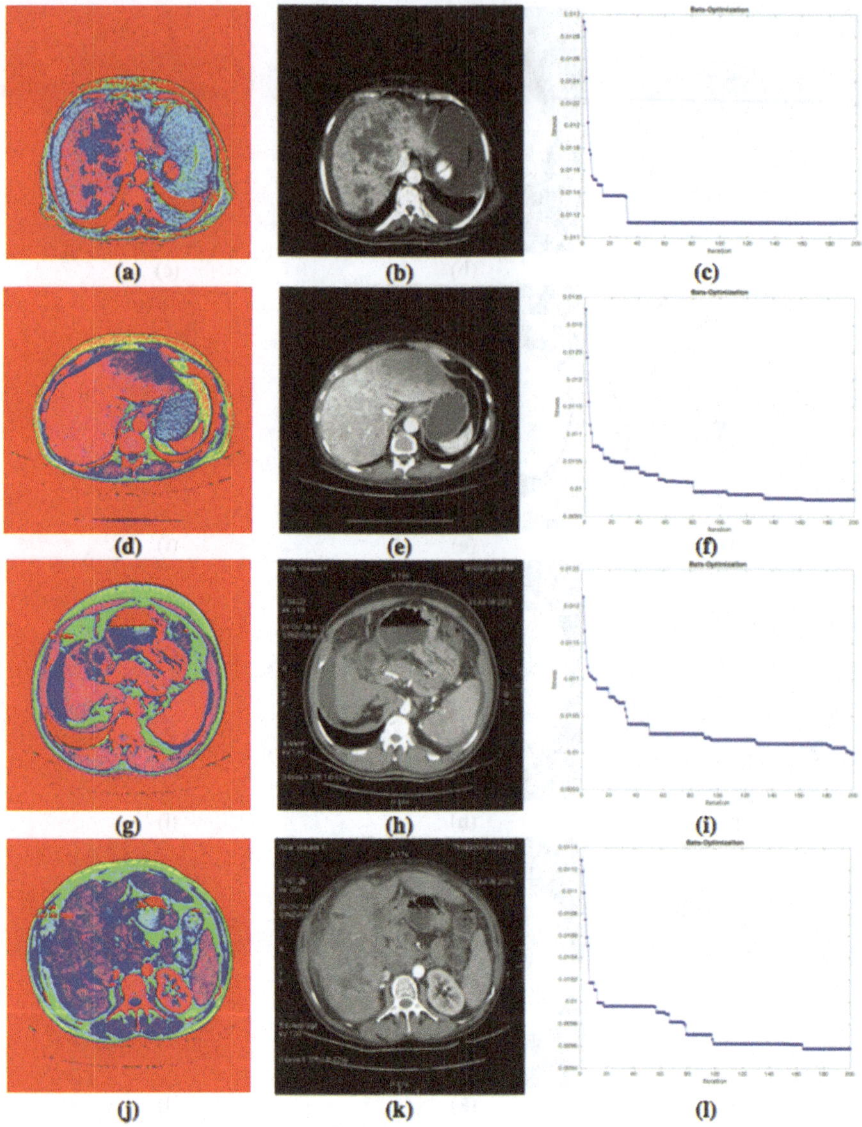

Fig. 22 Bat-VQ based on dynamic codebook design outputs (a-l) (ID1–ID4)

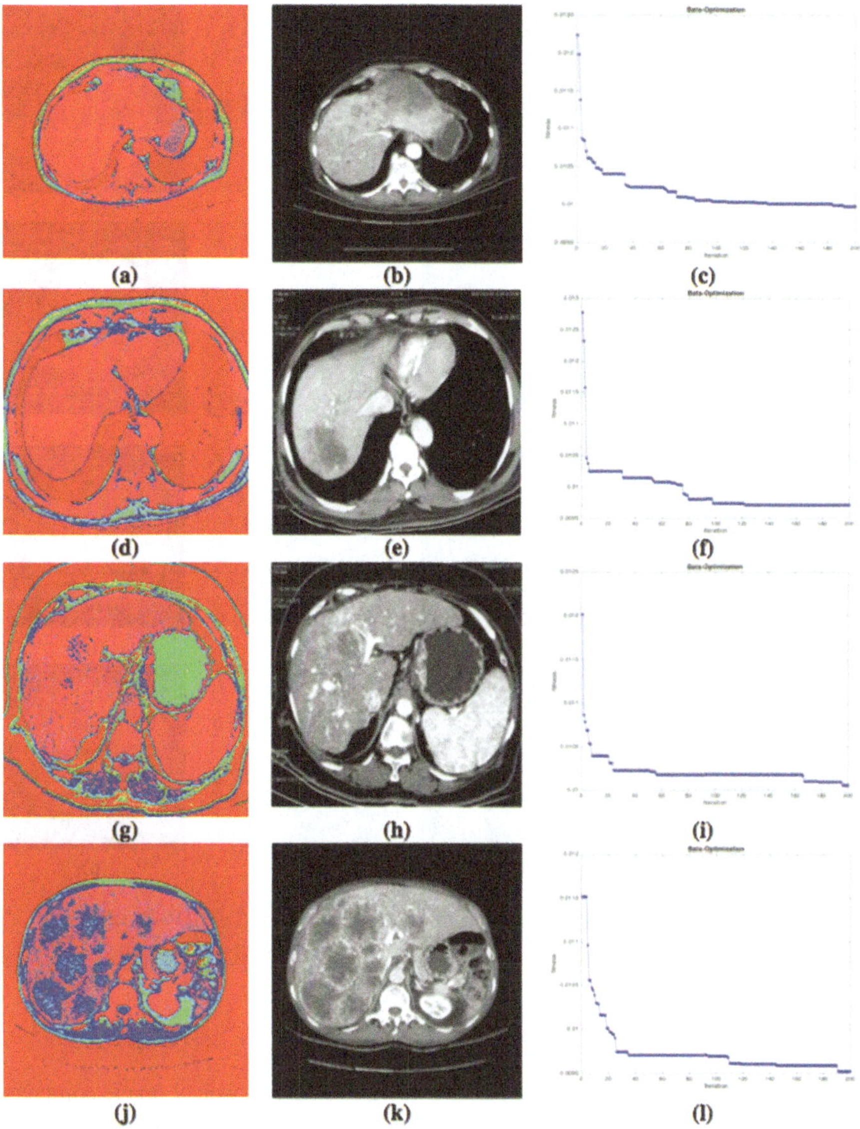

Fig. 23 Bat-VQ based on dynamic codebook design outputs (a-l) (ID5–ID8)

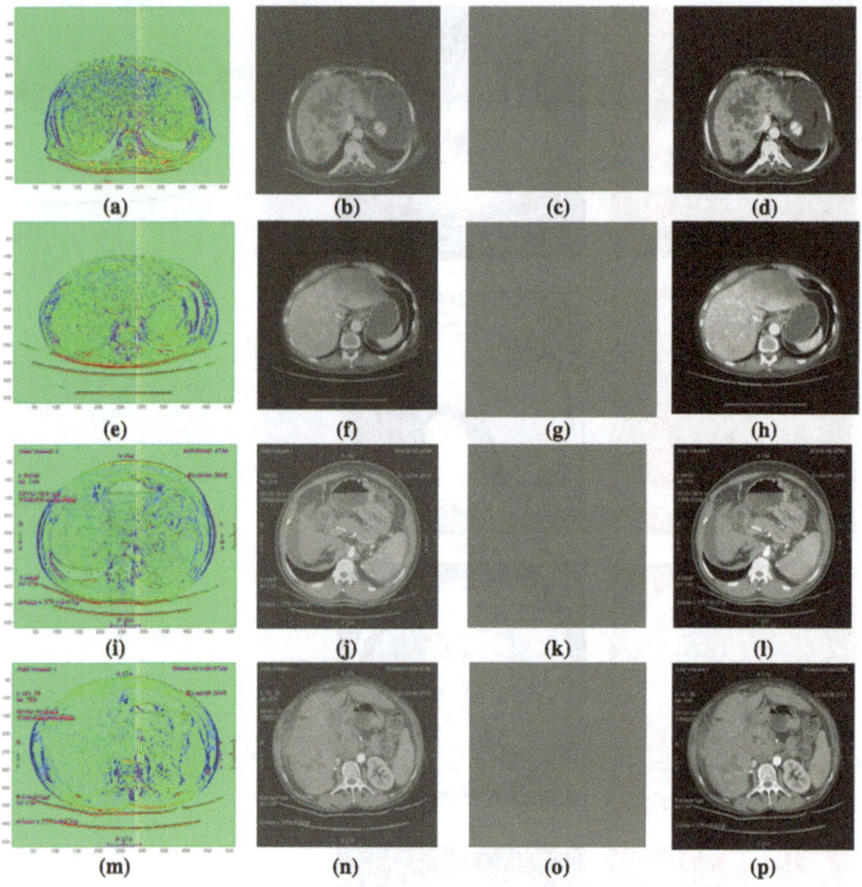

Fig. 24 Least square-based prediction lossless compression outputs (a-p) (ID1–ID4)

5 Future Works

Future work will emphasize on the super-resolution-based medical image compression. Super-resolution improves the resolution of image; the low-resolution images are converted into high-resolution images. The super-resolution eliminates the blur created by a low-resolution camera. This technique is employed to generate the original image with a high-resolution image for a given set of images with low resolution. For large datasets, super-resolution-based image compression generates a good compression ratio with optimum computation time. An autoregressive super-resolution network was employed in [34] for lossless image compression. The high-resolution images are preferred for medical image analysis since physicians are using disease diagnosis and therapeutic planning. In [35], feedback adaptive weighted dense network (FAWDN) is used for the medical image compression. The FAWDN

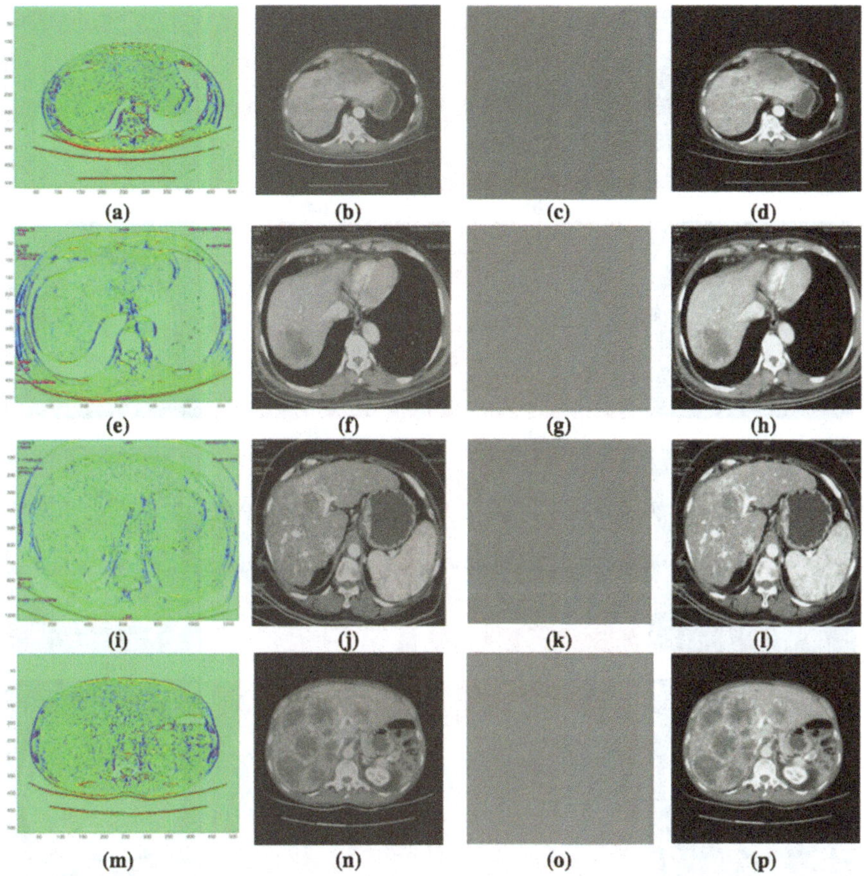

Fig. 25 Least square-based prediction lossless compression outputs (a-p) (ID5–ID8)

performance was found to be superior when compared with the classical super-resolution techniques. The deep residual neural network was employed in [36] for the efficient analysis of lesions in the diabetic retinopathy images. A hybrid lossless compression scheme comprising singular value decomposition, wavelet difference reduction, and discrete wavelet transform was used in [37], and the resolution enhancement was performed by discrete wavelet transform.

The medical images have a low signal-to-noise ratio (SNR) when compared with the natural images. The generative adversarial network (GAN)-based deep learning architecture plays a vital role in many computer vision applications, and in [38], it was employed for single image super-resolution. The lesion-focused super-resolution (LFSR) was coupled with the novel multi-scale GAN and generated superior results. The LFSR, coupled with multi-scale GAN, generates efficient results for MR brain images. The wavelet-based multi-channel and multi-scale

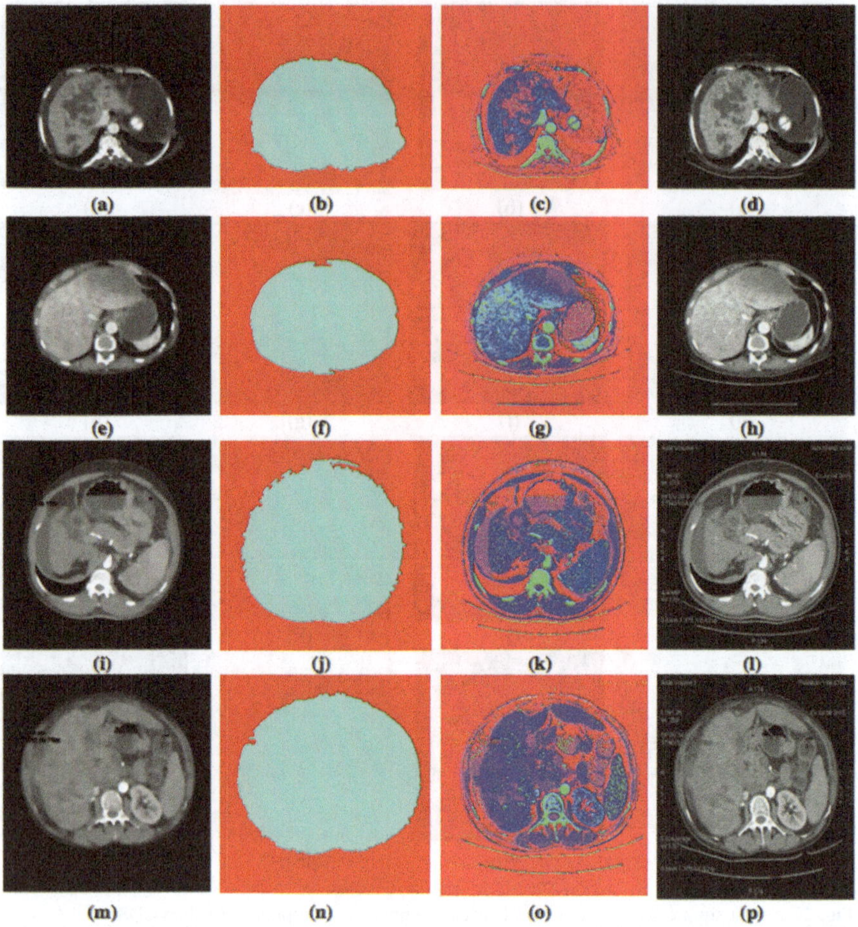

Fig. 26 CVQ compression outputs (a-p) (ID1–ID4)

cross-connected residual-in-dense grouped convolutional neural network (WCRDGCNN) was proposed for the super-resolution enhancement of medical images, and db9 wavelet was employed in this work. The edge preservation was efficient in this approach. PSNR and SSIM did performance metrics evaluation with the highest values of 36.18 and 0.9597 [39]. The progressive generative adversarial network (P-GAN) was proposed in [40] for super-resolution enhancement of MR and retinal images. The triplet loss function used along with the P-GAN improves the quality of the image and works well for cardiac LV segmentation. An improved GAN was proposed for the medical image super-resolution and generated efficient results compared with the existing deep learning models like EDSR, SRGAN, VDSR, and D-DBPN [41]. Novel medical image reconstruction was proposed

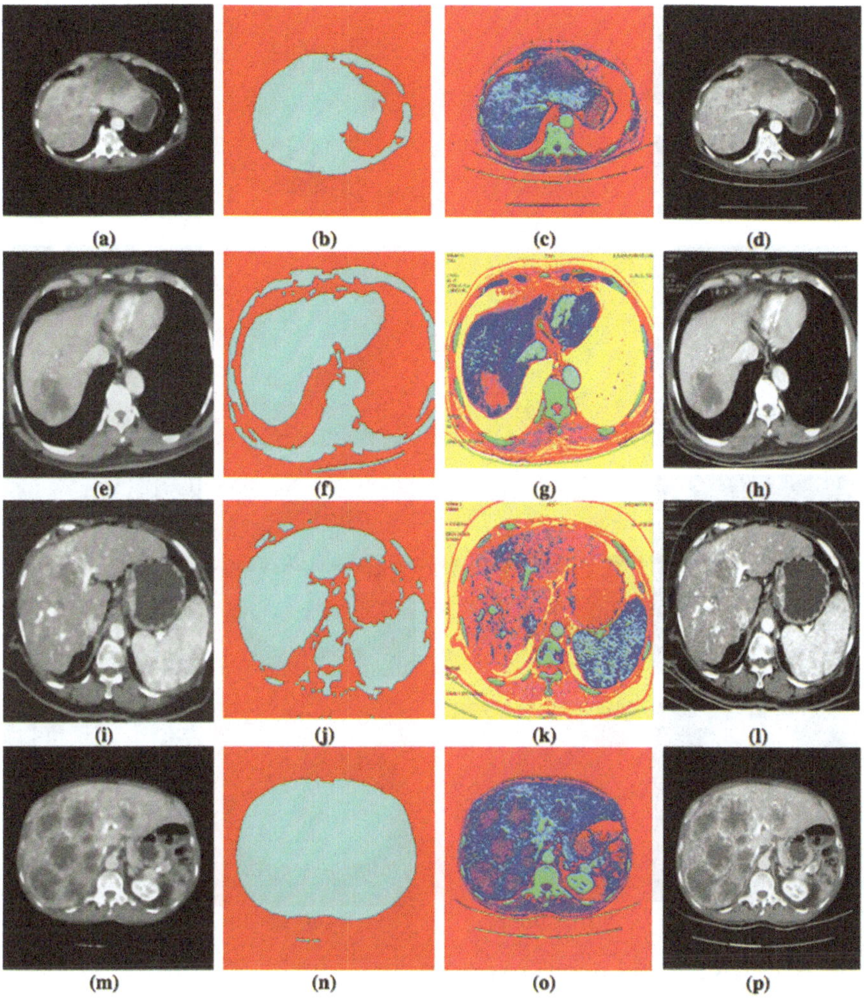

Fig. 27 CVQ compression outputs (a-p) (ID5–ID8)

based on the adaptive multi-dictionary learning and structural self-similarity for medical images [42]. The residual convolution neural network was submitted for the 3D MR images and generated efficient results compared with the classical techniques [43]. The role of the compression algorithm in the medical field was highlighted in [44], and different medical imaging modalities with classical compression models are also discussed.

Artifacts can also be improved with the help of metaheuristics [45, 46].

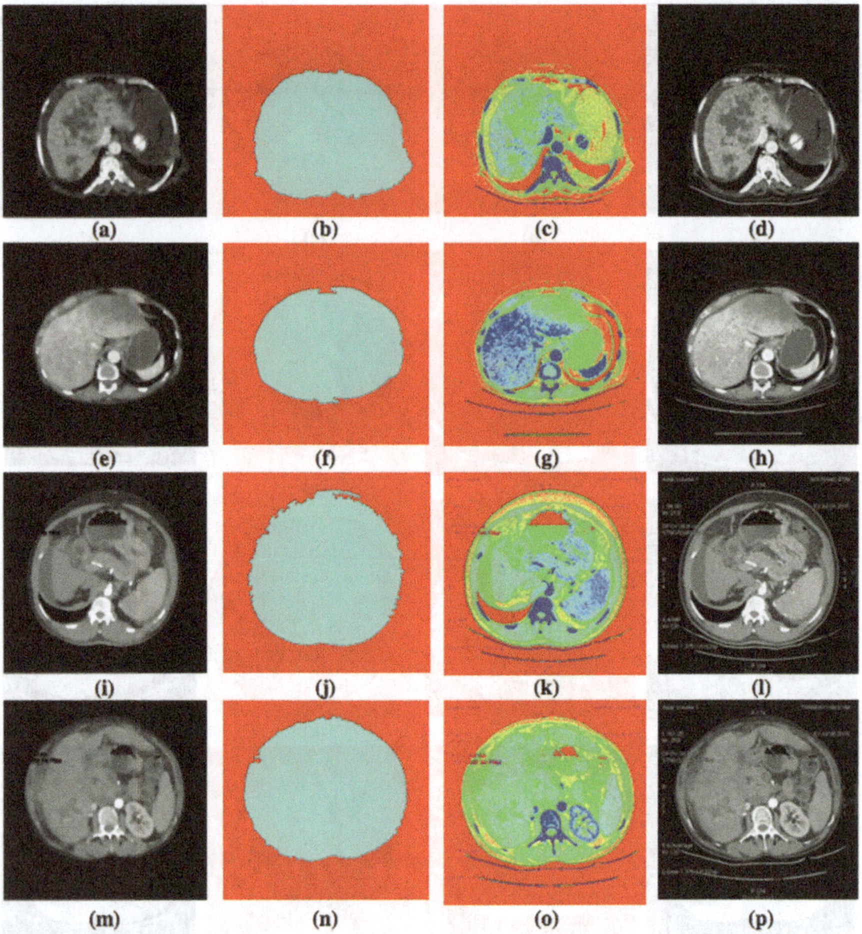

Fig. 28 CVQ-SA compression outputs (a-p) (ID1–ID4)

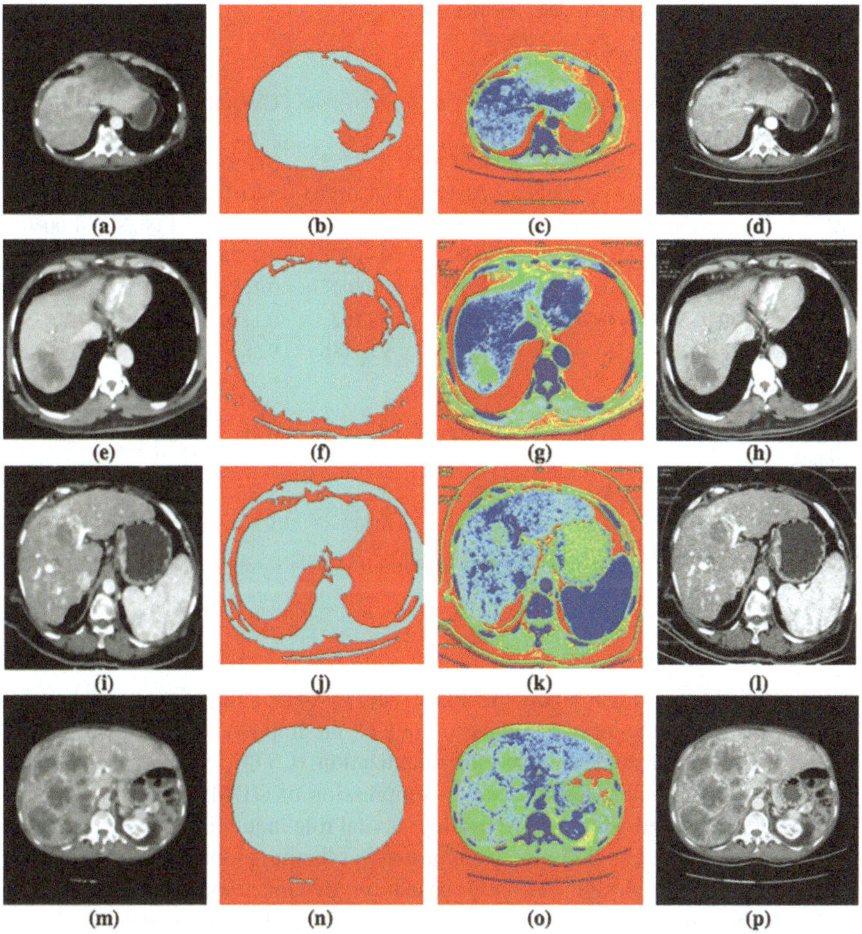

Fig. 29 CVQ-SA compression outputs (a-p) (ID5–ID8)

Table 2 Mean values of performance metrics

Mean value of metrics	Bat-VQ (Static codebook design)	Bat-VQ (Dynamic codebook design)	LS	CVQ	CVQ-SA	JPEG Lossy	JPEG Lossless
PSNR	37.3335	39.8467	45.8499	31.8295	31.797	30.6969	30.9187
AD	0.0002	−0.0042	0.0348	−0.01703	−0.0172	0.2240	0.1317
SC	1.0010	1.0056	1.0046	1.0054	1.0053	1.0028	1.0060
NK	0.9987	0.9966	0.9975	0.9946	0.9948	0.9954	0.9995
LMSE	0.20045	0.0616	0.0187	0.1402	0.1676	0.4123	0.3128
NAE	0.0329	0.0453	0.0107	0.1162	0.1179	0.0818	0.1035
Percentage of matching	92.3223	95.0664	94.295	93.3186	92.868	68.0389	83.4422
Space savings	0.70	0.70	0.81	0.83	0.81	0.70	0.67
nbits/pixel	5	5	3	3	3	3	3
CR	3.3	3.3	5.64	5.92	5.5	6.13	5.21

6 Conclusion

This research work focuses on the lossy and lossless compression models for medical images. In the lossy compression model, the improved vector quantization based on BOA and contextual vector quantization (CVQ) relying on simulated annealing yields efficient results for the compression of CT/MR images compared with the classical methods. The PSNR plays a vital role in judging the reconstructed image quality, hence out of the lossy compression schemes proposed; Bat-VQ-based on static codebook design, Bat-VQ based on dynamic codebook design and CVQ-SA, best results are given by Bat-VQ based on dynamic codebook design and CVQ-SA schemes. In terms of the compression ratio, under the lossy schemes proposed, the CVQ-SA worked efficiently. The LS-based prediction lossless algorithm was also proposed in this research work that yields robust compression results compared with the classical algorithms in terms of PSNR and the compression ratio. Also, metrics like structural similarity index, visual information fidelity, and visual SNR help validate compression algorithms.

Acknowledgments The authors would like to acknowledge the support from Metro Scans and Research Laboratory, Trivandrum, for providing the DICOM medical images in this research work. The author S.N Kumar would also like to acknowledge the Schmitt Centre for Biomedical Instrumentation (SCBMI) of Amal Jyothi College of Engineering for supporting this research work.

References

1. Ukrit, M. F., Umamageswari, A., & Suresh, G. R. (2011). A survey on lossless compression for medical images. *International Journal of Computers and Applications, 31*, 47–50.
2. Yue, L., Shen, H., Li, J., et al. (2016). Image super-resolution: The techniques, applications, and future. *Signal Processing, 128*, 389–408. https://doi.org/10.1016/j.sigpro.2016.05.002.
3. Deshpande, A., Patavardhan, P., Estrela, V. V., & Razmjooy, N. (2020). Deep learning as an alternative to super-resolution imaging in UAV systems. In V. V. Estrela, J. Hemanth, O. Saotome, G. Nikolakopoulos, & R. Sabatini (Eds.), *Imaging and sensing for unmanned aircraft systems* (Vol. 2, 9, pp. 177–212). London: IET. https://doi.org/10.1049/PBCE120G_ch9.
4. de Jesus, M. A., Estrela, V. V., Saotome, O., & Stutz, D. (2018). Super-resolution via particle swarm optimization variants. In J. Hemanth & V. Balas (Eds.), *Biologically rationalized computing techniques for image processing applications. Lecture notes in computational vision and biomechanics* (Vol. 25, pp. 317–337). Cham: Springer. https://doi.org/10.1007/978-3-319-61316-1_14.
5. Dong, C., Deng, Y., Loy, C. C., & Tang, X. (2015). Compression artifacts reduction by a deep convolutional network. In *2015 IEEE International Conference on Computer Vision*, pp. 576–584. https://doi.org/10.1109/ICCV.2015.73.
6. Shen, J. J., Yeh, C. H., & Jan, J. K. (2019). A new approach of lossy image compression based on hybrid image resizing techniques. *International Arab Journal of Information Technology, 16*, 226–235.
7. Cheng, Z., Sun, H., Takeuchi, M., & Katto, J. (2018). Performance comparison of convolutional autoencoders, generative adversarial networks and super-resolution for image compression. In *CVPR Workshops 2018*, pp. 2613–2616.
8. Sharma, U., Sood, M., & Puthooran, E. (2018). Lossless compression of medical image sequences using a resolution independent predictor and block adaptive encoding. *International Journal of Electrical and Computer Engineering, 9*, 69–79. https://doi.org/10.32985/ijeces.9.2.4.
9. Heindel, A., Wige, E., & Kaup, A. (2017). Low-complexity enhancement layer compression for scalable lossless video coding based on HEVC. *IEEE Transactions on Circuits and Systems for Video Technology, 27*, 1749–1760. https://doi.org/10.1109/TCSVT.2016.2556338.
10. Kiran, B., & Khedkar, A. S. D. (2016). Fingerprint compression using singular value decomposition. *International Journal of Science and Research, 5*, 884–887.
11. Coelho, A. M., & Estrela, V. V. (2012). EM-based mixture models applied to video event detection. In P. Sanguansat (Ed.), *Principal component analysis – engineering applications* (pp. 101–124). London: IntechOpen. https://doi.org/10.5772/38129.
12. Blinder, D., Ahar, A., Bettens, S., et al. (2019). Signal processing challenges for digital holographic video display systems. *Signal Processing: Image Communication, 70*, 114–130. https://doi.org/10.1016/j.image.2018.09.014.
13. Estrela, V. V., & Coelho, A. M. (2013). State-of-the-art motion estimation in the context of 3D TV. In R. A. Farrugia & C. J. Debono (Eds.), *Multimedia networking and coding* (pp. 148–173). Hershey: IGI Global. https://doi.org/10.4018/978-1-4666-2660-7.ch006.
14. Mentzer, F., Agustsson, E., Tschannen, M., et al. (2019). Practical full resolution learned lossless image compression. In *2019 IEEE/CVF Conference on Computer Vision and Pattern Recognition*, pp. 10621–10630. https://doi.org/10.1109/CVPR.2019.01088.
15. Bhagat, A. P., & Atique, M. (2012). Medical images: Formats, compression techniques, and DICOM image retrieval a survey. In *2012 International Conference on Devices, Circuits and Systems, ICDCS 2012*, pp. 172–176.
16. Masood, S., Sharif, M., Yasmin, M., et al. (2013). Brain image compression: A brief survey. *Research Journal of Applied Sciences, Engineering and Technology, 5*, 49–59. https://doi.org/10.19026/rjaset.5.5083.

17. Yang, X. S. (2010). A new metaheuristic bat-inspired algorithm. In *Nature inspired cooperative strategies for optimization (NICSO 2010). Studies in computational intelligence* (Vol. 284, pp. 65–74). Berlin, Heidelberg: Springer. https://doi.org/10.1007/978-3-642-12538-6_6.
18. Yang, X. S. (2011). Bat algorithm for multi-objective optimisation. *International Journal of Bio-Inspired Computation, 3*, 267–274. https://doi.org/10.1504/IJBIC.2011.042259.
19. Gálvez, A., Fister, I., Fister, I., et al. (2019). Computing rational border curves of melanoma and other skin lesions from medical images with bat algorithm. In *GECCO 2019 Companion – Proceedings of the Genetic and Evolutionary Computation Conference Companion*, pp. 1675–1682. https://doi.org/10.1145/3319619.3326873.
20. Gupta, D., Arora, J., Agrawal, U., et al. (2019). Optimized Binary Bat algorithm for classification of white blood cells. *Measurement. Journal of the International Measurement Confederation, 143*, 180–190. https://doi.org/10.1016/j.measurement.2019.01.002.
21. Lu, Q., Zhang, Z., & Yue, C. (2019). Image segmentation based on bat algorithm and pulse coupled neural network. In *Electronic Information Technology and Computer Engineering (EITCE) 2019*, pp. 4–7.
22. Lu, S., Wang, S. H., & Zhang, Y. D. (2020). Detection of abnormal brain in MRI via improved AlexNet and ELM optimized by chaotic bat algorithm. *Neural Computing and Applications*, 4. https://doi.org/10.1007/s00521-020-05082-4.
23. Estrela, V. V., & Galatsanos, N. P. (2000). Spatially adaptive regularized pel-recursive motion estimation based on the EM algorithm. In *Proceedings of the SPIE 3974, Image and Video Communications and Processing 2000* (19 April 2000). https://doi.org/10.1117/12.382969.
24. Wang, P., Lin, J., & Wang, M. (2015). An image reconstruction algorithm for electrical capacitance tomography based on simulated annealing particle swarm optimization. *Journal of Applied Research and Technology, 13*, 197–204.
25. Wang, P., Wang, M., & Lin, J. (2014). A novel image reconstruction algorithm for electrical capacitance tomography. In *2014 International Symposium on Computer, Consumer and Control*, pp. 227–230.
26. Xiang, G. E. (2008). JPEG 2000 still image compression/decompression system software realization. *JPEG 2000 Still Image Compression/Decompression System Software Realization 8*, 1.
27. Lawson, S., & Zhu, J. (2002). Image compression using wavelets and JPEG2000: A tutorial. *Electronics & Communication Engineering Journal, 14*, 112–121. https://doi.org/10.1049/ecej:20020303.
28. Equitz, W. H. (1989). A new vector quantization clustering algorithm. *IEEE Transactions on Acoustics, 37*, 1568–1575. https://doi.org/10.1109/29.35395.
29. Kumar, S. N., Fred, A. L., Kumar, H. A., & Varghese, P. S. (2019). Bat optimization-based vector quantization algorithm for compression of CT medical images. In B. Gulyás, P. Padmanabhan, A. Fred, T. Kumar, & S. Kumar (Eds.), *ICTMI 2017* (pp. 53–64). Singapore: Springer.
30. Karri, C., & Jena, U. (2016). Fast vector quantization using a Bat algorithm for image compression. *Engineering Science and Technology, an International Journal, 19*, 769–781. https://doi.org/10.1016/j.jestch.2015.11.003.
31. Askelöf, J., Carlander, M. L., & Christopoulos, C. (2002). Region of interest coding in JPEG 2000. *Signal Processing: Image Communication, 17*, 105–111. https://doi.org/10.1016/S0923-5965(01)00026-1.
32. Huang, H. C., Pan, J. S., Lu, Z. M., et al. (2001). Vector quantization based on genetic simulated annealing. *Signal Processing, 81*, 1513–1523. https://doi.org/10.1016/S0165-1684(01)00048-2.
33. Eskicioglu, A. M., & Fisher, P. (1995). Image quality measures and their performance. *IEEE Transactions on Communications, 43*, 2959–2965.
34. Cao, S., Wu, C.-Y., & Krähenbühl, P. (2020). Lossless image compression through super-resolution. arXiv:2004.02872.

35. Chen, L., Yang, X., Jeon, G., et al. (2020). A trusted medical image super-resolution method based on feedback adaptive weighted dense network. *Artificial Intelligence in Medicine, 106*, 101857. https://doi.org/10.1016/j.colsurfa.2020.124658.
36. Ren, S., Jain, D. K., Guo, K., et al. (2019). Towards efficient medical lesion image super-resolution based on deep residual networks. *Signal Processing: Image Communication, 75*, 1–10. https://doi.org/10.1016/j.image.2019.03.008.
37. Anbarjafari, G. (2017). Super resolution based image compression using wavelet transforms and singular value decomposition. *Tamap Journal of Engineering, 2017*, 1–9.
38. Mahapatra, D., & Bozorgtabar, B. (2019). Progressive generative adversarial networks for medical image super resolution. arXiv:1902.02144.
39. Amaranageswarao, G., Deivalakshmi, S., & Ko, S. B. (2020). Wavelet based medical image super resolution using cross connected residual-in-dense grouped convolutional neural network. *Journal of Visual Communication and Image Representation, 70*, 102819. https://doi.org/10.1016/j.jvcir.2020.102819.
40. Mahapatra, D., Bozorgtabar, B., & Garnavi, R. (2019). Image super-resolution using progressive generative adversarial networks for medical image analysis. *Computerized Medical Imaging and Graphics, 71*, 30–39. https://doi.org/10.1016/j.compmedimag.2018.10.
41. Zhao, J., An, P., Huang, X., Yang, C., & Shen, L. (2019). Light field image compression via CNN-based EPI super-resolution and decoder-side quality enhancement. *IEEE Access, 7*, 135982–135998.
42. Zhang, F., Wu, Y., Xiao, Z., et al. (2019). Super resolution reconstruction for medical image based on adaptive multi-dictionary learning and structural self-similarity. *Computer Assisted Surgery, 24*, 81–88. https://doi.org/10.1080/24699322.2018.1560092.
43. Du, J., He, Z., Wang, L., et al. (2020). Super-resolution reconstruction of single anisotropic 3D MR images using residual convolutional neural network. *Neurocomputing, 392*, 209–220. https://doi.org/10.1016/j.neucom.2018.10.102.
44. Liu, F., Hernandez-Cabronero, M., Sanchez, V., Marcellin, M. W., & Bilgin, A. (2017). The current role of image compression standards in medical imaging. *Information, 8*(4), 131.
45. Razmjooy, N., & Estrela, V. V. (2019). *Applications of image processing and soft computing systems in agriculture* (pp. 1–300). Hershey: IGI Global. https://doi.org/10.4018/978-1-5225-8027-0.
46. Razmjooy, N., Estrela, V. V., & Loschi, H. J. (2019). A study on metaheuristic-based neural networks for image segmentation purposes. In *Data science theory, analysis and applications*. Abingdon: Taylor and Francis.

Super-Resolution-Based Human-Computer Interaction System for Speech and Hearing Impaired Using Real-Time Hand Gesture Recognition System

Suriya Sundaramoorthy and Balaji Muthazhagan

1 Introduction

The interaction which happens between a human kind and a computer builds an environment known as gesture recognition system [1]. A mode of communication will be established by the recognized gestures between the system and the user. In the computer science research, the need for human-computer interaction (HCI) system has evolved as a promising and active field [2, 3]. This project aims to utilize hand gesture (input) to build a communication with a computer system, as a result to develop a hand gesture-based recognition system. Usability and functionality are the necessary characteristics involved in such a design as described in [4].

The meaning of functionality is defined as how well the recognition of gesture is established and how well they are stored in the computer system. Whereas, how well the user can interact with the system is defined as usability. To control the machine, hand gestures provide a flexible way [4]. Hand gesture recognition (HGR) can be static or dynamic. Less computational complexity is required for static gestures [5]. Whereas, the dynamic gestures (sequence of postures) are more complex and thus they are ideal for real-time environments [5, 6]. Methods are using additional hardware devices and others that rely on computer vision cues like the hand's appearance to segment hands and extract the necessary features [7]. However, the system's goal is to eliminate any external depth cameras to identify the gestures.

This chapter has been structured as follows: The first section introduces the need for such a system, the second section sheds some light on existing and recent

S. Sundaramoorthy (✉) · B. Muthazhagan
Department of Computer Science & Engineering, Velammal College of Engineering and Technology, Anna University, Madurai, Tamil Nadu, India

PSG College of Technology, Coimbatore, Tamil Nadu, India

A. Deshpande et al. (eds.), *Computational Intelligence Methods for Super-Resolution in Image Processing Applications*, https://doi.org/10.1007/978-3-030-67921-7_7

developments in this area, the third section expounds on the proposed approach with a special focus on super-resolution (SR), the fourth section brings in future work. Finally, the fifth section concludes with future trends in this area and observations derived from this work.

2 Literature Review

2.1 Input Methods for Gesture Recognition

2.1.1 Pen-Based

For a period, pen or a mouse was considered as 2-D input device to perceive gestures.

The early Sketchpad framework was using light-pen gestures in the 1960s. Since the 1970s, a small number of industrial systems have employed pen gestures. Document editing, air traffic control, and certain specific design tasks with editing splines are some of the examples of recognizing through gestures [8, 25]. To control a virtual environment from a pen-based gesture recognition to coordinate speech recognition, frameworks like the OGI Quick Set system have been useful [26]. Sixty-eight gestures of pen is acknowledged by Quick Set, the gestures such as editing gestures, map symbols, taps, and route and region indicators [8]. Quick Set acknowledges 68 pen gestures, map symbols, editing gestures, route and region indicators, and taps [8]. The utilization of external gadgets, for example, a pen, is quite complicated for physically challenged people, thereby preventing a gesture-based recognition system's main aim to be achieved [8].

2.1.2 Data Gloves

Depending on the position, the degree of accuracy, completeness, and precision of hand, configuration is estimated in many commercial devices. Instrumented hand-mounted gloves and exoskeleton are some of the examples of these devices which are usually referred as data gloves [8].

Some benefits of using the data gloves: They measure the hand and finger parameters directly, data provisioning, high sampling frequency, easy if used, line of sight, less cost version, and data translation independency feature [27]. However, there are couple of disadvantages with the usage of data glove which includes decrease in comfort and movement, challenges for calibration, inexpensive system, and accurate system's expensiveness. Specifically, the main disadvantage is, the data glove is expected to be worn by the user when each time the user collaborates with the system [8].

2.1.3 Bodysuits

By strategically placing the created small region of dots on the human body, the patterns can be discovered such as activities, identities, gestures, and other aspects of the body.

By calculating the three-dimensional location of the markers connected to the body, human postures and motion can be identified [8]. There are benefits and drawbacks in association with bodysuits. Body suits are expensive and are complicated, however, only at high sampling rate, they provide valid solutions [8].

2.1.4 Vision-Based Interfaces

The major drawback with respect to the tracker-based system is that the system gets complicated. This results the user to wear the unusual device, which is however, an important aspect in the tracker-based systems. To wear and to get adjusted to the device takes considerable efforts [8]. These limitations are faced by the optical systems with body markers. In the vision-based interfaces, camera of one or more could capture images. Ten frames per second or more are the capture rate of the images. To identify the gesture and a person's activity, the captured images are interpreted to build a visual feature. The position of the cameras usually is static in the environment.

A number of experiments have been carried out in the past 10 years in the field of computer vision community which includes identifying and discovering faces and their expressions, facial gesture, human action, gesture, and lip movement [8]. The major disadvantage with the vision-based system is the account of occlusions.

Some portion of the human body is blocked which results their being unrecognizable by the cameras. For example, when the camera is in the front, the back side of the human body is hidden. When part of an object overlaps itself and from the camera that part gets hidden is when the self-occlusion occurs. Complete look of fingers, hand, and body cannot be achieved by a single view. However, integration and correlation problems occur when more than one cameras are used [8].

Unless the effect of movement and a reliable model of object motion is available, complete body tracking is possible with respect to the occlusion problem. For gesture recognition all the body motion parameters may not be the preconditions for the retrieval [4]. By many of the computer vision systems cameras are being used for identification.

There is a necessity for a preprocessing step to enlarge the images, remove noises, and transform the scenes. Next, the feature extraction step comes to represent the main images' features. The elements could be either two- or three-dimensional. The properties are analyzed such as the parameters of the body and the statistical properties. If they are appropriate, they are then classified as a gesture [4]. For distinguishing hand gestures in the real-time scenario, this above approach is the most suitable method.

2.2 Recent Works Related to Gesture Recognition

To communicate with people or with each other, communication is the primary channel, according to [14]. There has been an increase in the patients of dumb and deaf because of possible accidents, oral diseases, and congenital disabilities. Visual communication is relied upon when a deaf or a mentally impaired person communicates with an average person. There are many spoken languages around the world to be understood and interpreted. It is often difficult for the people who have difficulty in speaking and hearing to understand the conversation with other person, respectively. Either from lip-reading or through sign language, sometimes people misinterpret the messages. Thus to bridge the gap between the specially challenged and the normal person, this project is conducted. Through synthesized speech, an electronic speaking system eases the problem for speech-impaired patients, which appears in [15].

Consisting of "bend sensitive resistance elements" multiple flex sensors are internally equipped for the data glove. The sensors produce a change in resistance of various elements for each gesture. The change in resistance is then fed into ARDUINO UNO microcontroller as an analog input. LCD with a switch containing audio commands relates to microcontroller. Recorded audio command is played through a speaker after displaying the command on the LCD, when the switch is pressed causing the signal conditioning to recognize the specific gesture. The voice will be sent to the server when the user will register his voice [16]. The speech to text is converted by the server. The sign will be generated after the text is classified. The generated sign will be sent to the deaf or dumb person's application. To communicate with the visually impaired people signs to speech can be developed. Using this system, communication can be made efficiently for the visually and hearing-impaired people. The research from [17] reminds one of a translator's need to understand what they speak and share with us. The translation of normal sign language to speech is how the sign language translation system works, hence, easing the communication between the normal and dumb. Therefore, the aim is to develop a system which establishes a communication with a normal person and a speech hearing impaired. Every part of human life has been influenced by informational and communication technology [18]. The influence has affected the way we modified, the way we work, travel, acknowledge, and convey. The utilization of ICT has enhanced deaf people's satisfaction by creating frameworks.

In an almost impaired group, gesture-based communication is a method of essential.

The issue emerges when hard of hearing individuals attempt to convey what needs be to other individuals with the assistance of these gesture-based communication language structures and bad habits a versa. The application gives hard of hearing individuals a method for getting more shut to cutting edge innovation by utilizing discourse to picture interpretation. This deaf individual can learn new advances by looking toward changing pictures to assimilate the discourse framework.

Reference [19] will give a proper demonstration of our system and address the problems that the less fortunate people and the normal people face while communicating. The modules in the present scheme include train patterns, pattern recognition, text-to-speech, and data dictionary. These modules will help us in achieving our objective.

2.3 *Necessary Literature for the Proposed System*

2.3.1 Color Models

Identifying the skin color is a significant challenge. To overcome this challenge, various color models for pixel-based skin detection were studied [9]. The color spaces used commonly in computer vision applications are as follows:

- Red, Green, and Blue (RGB) color space relies on three channels: Red (R), Green (G), and Blue (B). The advantage is the simplicity and ease of this color space. However, they are not uniform perceptually. There is a solid association between the RGB components and the differentiation of chrominance and luminance is not done [9].
- The Hue, Saturation, and Value (HSV) color space: Of a specified area, Hue refers in identifying the extensive color (Green, Red, Yellow, or Purple). Saturation is involved with the color's brightness. It is widely used in the computer vision systems, because the intensity is independent of the pixel information. Thus, HSV model provides poor results for low brightness [9].
- CIELAB color space highlights three major values or components as follows: L* Lightness is represented, a* Green-Red color component is represented, and b* Blue-yellow color components is represented. Same amount of change is expressed when the values change [9]. In manipulating the images, CIELAB is mostly used. For example, from the image gathered with a camera or a scanner, CIELAB removes or sharpens some featuresspi2;[9].

2.3.2 Convex Defect-Based Fingertip Detection

The contour will be the best method for analyzing fingertip detection since it identifies the fingertip correctly based on edge detection [10]. The least squares (LS) ellipse fitting is another method for detecting the fingertips [11, 28, 29]. Along with a good quality contour [12], these require ample amount of computation. The boundary of any given shape is what the contour represents. From Fig. 1, it can be noticed that the outline could be obtained for the hand image [22–25, 30]. Type of gesture can be revealed by contour analysis of the image [12].

Convex polygon (convex hull) surrounds contour region found in the analysis of contour. Of the specified hand image, the polygon will wrap all the

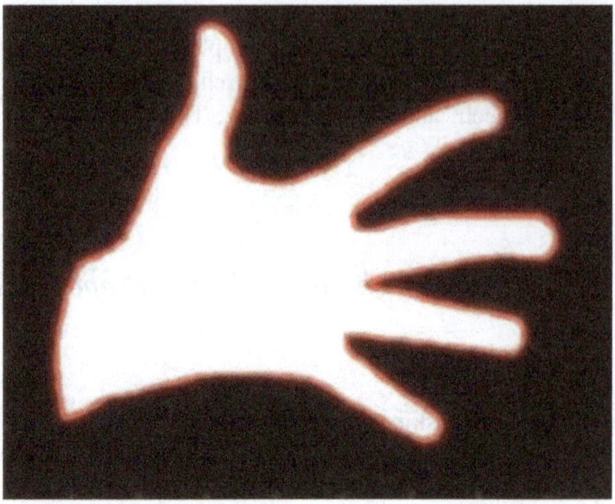

Fig. 1 Hand contour

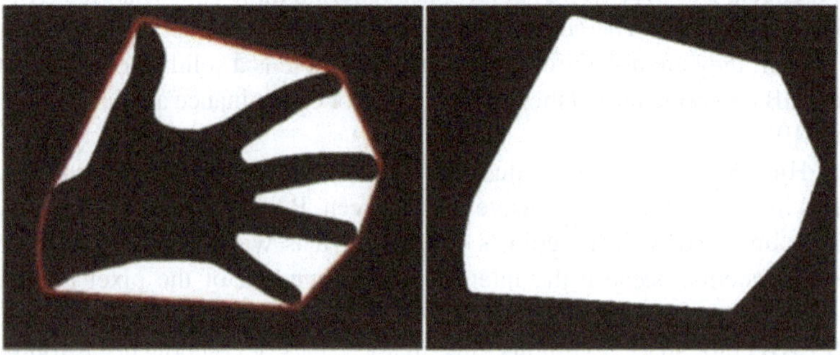

Fig. 2 Convex hull of hand

vertices. Figure 2 illustrates the hand gesture's convex hull in addition to the separated convex hull [12].

Convex defect is the difference between gesture contour and the convex hull. In Fig. 3, it can be seen that the areas numbered 1–6 are the convex defects. Convex defect point is identified by three components, that is, start, end, and concave contour point [12]. For convex defect 2, start is its P1, end is P2, and concave point is the P3, and d_2 (the distance represented by) is the depth of convex defect [12].

Gesture convexity, denoted by δ, concerning the convex hull, is the hand gesture's intensity of tightness. The area ratio of the gesture contour and the convex hull gives the following value:

$$\delta = \frac{\text{contour}_{\text{Area}}}{\text{hull}_{\text{Area}}}, \tag{1}$$

where $\text{hull}_{\text{Area}}$ signifies the area of the convex hull, $\text{contour}_{\text{Area}}$ describes the area of gesture contour, and δ is a value in the range (0,1), such that the bigger the value, the tighter it will be [12].

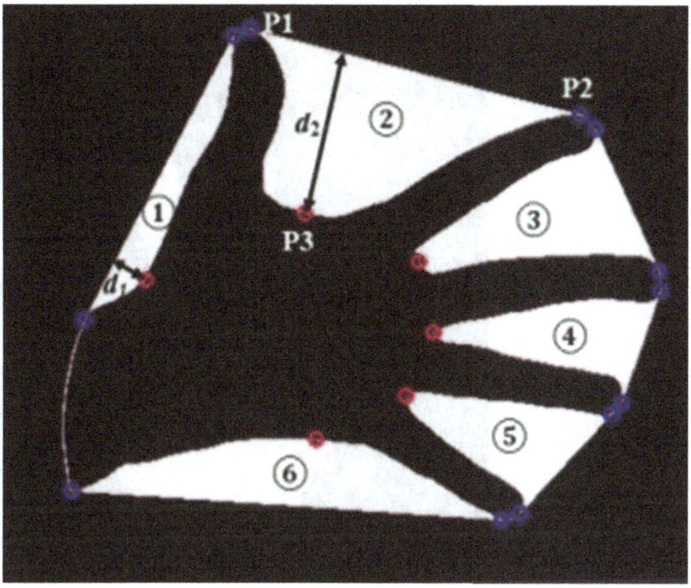

Fig. 3 Convex defects of hand

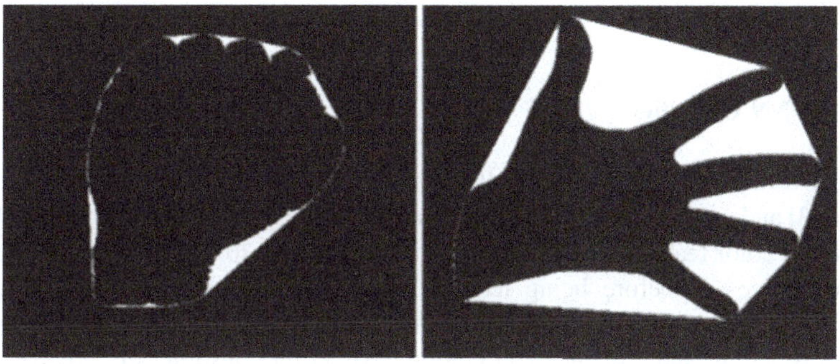

Fig. 4 Tightness of hand

The gesture contour to convex hull becomes tighter once $\text{hull}_{\text{Area}} > \text{contour}_{\text{Area}}$. Figure 4 shows an example for more clarification of the tightness of hand.

Different gestures can be distinguished depending on the relative position between the fingertips and the angle values of α and β,

$$\alpha = \theta_1 + \theta_2 + \cdots + \theta_{N-1} \tag{2}$$

$$\beta = \theta_{N-1} \tag{3}$$

where α is the summation of angles with centroid as vertex and β is the value of the angle with centroid as vertex as highlighted in Fig. 5, subsequently,

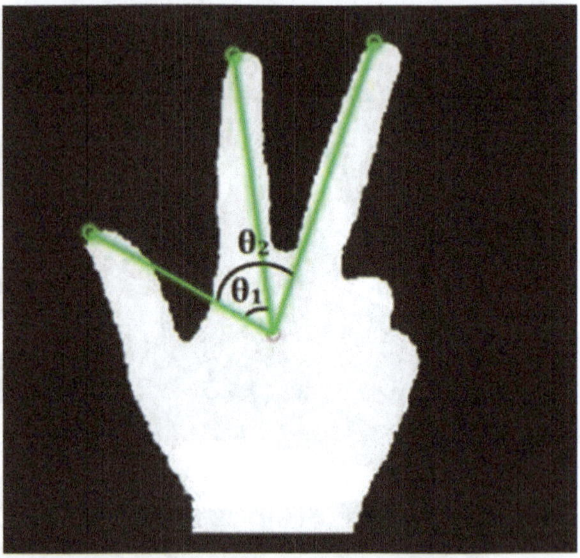

Fig. 5 Definition of relative position of fingers

where θ is the angle between the first fingertip and the other fingertip and N is the number of fingertips as shown in Fig. 5. For example, for $N = 3$, $\alpha = \theta_1 + \theta_2$, and $\beta = \theta_2$ (Figs. 6 and 7).

2.3.3 CNN Classifier

The convolutional neural network (CNN) model has been modified from the literature [13] and can distinguish gestures. The CNN processes a binary image such that the hand's color features do not impact on the classifier [26, 40]. The image needs to be first processed before being fed to the CNN classifier. Hand color filtering, Gaussian blurring, thresholding, Background subtraction, morphological transformation (opening and closing), and contour extraction constitute the preprocessing phase (Fig. 8).

By preprocessing the images, the detector works. By tracking the particular movement of the hands' particular point, the hand's contour region is extracted. Which in turn estimates the palm radius and hand center and can control the mouse cursor [13]. Finally, resizing the region of contour to a fixed size is carried out with respect to a constant aspect ratio, before being fed to the CNN classifier, and then put onto a canvas [13] (Fig. 9).

3 Proposed System

The system can be divided broadly into two parts:

Fig. 6 Relative position of different hand gestures

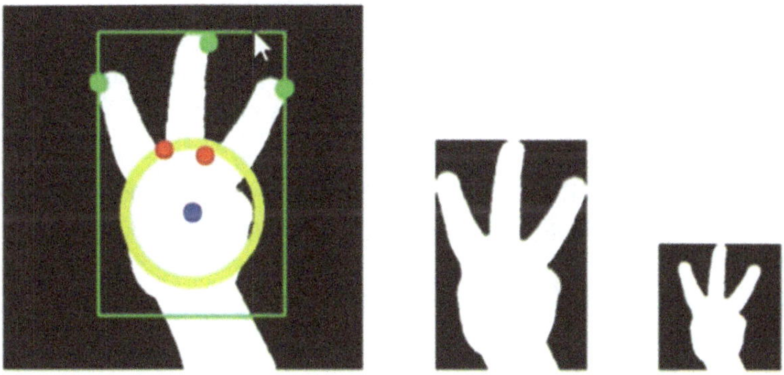

Fig. 7 Extraction of features

Fig. 8 Preprocessing steps

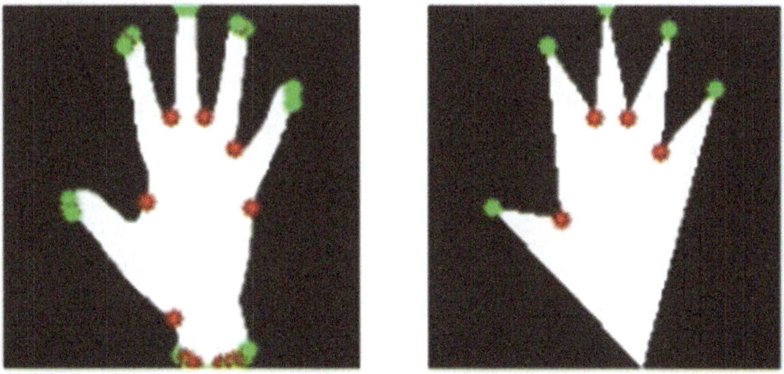

Fig. 9 Polygonal approximation

- Refinement of real-time data using super-resolution: Since the real-time data can be captured using low-resolution cameras, the image must be upscaled so that it can be fed into the prediction phase.
- Feeding refined image into the neural network to predict gestures: From a specified folder images, the CNN model is generated and trained. In an hdf5 format weight file, the obtained weights are locally stored. As an input to the system, the user interacts and gestures. Hence, the system produces real-time predictions depending on the gesture comparing with the locally stored weight file.

3.1 Super-Resolution in Gesture Recognition

Super-resolution is a technique to up-sample the low-resolution image [49–51]. The SR model relying on CNNs comes from [20] and is known as SRCNN. There are essentially three parts to this:

- Patch Extraction and Representation: The first phase of this model carries out the convolution in the image's patches. This can be represented as

$$F_1(x) = \max(0, W_1 * x + B_1),$$

  where $W_1 = n_1$ filters of support $c * s_1 * s_1$, here c is the number of channels in the input image, and s_1 is the filter's size, B_1 contains the biases of dimension n_1 and * represents the convolution operation. Following the convolution process, there is an n_1-dimensional vector on which we apply the Rectified Linear Unit (ReLU) max filter.
- Nonlinear Mapping: This step considers the size of the filter to be 1×1. Much like the previous phase, this can be formulated as

$$F_2(x) = \max(0, W_2 * F_1(x) + B_2),$$

  where $W_2 = n_2$ means a filter of support $n_1 * s_2 * s_2$ with n_1 as the depth and s_2 is the size of the filter, $B_2 =$ biases of dimension n_2 and * represents the convolution operation.
- Reconstruction: In this step, we try to average out to get a higher resolution image:

$$F_3(x) = W_{33} * F_2(x) + B,$$

  where $W_3 = g$ is a filter of size $n_2 * s_3 * s_3$, where n_2 is the depth and s_3 is the size of the filer, B_3 is the bias of dimension g.

Figure 10 shows a super-resolution version of a low-resolution hand image using SRCNN.

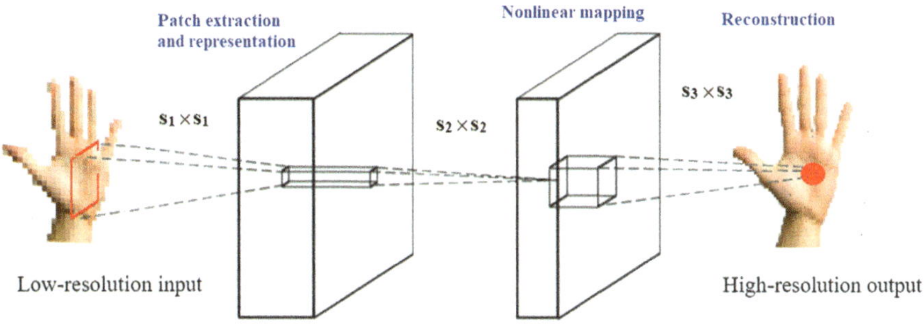

Fig. 10 Super-resolution version of a low-resolution hand image using SRCNN

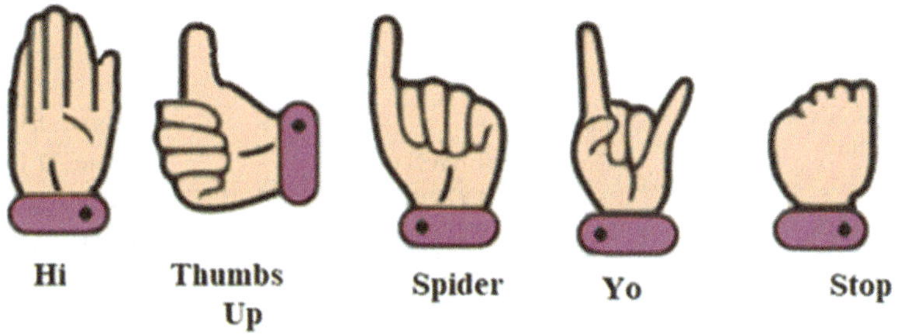

Fig. 11 Gestures in the proposed system

3.2 Feeding Refined Image into the Neural Network to Predict Gestures

3.2.1 Gestures

The gestures in the proposed system are shown in Fig. 11.

3.2.2 Training the Model

In the system, two CNN models were considered for training with the contour section of hand images as input and the gesture they represent (Fig. 12).

CNN model 1 consists of 15 layers (Table 1):

For model 1, the output illustrates the output terminal as obtained from the virtual machine. The cost function for cross-validation data is represented using val_loss, and the cost function for training data is the loss (Fig. 13).

CNN model 2 consists of 12 layers (Table 2):

For model 2, the below output illustrates the output as obtained on the virtual machine.

Fig. 12 Sample training binary image

Table 1 CNN model 1 architecture

Layer (type)	Output shape
Conv2d_1 (Conv2D)	(None,32,198,198)
Activation_1 (Activation)	(None,32,198,198)
Conv2d_2 (Conv2D)	(None,32,198,196)
Activation_2 (Activation)	(None,32,198,196)
max_pooling2d_1	(None,32, 98, 98)
Dropout_1 (Dropout)	(None,32, 98, 98)
Flatten_1 (Flatten)	(None,307328)
dense_1 (Dense)	(None,128)
Activation_3 (Activation)	(None,128)
Dropout_2 (Dropout)	(None,128)
dense_2 (Dense)	(None,64)
Activation_4 (Activation)	(None,64)
Dropout_3 (Dropout)	(None,64)
dense_3 (Dense)	(None,5)
Activation_5 (Activation)	(None,5)

The cost function for cross-validation data is represented using val_loss, and the cost function for training data is the loss (Fig. 14).

3.2.3 Prediction Phase

The prediction phase follows the application of the super-resolution of the real-time images followed by a skin mask filter and binary threshold filter to extract only the relevant details. Thereafter, it is fed into the proposed CNN models to predict the relevant gestures.

Figure 15 shows the skin mask filter: It scans the hand, which is our region of interest (ROI) and extracts the layout of the hand in a skin-like color. If the skin color

```
| Connected to the serial port of the VM.        |
| If no login prompt is displayed, press ENTER. |
+-----------------------------------------------+

3200/3200 [==============================] - 474s 148ms/step - loss: 0.0389 - acc: 0.9872 - val
_loss: 0.0194 - val_acc: 0.9938
Epoch 9/14
3200/3200 [==============================] - 475s 148ms/step - loss: 0.0378 - acc: 0.9869 - val
_loss: 0.0158 - val_acc: 0.9975
Epoch 10/14
3200/3200 [==============================] - 475s 148ms/step - loss: 0.0247 - acc: 0.9925 - val
_loss: 0.0251 - val_acc: 0.9962
Epoch 11/14
3200/3200 [==============================] - 474s 148ms/step - loss: 0.0227 - acc: 0.9916 - val
_loss: 0.0211 - val_acc: 0.9950
Epoch 12/14
3200/3200 [==============================] - 475s 148ms/step - loss: 0.0205 - acc: 0.9928 - val
_loss: 0.0207 - val_acc: 0.9962
Epoch 13/14
3200/3200 [==============================] - 474s 148ms/step - loss: 0.0178 - acc: 0.9922 - val
_loss: 0.0198 - val_acc: 0.9975
Epoch 14/14
3200/3200 [==============================] - 474s 148ms/step - loss: 0.0178 - acc: 0.9950 - val
_loss: 0.0239 - val_acc: 0.9975
Saving trained weightsL
Enter file namenewtestweight50002
Press any key to continue
```

Fig. 13 Training output of model 1

Table 2 CNN model 2 architecture

Layer (type)	Output shape
Conv2d_1 (Conv2D)	(None,32,198,198)
Activation_1 (Activation)	(None,32,198,198)
Conv2d_2 (Conv2D)	(None,32,198,196)
Activation_2 (Activation)	(None,32,198,196)
max_pooling2d_1	(None,32, 98, 98)
Dropout_1 (Dropout)	(None,32, 98, 98)
Flatten_1 (Flatten)	(None,307328)
dense_1 (Dense)	(None,128)
Activation_3 (Activation)	(None,128)
Dropout_2 (Dropout)	(None,128)
dense_2 (Dense)	(None,5)
Activation_4 (Activation)	(None,5)

matches with the background, then it fails in some cases. This leads to errors because it confuses the system and the region of interest could be inaccurate.

Figure 16 shows the optimal way to obtain the contour of the hand, which is our ROI. Regardless of the factors like background, nail polish, etc., it provides a black border with a white background which helps to focus and provide accurate results.

Figure 17 depicts the prediction of gestures pause/play. On VLC Media player, this has been implemented. To perform the functions on it, the model has been trained to specific gestures as input. Based on the trained data, when a gesture is "stop," then it pauses the video, based on the trained data. In addition, a gesture for playing the video exists.

```
Epoch 7/14
3200/3200 [==============================] - 474s 148ms/step - loss: 0.0023 - acc: 0.9997 - val
_loss: 0.0247 - val_acc: 0.9963
Epoch 8/14
3200/3200 [==============================] - 475s 149ms/step - loss: 0.0095 - acc: 0.9969 - val
_loss: 0.0274 - val_acc: 0.9950
Epoch 9/14
3200/3200 [==============================] - 474s 148ms/step - loss: 0.0019 - acc: 0.9994 - val
_loss: 0.0267 - val_acc: 0.9963
Epoch 10/14
3200/3200 [==============================] - 475s 148ms/step - loss: 0.0023 - acc: 0.9997 - val
_loss: 0.0221 - val_acc: 0.9975
Epoch 11/14
3200/3200 [==============================] - 475s 148ms/step - loss: 0.0014 - acc: 0.9997 - val
_loss: 0.0250 - val_acc: 0.9975
Epoch 12/14
3200/3200 [==============================] - 474s 148ms/step - loss: 0.0019 - acc: 0.9997 - val
_loss: 0.0245 - val_acc: 0.9975
Epoch 13/14
3200/3200 [==============================] - 476s 149ms/step - loss: 0.0010 - acc: 0.9997 - val
_loss: 0.0281 - val_acc: 0.9975
Epoch 14/14
3200/3200 [==============================] - 475s 148ms/step - loss: 0.0020 - acc: 0.9991 - val
_loss: 0.0241 - val_acc: 0.9975
Saving trained weights1
Enter file namenewtestweight5003
Press any key to continue
```

Fig. 14 Training output of model 2

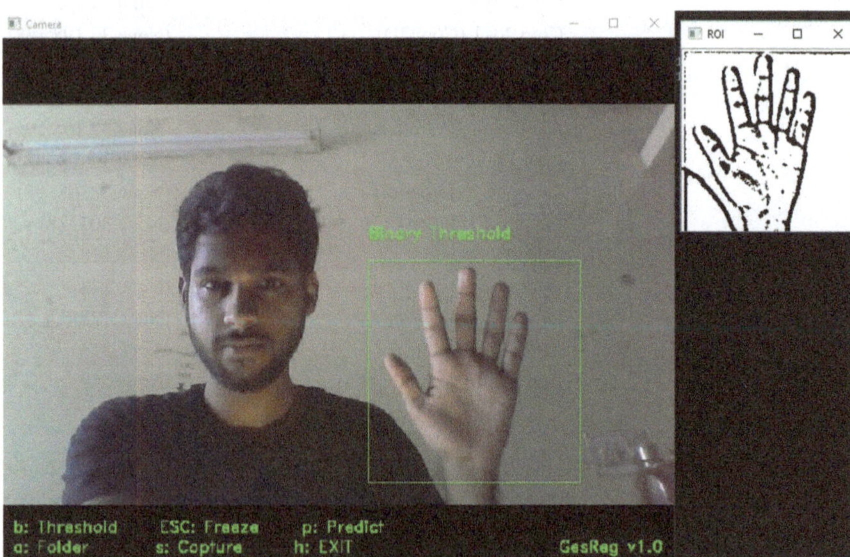

Fig. 15 Extraction of the region of interest with binary threshold filter active

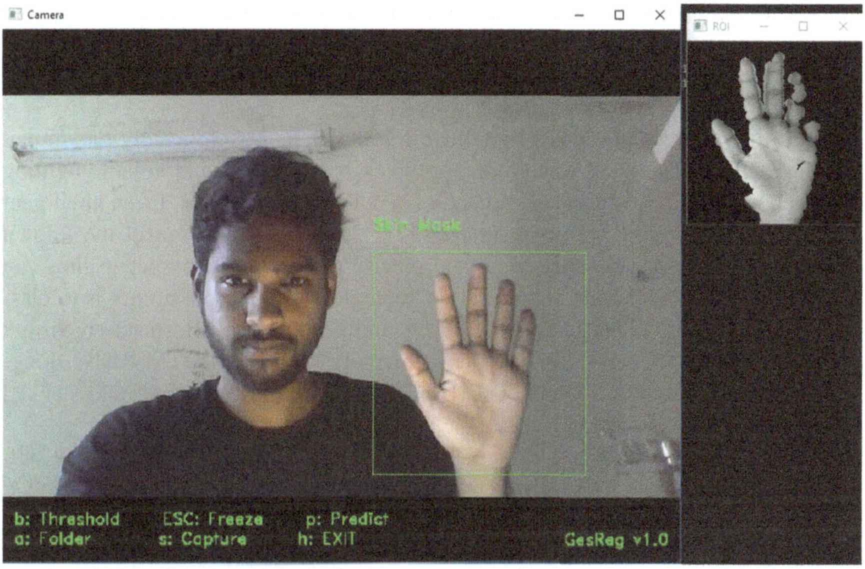

Fig. 16 Extraction of the region of interest with skin mask filter active

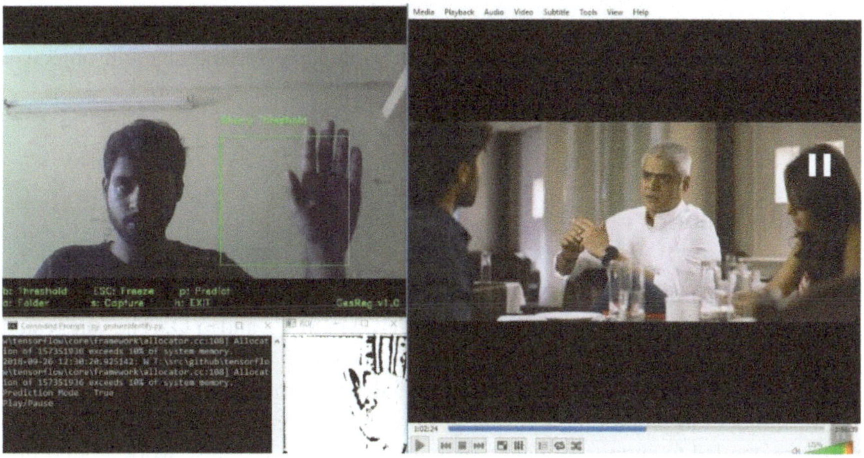

Fig. 17 Prediction of gesture pause/play

4 Future Work

HGR can answer a call or control the radio-based systems while operating the equipment as it is the case with controlling a drone. Nowadays, this type of functionality is an established feature in more expensive and automated devices [22–25]. High-resolution (HR) time-of-flight (ToF) cameras and powerful processing platforms can usually form the core of these systems albeit with a high price tag. Hence, the possibility to design an algorithm to predict hand gestures with low latency via, for instance, an inexpensive low-resolution thermal camera is worth investigation.

With the progress in artificial intelligence, HGR has a significantly paramount role in HCI applications. Furthermore, complex backgrounds and volatile lighting still challenges hand gestures' identification in real-world scenarios. Fresh intelligent fusion schemes relying on depth images [22, 26, 27, 30] to unravel the known caveats and attain real-time dynamic and precise recognition. New algorithms can utilize better statistical hand segmentation strategies. Another possibility is to classify static hand pose via different types of DL architecture [31, 32], such as recurrent neural networks (RNNs), long short-term memory (LSTM), LDTM-RNNs, long-term recurrent convolution networks (LRCN) [33], CNN-RNN [34], and stacked denoising autoencoder (SDAE) [35].

Improved model matching strategies can deliver remarkable experimental outcomes with outstanding performance in complex real-world scenarios with lower computational complexity and better applicability [36].

Human action recognition can also use metaheuristics to improve performance. Features can be better handled and optimized with techniques like particle swarm optimization, SVM, and genetic algorithms among others. These feature vectors can be assembled using Harris corner points, histograms, and wavelet coefficients and other extracted features from the spatiotemporal video sequence. The system's computational complexity and the feature space can be lowered by a multi-objective particle swarm optimization tactic [37–48].

5 Conclusion

The hand gesture recognition system was successfully built and predicted the following gestures: HI, SPIDER, STOP, THUMBSUP, and YO. The suggested system used computational intelligence super-resolution to upscale the input image being fed. Later, the processed image was fed into a trained convolutional neural network model to predict the relevant gesture. The system was scalable and can be further extended to support a wide variety of gestures. The future work around this should focus on improving the dataset for the gestures. The dataset should be handcrafted, encompassing various lighting conditions. This will ensure that the observations will not be biased and can increase the accuracy of the system. The super-resolution algorithm can be tweaked further to encompass further robustness.

References

1. Murthy, G. R. S., & Jadon, R. S. (2009). A review of vision based hand gestures recognition. *International Journal of Information Technology and Knowledge Management, 2*(2), 405–410.
2. Shah, K. N., Rathod, K. R., & Agravat, S. J. (2014). A survey on human computer interaction mechanism using finger tracking. arXiv preprint arXiv:1402.0693.
3. Garg, P., Aggarwal, N., & Sofat, S. (2009). Vision based hand gesture recognition. *World Academy of Science, Engineering and Technology, 49*, 972–977.
4. Karray, F., Alemzadeh, M., Saleh, J. A., & Arab, M. N. (2008). Human computer interaction: Overview on state of the art. *International Journal on Smart Sensing and Intelligent Systems, 1* (1), 137–159.
5. Rautaray, S. S., & Agrawal, A. (2015). Vision based hand gesture recognition for human computer interaction: A survey. *Artificial Intelligence Review, 43*, 1–54.
6. Li, X. (2008). *Gesture recognition based on fuzzy C-means clustering algorithm*. Department of Computer Science, The University of Tennessee Knoxville.
7. Mitra, S., & Acharya, T. (2007). Gesture recognition: A survey. *IEEE Transactions on Systems, Man and Cybernetics, Part C: Applications and Reviews, 37*(3), 311–324. https://doi.org/10.1109/TSMCC.2007.893280.
8. Wysoski, S. G., Lamar, M. V., Kuroyanagi, S., & Iwata, A. (2012). A rotation invariant approach on static-gesture recognition using boundary histograms. *International Journal of Artificial Intelligence & Applications (IJAIA), 3*(4), 173.
9. Rivera, L. A., Estrela, V. V., Carvalho, P. C. P., & Velho, L. (2004). Oriented bounding boxes based on multiresolution contours, Journal of WSCG. In *Proceedings of the 12-th International Conference in Central Europe on Computer Graphics, Visualization and Computer Vision'2004, WSCG 2004*, University of West Bohemia, Campus Bory, Plzen-Bory, Czech Republic, February 2–6, 2004 (Short Papers), 219–212.
10. Stanney, K. M. (2002). Handbook of virtual environments design, implementation, and applications, Gesture recognition Chapter #10 by Matthew Turk.
11. Vezhnevets, V., Sazonov, V., & Andreeva, A. (2003). A survey on pixel-based skin color detection techniques. In *International Conference GraphiCon 2003*, Moscow, Russia.
12. Argyros, A. A., & Lourakis, M. I. A. (2006). Vision-based interpretation of hand gestures for remote control of a computer mouse [C]. In *Proceeding of the International Conference on Computer Vision in Human-Computer Interaction*.
13. Lee, T., Hollerer, T., & Handy, A. R. (2007). Markerless inspection of augmented reality objects using fingertip tracking [C]. In *11th IEEE International Symposium on Wearable Computers*.
14. Xu, Y., Park, D.-W., & Pok, G. C. (2017). Hand gesture recognition based on convex defect detection. *International Journal of Applied Engineering Research, 12*(18), 7075–7079.
15. Towards Data Science. (2018). *A simple 2D CNN for MNIST digit recognition – Towards Data Science*. [online] Available at: https://towardsdatascience.com/a-simple-2d-cnn-for-mnist-digitrecognition-a998dbc1e79a. Accessed 12 Sept. 2018.
16. Prajapati, R., Pandey, V., Jamindar, N., Yadav, N., & Phadnis, N. (2018). Hand gesture recognition and voice conversion for deaf and dumb. *International Research Journal of Engineering and Technology (IRJET), 5*(4), 1373–1376.
17. Hussain, M., & Ravinder, K. (2018). Interactive communication interpreter for deaf dumb and blind people. *International Journal of Scientific Engineering and Technology Research, 7*(2), 0208–0211.
18. Narute, P., Pote, A., Poman, A., & Pawar, S. (2018). An efficient communication system for blind, dumb and deaf people. *International Research Journal of Engineering and Technology (IRJET), 5*(1), 1561–1563.
19. Sontakke, D., Irkhede, T., Gawande, A., Waikar, J., Nikore, N., & Rahangdale, S. (2017). System for effective communication with deaf and mute people. *International Journal of Engineering Science and Computing, 7*(2), 4375–4376.

20. Kawale, N., Hiranwar, D., & Bomewar, M. (2017). An android messenger application for dumb and deaf people. *International Journal of Scientific Research in Science and Technology, 3*, 98–102.
21. Shaikh, S. I., Memon, I. M., Shetty, S. J., Vakanerwala, A. S., & Pawar, S. E. (2016). Communication system to help deaf and dumb communicate with normal people. *International Research Journal of Engineering and Technology (IRJET), 3*(4), 1793–1799.
22. Donahue, J., Hendricks, L. A., Rohrbach, M., Venugopalan, S., Guadarrama, S., Saenko, K., & Darrell, T. (2015). Long-term recurrent convolutional networks for visual recognition and description. In *2015 IEEE Conference on Computer Vision and Pattern Recognition (CVPR)*, 2625–2634.
23. de Jesus, M. A., Estrela, V. V., Huacasi, W. D., Razmjooy, N., Plaza, P., & Peixoto, A. B. M. (2020). Using transmedia approaches in STEM. In *2020 IEEE Global Engineering Education Conference (EDUCON)*, 1013–1016. https://doi.org/10.1109/EDUCON45650.2020.9125239.
24. Arshaghi, A., Razmjooy, N., Estrela, V. V., Burdziakowski, P., Nascimento, D. A., Deshpande, A., & Patavardhan, P. P. (2020). Image transmission in UAV MIMO UWB-OSTBC system over Rayleigh channel using multiple description coding (MDC). In *Imaging and sensing for unmanned aircraft systems: Volume 2: Deployment and applications*. Stevenage: IET.
25. Estrela, V. V., et al. (2019). Why software-defined radio (SDR) matters in healthcare? *Medical Technologies Journal, 3*(3), 421–429.
26. Aroma, R. J., Raimond, K., Razmjooy, N., Estrela, V. V., & Hemanth, J. (2020). Multispectral vs. hyperspectral imaging for unmanned aerial vehicles: Current and prospective state of affairs. In V. V. Estrela, J. Hemanth, O. Saotome, G. Nikolakopoulos, & R. Sabatini (Eds.), *Imaging and sensing for unmanned aircraft systems* (Vol. 2, pp. 133–156). London: IET. https://doi.org/10.1049/PBCE120G_ch7.
27. Deshpande, A., Patavardhan, P., Estrela, V. V., & Razmjooy, N. (2020). Deep learning as an alternative to super-resolution imaging in UAV systems. In V. V. Estrela, J. Hemanth, O. Saotome, G. Nikolakopoulos, & R. Sabatini (Eds.), *Imaging and sensing for unmanned aircraft systems* (Vol. 2, pp. 177–212). London: IET. https://doi.org/10.1049/PBCE120G_ch9.
28. Estrela, V. V., Rivera, L. A., Beggio, P. C., & Lopes, R. T. (2003). Regularized pel-recursive motion estimation using generalized cross-validation and spatial adaptation. In *Proceedings of the XVI Brazilian Symposium on Computer Graphics and Image Processing (SIBGRAPI 2003)*. https://doi.org/10.1109/SIBGRA.2003.1241027.
29. de Jesus, M. A., & Estrela, V. V. (2017). Optical flow estimation using total least squares variants. *Oriental Journal of Computer Science and Technology (OJCST), 10*, 563–579. https://doi.org/10.13005/ojcst/10.03.03.
30. Wang, W., Ying, R., Qian, J., Ge, H., Wang, J., & Liu, P. (2017). Real-time hand gesture recognition based on a fusion learning method. In *2017 International Conference on Computational Science and Computational Intelligence (CSCI)*, 535–540.
31. Obaid, F., Babadi, A., & Yoosofan, A. (2020). Hand gesture recognition in video sequences using deep convolutional and recurrent neural networks. *Applied Computer Systems, 25*, 57–61.
32. Guo, H., Yang, Y., & Cai, H. (2019). Exploiting LSTM-RNNs and 3D skeleton features for hand gesture recognition. In *2019 WRC Symposium on Advanced Robotics and Automation (WRC SARA)*, 322–327.
33. John, V., Boyali, A., Mita, S., Imanishi, M., & Sanma, N. (2016). Deep learning-based fast hand gesture recognition using representative frames. In *2016 International Conference on Digital Image Computing: Techniques and Applications (DICTA)*, 1–8.
34. Lai, K., & Yanushkevich, S. (2018). CNN+RNN depth and skeleton based dynamic hand gesture recognition. In *2018 24th International Conference on Pattern Recognition (ICPR)*, 3451–3456. https://doi.org/10.1109/ICPR.2018.8545718.
35. Ma, M., Gao, Z., Wu, J., Chen, Y., & Zhu, Q. (2018). A recognition method of hand gesture based on stacked denoising autoencoder. In *Proceedings of the fifth Euro-China conference on intelligent data analysis and applications, advances in intelligent systems and computing* (Vol. 891, pp. 736–744). Cham: Springer. https://doi.org/10.1007/978-3-030-03766-6_83.

36. Min, X., Zhang, W., Sun, S., Zhao, N., Tang, S., & Zhuang, Y. (2019). VPModel: High-fidelity product simulation in a virtual-physical environment. *IEEE Transactions on Visualization and Computer Graphics, 25*, 3083–3093.
37. Razmjooy, N., Estrela, V. V., & Loschi, H. J. (2019). A study on metaheuristic-based neural networks for image segmentation purposes. In *Data science* (pp. 25–49). CRC Press.
38. Razmjooy, N., Ashourian, M., Karimifard, M., Estrela, V. V., Loschi, H. J., do Nascimento, D., França, R. P., & Vishnevski, M. (2020). Computer-aided diagnosis of skin cancer: A review. *Current Medical Imaging, 16*(7), 781–793.
39. Berlin, S. J., & John, M. (2020). Particle swarm optimization with deep learning for human action recognition. *Multimedia Tools and Applications, 79*, 17349–17371. https://doi.org/10.1007/s11042-020-08704-0.
40. Al-Berry, M. N., Ebied, H. M., Hussein, A. S., & Tolba, M. F. (2014). Human action recognition via multi-scale 3D stationary wavelet analysis. In *14th international conference on hybrid intelligent systems* (pp. 254–259). Kuwait: IEEE. https://doi.org/10.1109/HIS.2014.7086208.
41. Han, Y., Zhang, P., Zhuo, T., Huang, W., & Zhang, Y. (2018). Going deeper with two-stream ConvNets for action recognition in video surveillance. *Pattern Recognition Letters, 107*, 83–90. https://doi.org/10.1016/j.patrec.2017.08.015.
42. Ji, X., Cheng, J., Feng, W., & Tao, D. (2018). Skeleton embedded motion body partition for human action recognition using depth sequences. *Signal Processing, 143*, 56–68. https://doi.org/10.1016/j.sigpro.2017.08.016.
43. Kumar, S. U., & Inbarani, H. H. (2017). PSO-based feature selection and neighborhood rough set-based classification for BCI multiclass motor imagery task. *Neural Computing and Applications, 28*(11), 3239–3258. https://doi.org/10.1007/s00521-016-2236-5.
44. Huang, C. L., & Dun, J. F. (2008). A distributed PSO-SVM hybrid system with feature selection and parameter optimization. *Applied Soft Computing, 8*(4), 1381–1391. https://doi.org/10.1016/j.asoc.2007.10.007.
45. Huynh-The, T., Banos, O., Le, B. V., Bui, D. M., Lee, S., Yoon, Y., & Le-Tien, T. (2015). PAM-based flexible generative topic model for 3D interactive activity recognition. In *Proceedings of the International Conference on Advanced Technologies for Communications*, Vietnam, 117–122. https://doi.org/10.1109/ATC.2015.7388302.
46. Golash, R., & Jain, Y. K. (2017). Motion estimation and tracking of hand using Harris-Laplace feature based approach. *Biometrics and Bioinformatics, 9*, 157–163.
47. Guan, T., Han, F., & Han, H. (2019). A modified multi-objective particle swarm optimization based on levy flight and double-archive mechanism. *IEEE Access, 7*, 183444–183467.
48. El-Sawya, A. A., Zakib, E. M., & Rizk-Allhb, R. M. (2013). A novel hybrid ant colony optimization and firefly algorithm for multi-objective optimization problems. *International Journal of Mathematical Archive, 6*(1), 1–22.
49. Deshpande, A., & Patavardhan, P. (2017). Super resolution of long range captured multiframe iris polar images. *IET Biometrics, 6*(5), 360–368.
50. Deshpande, A., & Patavardhan, P. (2017). Multiframe super-resolution for long range captured iris polar image. *IET Biometrics, 6*(2), 108–116.
51. Deshpande, A., Patavardhan, P., & Rao, D. H. (2015). Iterated back projection based super-resolution for iris feature extraction. *Elsevier Procedia Computer Science, 48*, 269–275.

Lossy Compression of Noisy Images Using Autoencoders for Computer Vision Applications

Dorsaf Sebai and Faouzi Ghorbel

1 Introduction

Computer vision (CV) is imperative today as it enhances the understanding of the information within images. A growing number of sensors that capture a handful of wavelengths (e.g., cameras) intensify the demand for CV. These devices replace the human eye with image processing algorithms substituting or complementing the human brain. In this context, the capture of environmental evidence through sensors poses a significant challenge. This obstacle leads to voluminous images and videos containing a large amount of information from different spectral bands.

The need to handle and fuse different types of imageries with different or deficient resolutions call for improvements in image quality. One of the most popular schemes is the super-resolution (SR), which ameliorates image quality by delivering high-resolution images and results in substantial space storages and other computational needs [1].

Autoencoders (AEs) perform unsupervised feature extraction, which can be particularly useful for computer vision algorithms. Such observation stems from the fact they can address the following:

1. The compression of images to reduce transmission delay between sensors and processing units
2. The feature extraction to not lose details that cannot be seen by the human eye but are picked up by CNNs

AEs are normally trained, assuming artifact-free input images. This is not the case in real CV application scenarios, where there exist several types of noise. However, AEs perform image recovery effectively, and this will be explored in this

D. Sebai (✉) · F. Ghorbel
Cristal Laboratory, National School of Computer Science (ENSI), Manouba, Tunisia

A. Deshpande et al. (eds.), *Computational Intelligence Methods for Super-Resolution in Image Processing Applications*, https://doi.org/10.1007/978-3-030-67921-7_8

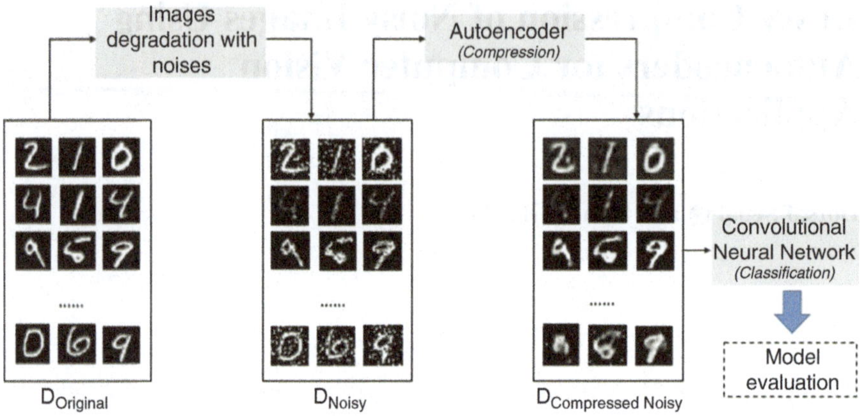

Fig. 1 Steps of the learning architecture: MNIST is an example of a benchmark dataset

research effort. Hence, noise removal of the decoder's desired output is fixed to be similar to the original images.

Nevertheless, this cannot be exploited if the noise arrives during image acquisition. In this case, we do not have pristine (i.e., artifact-free, clear, noiseless, and undistorted) images to use them as the desired AE output. This perfect image is also known as the ground truth (GT) and is clear, noiseless, and undistorted. The output is instead fixed to be similar to noisy images. This is indeed the quintessential case of real-world problems where sensors are often placed in noisy environments such as factories, trains, and military equipment types.

Since classification presents the most common step of almost all computer vision applications, this work aims to study the impact of image compression in the presence of noise using AE on CNN classification performances. This issue is particularly vital in this research effort [2–4] as it only studies the impact of standard compression techniques, for example, JPEG and JPEG 2000. However, artifacts of auto encoder-based compression methods have different characteristics from classical codecs [5].

Figure 1 depicts the steps of our learning architecture applied to different benchmark deep learning datasets

1. The constructed dataset D_{Noisy} includes distorted images by applying different common types of noise to original images of the D_{Original} dataset.
2. We construct the $D_{\text{Compressed Noisy}}$ dataset that includes the compressed images of the D_{Noisy} dataset using different AE configurations.
3. Images from the $D_{\text{Compressed Noisy}}$ dataset pass through a state-of-the-art CNN to classify them.
4. According to the results of step 3, the framework analyzes how AEs can deal with different types of noise.
5. Identify which of the noise types, once compressed, are more likely to generate classification errors.

The rest of this chapter is organized as follows. Section 2 introduces the general working principle of AEs and CNNs. In Sect. 3, we detail our approach to building the D_{Noisy} and $D_{\text{Compressed Noisy}}$ datasets as well as our learning architecture. Section 4 presents discussions about the CNN performance on unspoiled and compressed noisy images. In Sect. 5, this performance can be enhanced using a super-resolution-based learning architecture. Finally, Section 6 concludes this chapter.

2 Background

Machine learning techniques based on deep neural networks (DNNs) are more and more leveraged to address several issues and challenges of our daily life, such as image recognition [6], handwriting recognition [7], and autonomous driving vehicles [8]. Here, we will only introduce two of the well-known DNNs that the authors will use later, namely autoencoders (AEs) and CNNs.

2.1 Autoencoder Neural Network

An autoencoder (AE) is an unsupervised learning algorithm that will find patterns in a dataset by detecting key features [9]. The AE analyzes all of the images in the dataset. Then, it extracts some useful features automatically so that it can distinguish images using those features. Ordinarily speaking, AEs excel in tasks that imply dimensionality reduction, feature extraction, and data compression. Most AEs are shallow networks with an input layer (X), one or a few hidden layers (HLs), and an output layer (Y). An AE is mainly composed of two parts, the encoder and the decoder. The encoder selects the most critical features of the input to yield a compressed representation. In the example of Fig. 2, the authors will compress the digit image, moving from 784 dimensions of data to only 196 dimensions. The decoder is a reflection of the encoder network. It works to recover the data from the code and recreate the input as accurately as it can.

As one can see, there is no need to care about the reconstructed image in this network. What we do care about, though, is the code layer values, which represent the input layer of the decoder. If the network has learned enough, it can generate the replica of input images only based on the code layer value. Then, this vector of size 196 is good enough to represent the input image feature set.

2.2 Convolutional Neural Network

The advancements in computer vision with deep learning have been essentially perfected with time, thanks to CNNs, also known as ConvNets [10]. They are

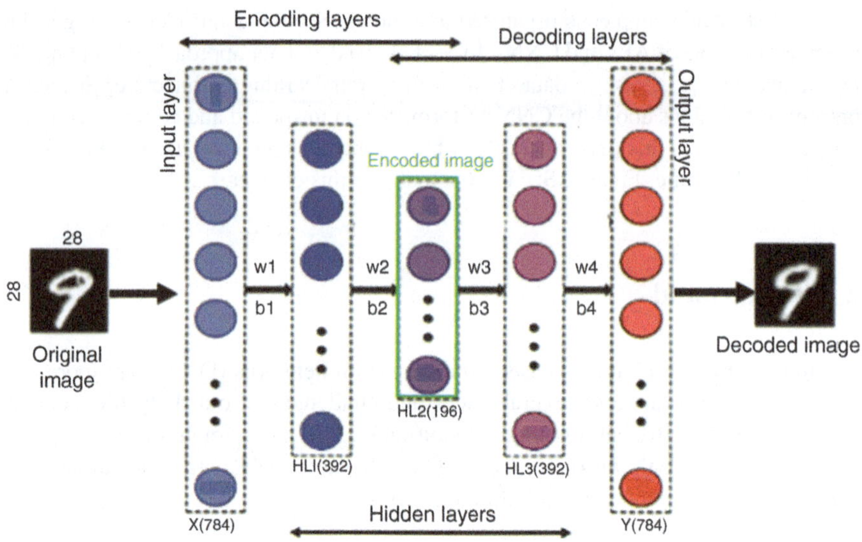

Fig. 2 Autoencoder architecture. An example of an encoder of depth 2: 1 intermediate layer of size 392 and 1 code layer of size 196

artificial neural networks inspired by the human visual cortex, especially fitted for image classification. Their primary purpose is to recognize and classify shapes, for example, objects and faces, in input images. They can carry out image classification by firstly extracting common low-level features. Secondly, they construct less elementary features, thanks to consecutive convolutional layers.

A CNN takes the input image, passes it along to convolutional, nonlinear, pooling, and fully connected layers to obtain an output. As their names indicate, neurons in convolutional layers generate images while convolving the input image with extracted features. The pooling layers downsample images of the previous layers that are mixed using a nonlinear function. Finally, the last pooling layer is attached to one or more fully connected layers that output a vector of dimension N, knowing that N is the number of classes to choose from.

3 Approach and Datasets

First, the D_{Original} datasets employed to generate the distorted D_{Noisy} ones are introduced. Second, we detail the settings of the AE used to construct the $D_{\text{Compressed Noisy}}$ dataset. Third, we specify the CNNs that we use to classify the so generated $D_{\text{Compressed Noisy}}$ images.

3.1 Datasets Description

As shown in Fig. 3, the learning architecture shown in Fig. 1 is evaluated on three well-known D_{Original} benchmark datasets:

- *Mixed National Institute of Standards and Technology (MNIST)* [11]: MNIST is a dataset of 28 × 28 size-normalized and centered images of handwritten digits. MNIST is composed of a training set of 60,000 samples and a test set of 10.000 samples labeled for 10 classes, one per each digit.
- *Fashion MNIST* [12]: This benchmark dataset consists of 60,000 and 10,000 training and test sets. Examples are 28 × 28 grayscale images belonging to 10 classes, namely t-shirt, trouser, dress, pullover, sandal, coat, shirt, sneaker, ankle, bag, and boot.
- *Canadian Institute for Advanced Research (CIFAR-10)* [13]: It has 60,000 color images of size 32 × 32 belonging to 10 mutually exclusive classes, with 6000 images for each category. It contains 50,000 and 10,000 training and testing samples. The classes are airplane, bird, cat, automobile, dog, deer, horse, ship, frog, and truck.

For each of the abovementioned datasets, the authors construct the D_{Noisy} training data by injecting several types and noise levels into the GT training images. The framework seeks five kinds of commonly faced noises, namely the Gaussian as an amplifier noise, Poisson is shot noise, Uniform is a quantization noise, Speckle is a multiplicative noise, and Salt & Pepper is an impulse noise. All these noise types are fairly presented in the D_{Noisy} dataset, that is, each of them shows 20% of the total samples. Furthermore, three levels, corresponding to variance values of 0.01 (level 1), 0.05 (level 2), and 0.1 (level 3), are considered for each noise. This training on a mixture of samples with multiple types of noises at different variances, rather than a certain type and variance, will provide the DNNs models, AE, and CNN for the case studied in this paper with stronger generalizing ability.

Fig. 3 Benchmark datasets *(from left to right)*: MNIST, Fashion MNIST, and CIFAR-10

3.2 Autoencoder Settings

The so obtained D_{Noisy} dataset is the entry of an AE to generate the $D_{\text{Compressed Noisy}}$ dataset from Fig. 1. Typically, training utilizes 70% of the noisy images, and the remaining 30% is for tests. Two different settings of the AE are considered. The former presents a vanilla AE with a fully connected code layer of size 32. The latter offers a deeper AE with 3 encoding fully connected layers of respective sizes equal to 128, 64, and 32. We note that an AE with fully connected layers based effectively operates for MNIST and Fashion MNIST, but not for CIFAR-10. Thus, the authors opted for a CNN-AE for this latter. Unlike simple handwritten digits and clothing items, there are many more features and details to extract from each CIFAR-10 image. To ensure the above-mentioned settings, we use a CNN-AE with 1 and 3 encoding convolutional layers.

These two settings are trained using Adam optimizer with a batch size of 128 for 100 epochs. The experiments are carried out using an initial learning rate of 0.1, decaying after every 20 epochs with an exponential rate of 0.1. During the AE back-propagation process, the metric leveraged, for both of our settings, to evaluate the quality of the output, concerning the input is the mean squared error (MSE) loss function. Here, we remind that the reconstructed images must be as close as possible to noisy ones since, as already mentioned, no noise-free pictures are available. The AE can be a denoiser when the reconstructed images must be similar to GT ones. Nevertheless, this cannot be our case as the noise occurs at the time of image acquisition and hence we do not own clean imageries of the scene.

3.3 CNNs for Classification

The $D_{\text{Compressed Noisy}}$ dataset (the AE output) is then provided as the input data to learn the CNN model. As for the AE, 70% of the $D_{\text{Compressed Noisy}}$ samples feed each CNN training epoch. Testing employs the remaining 30%. To achieve a fair comparison, we look for the CNN model that fits each benchmark dataset. If using the same CNN for all datasets, it would produce unbiased results as the CNN architecture is optimized for one dataset and not for the other. Thus, the authors opted for three CNN models optimized for each of our datasets, resulting in respective accuracy of 99.8%, 93%, and 90%. These models are largely detailed in [14–16].

4 Results and Discussion

Eight different training approaches are implemented to evaluate the impact of the noisy auto encoder-compressed images on one of the leading computer vision applications: classification. All these eight approaches share the same architecture

Table 1 Training strategies

Strategies	Training images	Noise variances	AE depth
1	GT	–	1
2	GT	–	3
3	Noisy (Gaussian, Poisson, Uniform, Speckle, and S&P)	0.01	1
4	Noisy (Gaussian, Poisson, Uniform, Speckle, and S&P)	0.01	3
5	Noisy (Gaussian, Poisson, Uniform, Speckle, and S&P)	0.05	1
6	Noisy (Gaussian, Poisson, Uniform, Speckle, and S&P)	0.05	3
7	Noisy (Gaussian, Poisson, Uniform, Speckle, and S&P)	0.1	1
8	Noisy (Gaussian, Poisson, Uniform, Speckle, and S&P)	0.1	3

of Fig. 1 but have different training strategies. As detailed in Table 1, these training strategies present different combinations of training images, noise variances, and AE depths. Their performances will then be analyzed via several classification metrics for MNIST, Fashion MNIST, and CIFAR-10.

4.1 Compressed GT Versus Compressed, Noisy Images

Noise-free and noisy images with lossy compression differ from each other. The authors compare CNN's classification results when it is trained on GT and compressed noisy images, respectively. Both of them are compressed using the two AE settings, namely depth 1 and depth 3 (*c.f.* Sect. 3.2). As a shorthand, we refer to these settings as AE_D1 and AE_D3. Their classification performances are evaluated in Tables 2 and 3, respectively, in terms of accuracy as well as macro and weighted averaged precision (*P*), recall (*R*), and F1-score (*F1*), expressed as percentages:

- *Accuracy metric*: It designates the proportion of correctly classified samples in their actual classes by the CNN. As shown in Eq. (1), the accuracy *A* is expressed as a function of true positives (*TP*), true negatives (*TN*), false positives (*FP*), and false negatives (*FN*) by

$$A = (TP + TN)/(TP + TN + FP + FN). \qquad (1)$$

- *Macro averaged P, R, and F1 metrics*: They refer to the average of the per-class precision, recall, and F1-score.
- *Weighted averaged P, R, and F1 metrics*: The average of the per-class *P*, *R*, and *F1*-score weighing each class's metric by the number of samples from that class.

Table 2 Classification performances of GT and compressed, noisy images considering *AE_D1* setting for *MNIST*, *Fashion MNIST*, and *CIFAR-10* datasets

		MNIST			
		GT images	**Noisy images**		
			ν = 0.01	**ν = 0.05**	**ν = 0.1**
Accuracy		97.73	96.30 (−1.43)	94.57 (−3.16)	93.57 (−4.16)
Macro Avg	**P**	97.74	96.28 (**−1.46**)	94.67 (−3.07)	93.79 (−3.95)
	R	97.72	96.26 (−1.46)	94.51 (−3.21)	93.47 (**−4.25**)
	F1	97.71	96.26 (−1.45)	94.49 (**−3.22**)	93.50 (−4.21)
Weighted Avg	**P**	97.76	96.32 (−1.44)	94.77 (−2.99)	93.84 (−3.92)
	R	97.73	96.30 (−1.43)	94.57 (−3.16)	93.57 (−4.16)
	F1	97.73	96.30 (−1.43)	94.57 (−3.16)	93.58 (−4.15)
		Fashion MNIST			
		GT images	**Noisy images**		
			ν = 0.01	**ν = 0.05**	**ν = 0.1**
Accuracy		81.77	81.67 (−0.10)	80.27 (−1.50)	78.47 (−3.30)
Macro Avg	**P**	81.53	81.25 (−0.28)	79.79 (−1.74)	77.52 (−4.01)
	R	81.69	81.25 (−0.44)	79.83 (−1.86)	78.01 (−3.68)
	F1	81.51	81.05 (**−0.46**)	79.45 (**−2.06**)	77.29 (**−4.22**)
Weighted Avg	**P**	81.78	81.62 (−0.16)	80.18 (−1.6)	78.01 (−3.77)
	R	81.77	81.67 (−0.10)	80.27 (−1.50)	78.47 (−3.30)
	F1	81.67	81.45 (−0.22)	79.87 (−1.80)	77.77 (−3.90)
		CIFAR-10			
		GT images	**Noisy images**		
			ν = 0.01	**ν = 0.05**	**ν = 0.1**
Accuracy		76.47	70.93 (−5.54)	64.20 (−12.27)	60.80 (−15.67)
Macro Avg	**P**	76.69	71.76 (−4.93)	64.68 (−12.01)	61.17 (−15.52)
	R	76.35	70.91 (−5.44)	64.22 (−12.13)	60.78 (−15.57)
	F1	75.76	70.04 (**−5.90**)	63.20 (**−12.75**)	59.65 (**−16.30**)
Weighted Avg	**P**	76.84	71.69 (−5.72)	64.62 (−12.56)	61.04 (−16.11)
	R	76.47	70.93 (−5.15)	64.20 (−12.22)	60.80 (−15.8)
	F1	75.89	69.99 (−5.54)	63.14 (−12.27)	59.59 (−15.67)

The above metrics can be employed, knowing the following definitions.

- *Precision (P) metric*: It is the proportion of predicted positives that is truly positive according to

$$P = TP/(TP + FP). \tag{2}$$

- *Recall (R) metric*: It is the proportion of actual positive outcomes that are correctly classified as follows:

$$R = TP/(TP + FN). \tag{3}$$

Table 3 Classification performances of GT and compressed, noisy images considering *AE_D3* setting for *MNIST*, *Fashion MNIST*, and *CIFAR-10* datasets

		MNIST			
		GT images	**Noisy images**		
			ν **= 0.01**	ν **= 0.05**	ν **= 0.1**
Accuracy		97.43	94.53 (−2.19)	93.47 (−3.96)	90.37 (−7.06)
Macro Avg	**P**	97.44	94.68 (−2.76)	93.58 (−3.86)	91.03 (−6.41)
	R	97.37	94.42 (**−2.95**)	93.41 (−3.96)	90.22 (**−7.15**)
	F1	97.39	94.50 (−2.89)	93.41 (**−3.98**)	90.28 (−7.11)
Weighted Avg	**P**	97.46	94.63 (−2.03)	93.63 (−3.83)	91.07 (−6.39)
	R	97.43	94.53 (−2.90)	93.47 (−3.96)	90.37 (−7.06)
	F1	97.43	94.53 (−2.90)	93.46 (−3.97)	90.38 (−7.05)
		Fashion MNIST			
		GT images	**Noisy images**		
			ν **= 0.01**	ν **= 0.05**	ν **= 0.1**
Accuracy		80.93	80.10 (−0.83)	79.97 (−0.96)	78.10 (−2.83)
Macro Avg	**P**	80.77	79.61 (−1.16)	79.45 (**−1.32**)	77.51 (−3.26)
	R	80.79	79.66 (−1.13)	79.62 (−1.17)	77.72 (−3.07)
	F1	80.53	79.33 (**−1.20**)	79.36 (−1.17)	77.25 (**−3.28**)
Weighted Avg	**P**	80.94	80.00 (−0.94)	79.85 (−1.09)	77.95 (−2.99)
	R	80.93	80.10 (−0.83)	79.97 (−0.96)	78.10 (−2.83)
	F1	80.69	79.75 (−0.94)	79.73 (−0.96)	77.66 (−3.03)
		CIFAR-10			
		GT images	**Noisy images**		
			ν **= 0.01**	ν **= 0.05**	ν **= 0.1**
Accuracy		60.20	53.30 (−6.90)	51.17 (−9.03)	47.87 (−12.33)
Macro Avg	**P**	60.05	52.81 (−7.24)	50.56 (−9.49)	47.37 (−12.68)
	R	60.10	53.29 (−6.81)	51.25 (−8.85)	47.93 (−12.17)
	F1	59.91	52.58 (**−7.64**)	50.52 (**−9.74**)	47.24 (**−13.01**)
Weighted Avg	**P**	60.17	52.77 (−7.33)	50.51 (−9.39)	47.34 (−12.67)
	R	60.20	53.30 (−7.40)	51.17 (−9.66)	47.87 (−12.83)
	F1	60.02	52.56 (−6.90)	50.45 (−9.03)	47.19 (−12.33)

- *F1-score (F1) metric*: It is the combination of *precision* and *recall* into a single number using a harmonic mean given by

$$\text{F1-score} = (2 \times P \times R)/(P + R). \tag{4}$$

The numbers in brackets in Tables 2 and 3 correspond to the difference between the metric values obtained for compressed, noisy images, and GT ones. Numbers that are in bold correspond to the largest ones per variance and AE setting.

From Tables 2 and 3, one can see that the CNN model, trained on noisy, compressed images, is sensitive to all the noise variances. Indeed, CNN's performance decreases as the noise level increases. Compared to GT images, the macro averaged recall of the CNN for compressed noisy MNIST images decreases, at a variance value of 0.01, by 1.46% for AE_D1 and 2.95% for AE_D3. The decrease is about 3% and 4% for variance 0.05 and 4% and 7% for variance 0.1. For the CIFAR-10, CNN's accuracy for noisy, compressed images decreases compared to GT images at variance 0.01, by 5.54% and 6.9% for the respective AE_D1 and AE_D3. Even with the noise level 0.1, where the variance value is still moderate, the accuracy drops more than 15% for AE_D1 and 12% for AE_D3.

The results from Tables 2 and 3 also exhibit that training strategies with AE_D1 allow better classification performance than AE_D3 on both GT and noisy images. For instance, the weighted averaged F1-score obtained for the AE_D3 policy, that is, 2, 4, 6, and 8 of Table 1, drops about 3% compared to the AE_D1 strategy, that is, 1, 3, 5, and 7, for Fashion MNIST. For the CIFAR-10, it drops more than 14%. It is known that if the data are highly nonlinear, then a deep AE should do well. This is not the case of our datasets where a vanilla AE is efficient, and no more hidden layers are required to produce a compact representation of MNIST, Fashion MNIST, and CIFAR-10 images. Visual evaluation of Fig. 4 is in support of the above interpretations. It provides samples of the MNIST D_{Original}, D_{Noisy} at variances 0.01, 0.05, and 0.1. The corresponding results for $D_{\text{Compressed Noisy}}$ are related to the AE_D1 and AE_D3 for the following kinds of noise: Gaussian, Poisson, Uniform, Speckle, and Salt & Pepper. As we can observe in some cases, the classification would not be robust for AE_D3, compared to AE_D1, as digits are more altered. For instance, the digit in the last row and second column of Fig. 4b can be interpreted as 6, even though it is a 5. On the other hand, the digit 3 in the seventh row and fifth column of Fig. 4c is interpreted differently. The CNN can misclassify it as an 8. For more results, refer to the appendix.

4.2 Results per Noise Type

Until now, our evaluation considered all types of noise together gathered in the test set. As already mentioned, the noise types are fairly presented in D_{Noisy} dataset, where each of them presents 20% of the total samples. This present section assesses the CNN performances while being trained on the mixture of the noise types but tested on every noise separately. Hence, one can study how damaging an AE compresses a specific kind of noise. The corresponding results appear in Fig. 5 in terms of accuracy, macro averaged F1-score, and weighted averaged F1-score for MNIST, Fashion MNIST, and CIFAR-10 with AE_D1.

Tables 2 and 3 reveal that the classification performance for noisy, compressed images decays when the noise level rises from level 1 to level 3. Although expected, we could not predict how much this decay could be if the authors have not carried out these experiments. As an example, the accuracy decays from 81.67% to 78.47%

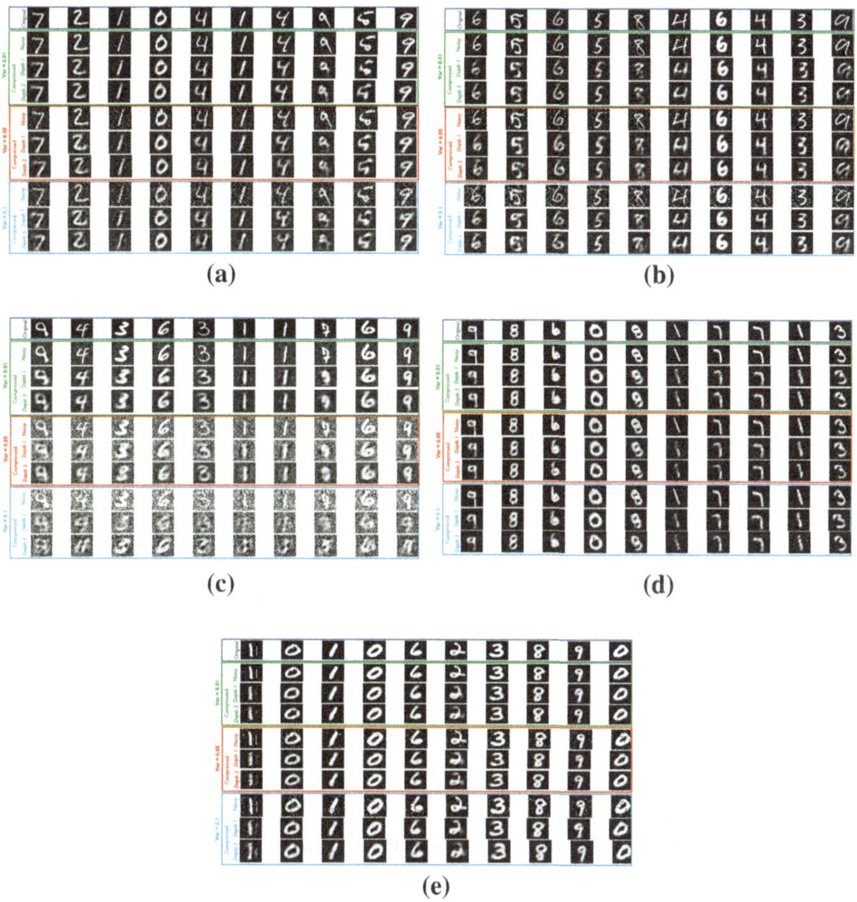

Fig. 4 Samples of the *MNIST* D_{Original} (blue rectangle), D_{Noisy} and its corresponding $D_{\text{Compressed Noisy}}$ at variances 0.01 (green rectangle), 0.05 (red rectangle), and 0.1 (cyan rectangle) using AE_D1 and AE_D3 for the following types of noise: (**a**) *Gaussian*, (**b**) *Poisson*, (**c**) *Uniform*, (**d**) *Speckle*, and (**e**) *Salt & Pepper*

and from 70.93% to 60.80% for the respective Fashion MNIST and CIFAR-10 datasets at AE_D1. The accuracy of MNIST is 96.3%, with a variance of 0.01 and drops to 93.57% for a variance of 0.1. Referring to Fig. 5, Gaussian and Uniform noises affect the most the classification performance as they present steep slope when passing from one noise level to another. This can be intuitively observed in Fig. 4a and c, where the compressed noisy images' quality is rapidly distorted when the noise variance rises from 0.01 to 0.05, then to 0.1. This decreasing speed is of slower order of magnitude for Poisson, Speckle, and Salt & Pepper noises. Furthermore, it is noticed that Uniform noise is the most damaging distortion since it has the worst classification performances for all metrics and datasets in Fig. 5. This can also be observed in Fig. 4c as well as in Figs. 10 and 15 in the Appendix.

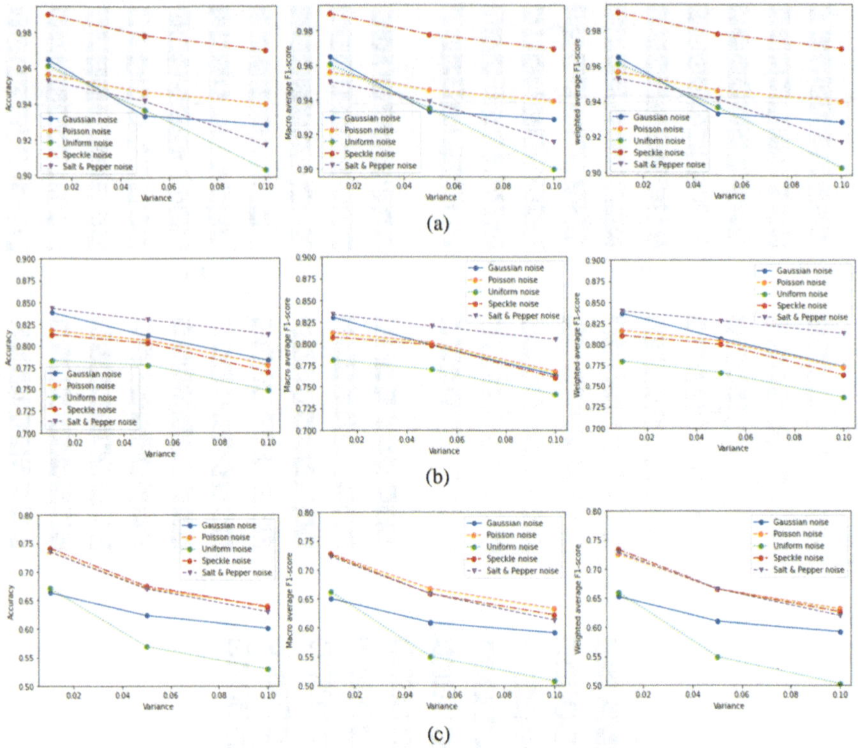

Fig. 5 Classification performance of the CNN model tested separately on each type of noise in terms of accuracy (first column), macro averaged F1-score (second column), and weighted averaged F1-score (third column)for (**a**) *MNIST*, (**b**) *Fashion MNIST*, and (**c**) *CIFAR-10*

5 Super-Resolution-Based Learning Architecture

So far, the authors have studied the noisy image AE-based compression impact on classification. In this section, the authors propose modifying the basic learning architecture of Fig. 1 to enhance CNN performances for compressed noisy image classification. Compared to the architecture of Fig. 1, the authors include a new step, marked in orange in Fig. 6, as a part of the enhanced learning architecture. As its name indicates, the SR converts low-resolution images from the AE to their corresponding high-resolution versions D_{SR}. Unlike the learning architecture of Fig. 1, the D_{Noisy} dataset passes through an AE that trains to output downscaled images by a factor of 4. The so generated images of the $D_{\mathrm{Compressed\ Noisy}}$ dataset are no longer images of the same resolution as D_{Noisy} ones. The first asset of such downsampling is an even higher reduction of the amount of data to transmit from sensors to processing units where classification is performed. The second asset consists of a noise reduction as noise decreases upon downsampling. Once in the processing unit and to recover the initial size of images before the downscaling, a

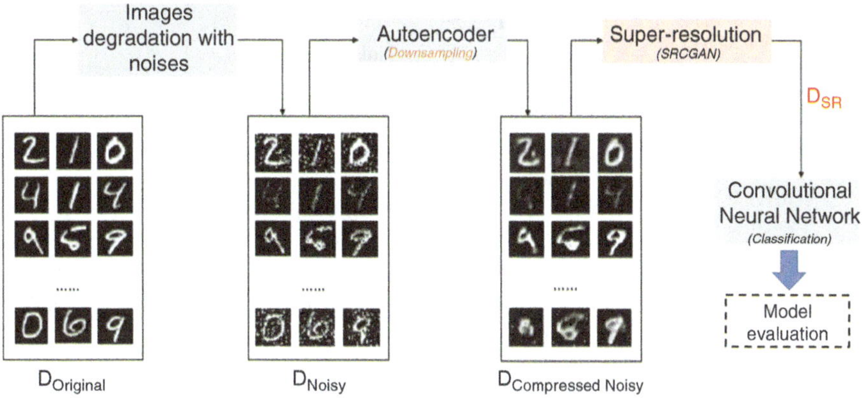

Fig. 6 Steps of the super-resolution-based learning architecture

super-resolution procedure is applied to images of the $D_{\text{Compressed Noisy}}$ dataset. Typically, one makes use of the SR conditional generative adversarial network (SRCGAN) as described in [17]. This latter employs a deep residual network as the generator to retrieve the original size from downscaled images. The downsampling step lessens noise, and images treated by SRCGAN will provide refined details that help the CNN to classify objects within their corresponding classes correctly.

Performances of the basic and enhanced learning architectures appear accurately in Table 4 to classify compressed, noisy images from the MNIST, Fashion MNIST, and CIFAR-10 datasets. As a shorthand, we refer to the basic learning architecture as learning architecture without/minus super resolution (LA − SR) and to the enhanced one as learning architecture with/plus super resolution (LA + SR). Comparisons consider the AE_D1 learning strategy since it presents better results than the AE_D3 one (cf. Sect. 4.1). Table 4 lists results where LA + SR enhances the classification performances for all the variance values and benchmark datasets. For the MNIST dataset, for example, the proposed LA + SE enables the CNN to reach an accuracy of 99.12%, 98.67%, and 96.54% for the variances equal to 0.01, 0.05, and 0.1, respectively. This element means an enhancement of 2.82%, 4.1%, and 2.97% compared to the LA-SR classification performances. A more vital enhancement stems from the Fashion MNIST and CIFAR-10 datasets. This enhancement is due to the super-resolution task carried out on images of $D_{\text{Compressed Noisy}}$ images where noise is already reduced, thanks to downsampling. Thus, the coarse details are reconstructed with a better and refined visual quality.

As shown in Table 5, LA + SR also provides better performances than LA − SR in terms of compression ratio. This can be justified by the downscaling performed by the AE to noisy images of the D_{Noisy} dataset. This resolution reduction highly decreases the size of D_{Noisy}, far more than the reduction produced by the LA − SR, where the AE reconstructs output images of the same resolution as the input ones.

Table 4 Classification *accuracy* of compressed, noisy images obtained by learning architectures *without and with super-resolution* for MNIST, Fashion MNIST, and CIFAR-10 datasets

MNIST			
	Noisy images		
	$\nu = 0.01$	$\nu = 0.05$	$\nu = 0.1$
LA − SR	96.30	94.57	93.57
LA + SR	99.12	98.67	96.54
Fashion MNIST			
	Noisy images		
	$\nu = 0.01$	$\nu = 0.05$	$\nu = 0.1$
LA − SR	81.67	80.27	78.47
LA + SR	89.61	87.99	85.45
CIFAR-10			
	Noisy images		
	$\nu = 0.01$	$\nu = 0.05$	$\nu = 0.1$
LA − SR	70.93	64.20	60.80
LA + SR	76.50	74.55	70.12

Table 5 *Compression ratio* obtained by learning architectures *without and with super-resolution* for MNIST, Fashion MNIST, and CIFAR-10

	MNIST	Fashion MNIST	CIFAR-10
LA − SR	123.07	89.46	83.13
LA + SR	162.81 (↑ 39.74)	132.81 (↑ 43.35)	157.39 (↑ 74.26)

6 Conclusion

For almost all computer vision applications, the deep neural networks, mainly the CNNs for classification, are trained and tested, considering that the images to process are distortion-free. Nevertheless, images can be subject to several distortions, particularly noise, and GT images do not exist in real applications. Furthermore, images are voluminous, and compression is a must to send them to the processing units. Noise, though, causes the degradation of all image compression algorithms, even the most robust ones. This chapter aims to address the lossy compression of noisy images, such as remote sensing ones. We aim to study the impact of noise on lossy image compression using vanilla and deep AE. Typically, the effect of Gaussian, Poisson, Uniform, Speckle, and Salt & Pepper noises on AE-based compression are studied. We first inject them into the training and test sets of MNIST, Fashion MNIST, and CIFAR-10 benchmark datasets. The so distorted images are then compressed using AEs of depth 1 and 3. A CNN is finally leveraged to classify the so compressed noisy images. The results show that the classification performance is susceptible to all noise types, albeit to varying degrees. Gaussian and Uniform, which are amplifier and quantization noises, affect CNN's resilience when classifying compressed, noisy images more than Poisson, Speckle, and Salt & Pepper ones, that are shot, multiplicative, and impulse noises.

Moreover, it is concluded that compressed, noisy images lead to lower classification performance than the GT images across different noises and levels. This decay becomes more critical and dramatically increases as images include more complex objects and textures. A maximum decay of 16.3% is noticed for CIFAR-10, whereas maximum decays of almost 7% and 4% are perceived for the simpler MNIST and Fashion MNIST datasets. The reason behind this is not the three-color channels or the slightly larger pixel count of the CIFAR-10 images, but rather their internal complexity.

Next, the primary learning architecture used for the above study is modified to extract a more useful signal for classification. Typically, the noisy, compressed images are no more passed, as they are, along with the CNN for classification. They are first downsampled so that we reduce noise while gaining in terms of compression ratio. Second, downsampled images are reconstructed using the SRCGAN super-resolution technique to retrieve details sufficient to enhance the basic learning architecture's classification performances. Experimental results have shown that better accuracy and compression ratio can be obtained. These promising results make the proposed LA + SR architecture potentially useful for computer vision systems. In this paper, we used SRCGAN as a super-resolution method. As future work, we aim to produce a complete study that assesses the ability of existing SR techniques to reduce noise. This study will enable us to distinguish which of the well-known SR approaches would be more efficient to produce clearer images so that CNNs could achieve better classification performances. The authors aim to evaluate SR techniques based on deep learning. They will be heterogeneously chosen because they belong to different categories, including linear, multi-branch, attention-based, residual, recursive, progressive, and adversarial designs [1]. The emphasis should also be on confronting the SR techniques to the ordinary denoising filters, for example, Gaussian, median, and bilateral, as well as the dominant efforts proposed in this context, for example, [18–22].

Appendix: Additional Results

Figure 7 depicts original samples from the MNIST, Fashion MNIST, and CIFAR-10 datasets that are compressed using the two settings of the AE. The aim is to represent the results of the AE while compressing noise-free images. To point out the AE's performance compressing noisy images, the results of Fig. 7 can be compared to those of Figs. 8, 9, 10, 11, 12, 13, 14, 15, 16, and 17. These show original samples of MNIST, Fashion MNIST, and CIFAR-10 datasets to which the five noises are respectively added at the three different variances. The figures also show these noisy samples compressed by two-auto encoder settings. Section 3 depicts the kinds of noise considered, variances, and AE settings.

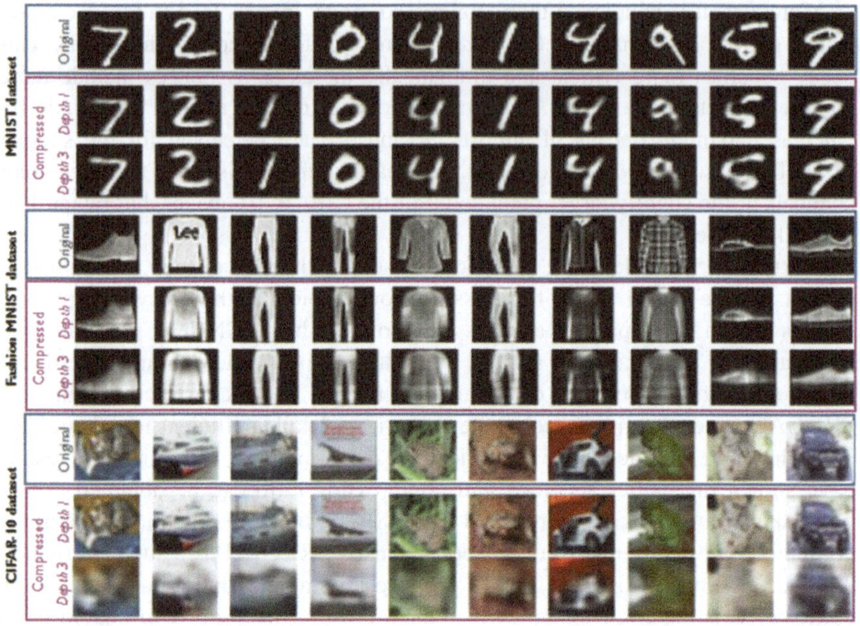

Fig. 7 GT samples from the *MNIST*, *Fashion MNIST*, and *CIFAR-10* compressed using the two AE settings

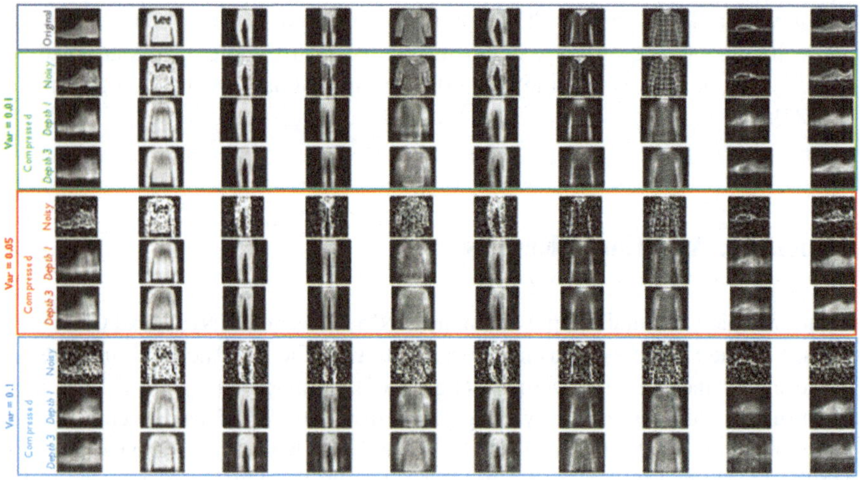

Fig. 8 Samples of the *Fashion MNIST* D_{Original}, D_{Noisy}, and $D_{\text{Compressed Noisy}}$ for the *Gaussian* noise

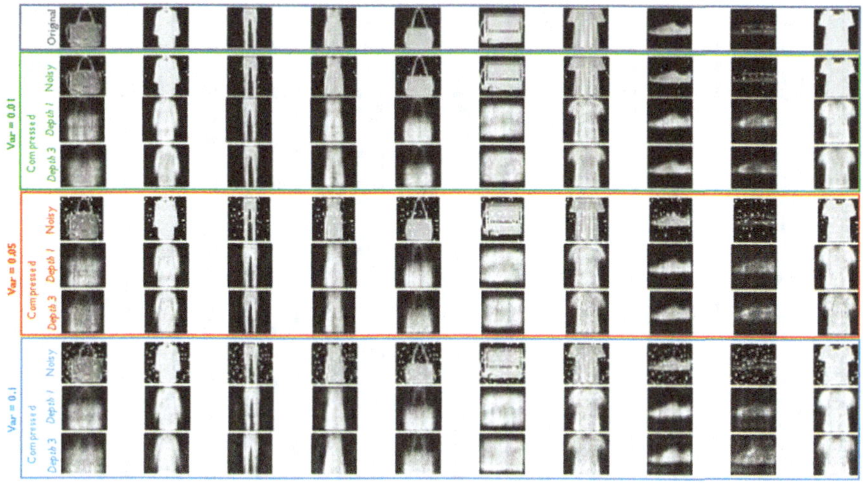

Fig. 9 Samples of the *Fashion MNIST* D_{Original}, D_{Noisy}, and $D_{\text{Compressed Noisy}}$ for the *Poisson* noise

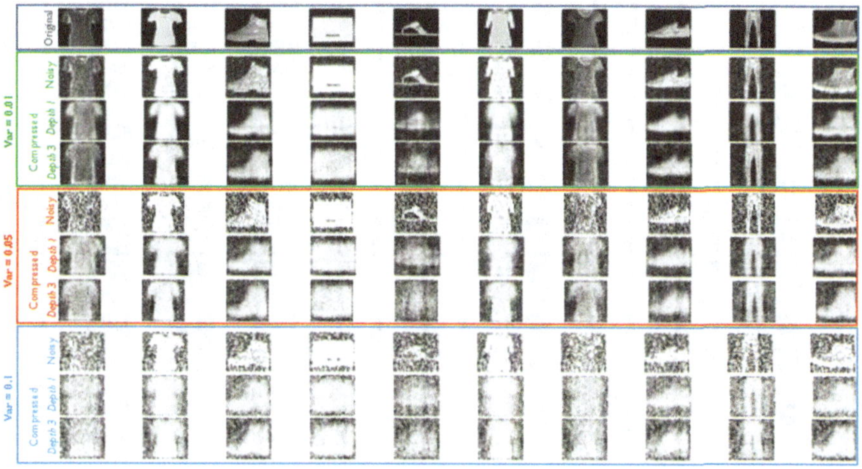

Fig. 10 Samples of the *Fashion MNIST* D_{Original}, D_{Noisy}, and $D_{\text{Compressed Noisy}}$ for the *Uniform* noise

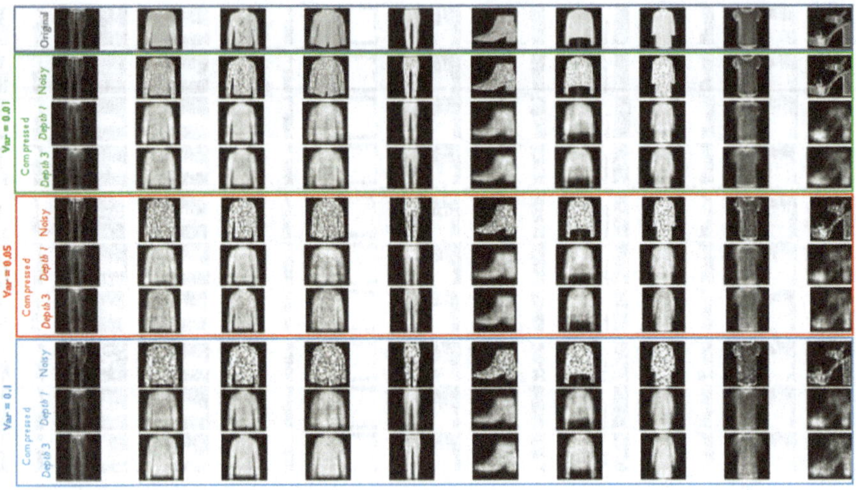

Fig. 11 Samples of the *Fashion MNIST* D_{Original}, D_{Noisy}, and $D_{\text{Compressed Noisy}}$ for the *Speckle* noise

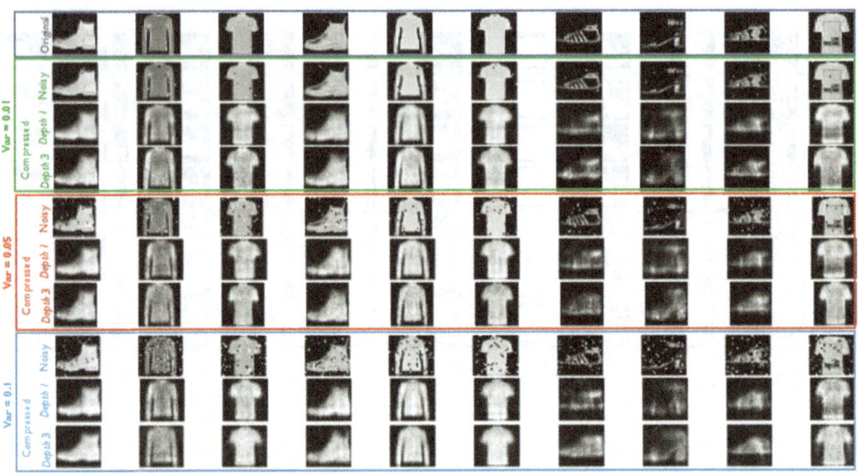

Fig. 12 Samples of the *Fashion MNIST* D_{Original}, D_{Noisy}, and $D_{\text{Compressed Noisy}}$ for the *Salt & Pepper* noise

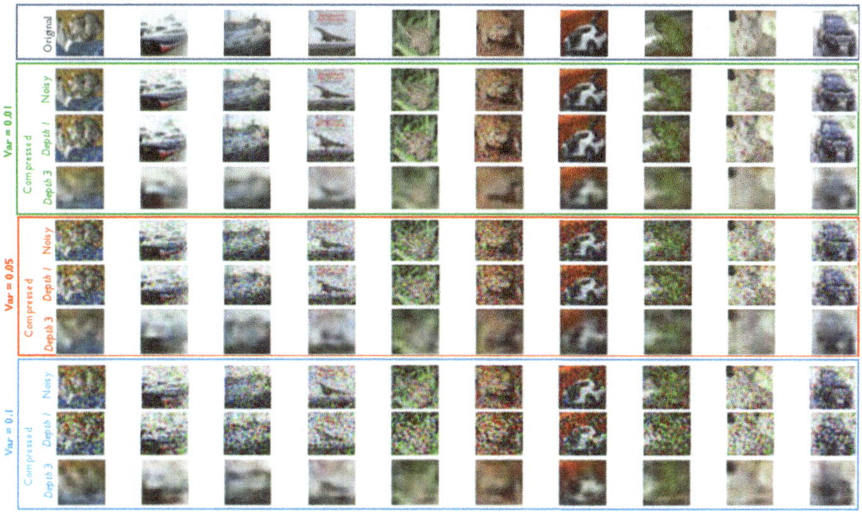

Fig. 13 Samples of the *CIFAR-10* D_{Original}, D_{Noisy}, and $D_{\text{Compressed Noisy}}$ for the *Gaussian* noise

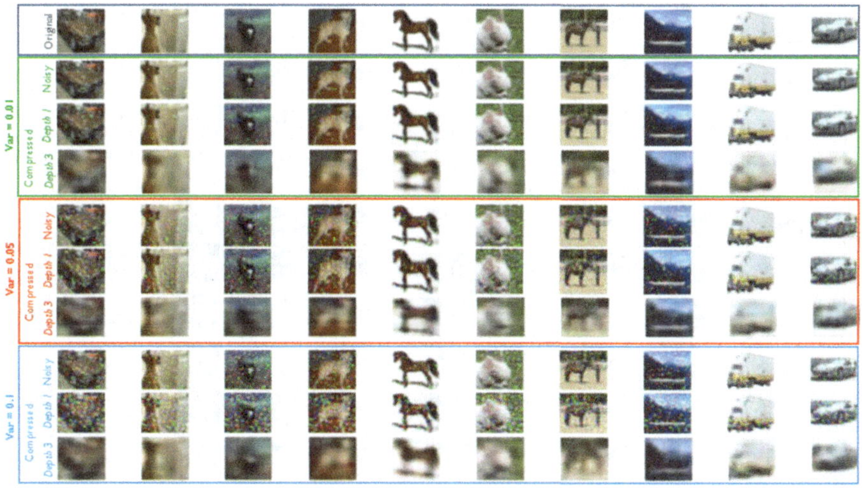

Fig. 14 Samples of the *CIFAR-10* D_{Original}, D_{Noisy}, and $D_{\text{Compressed Noisy}}$ for the *Poisson* noise

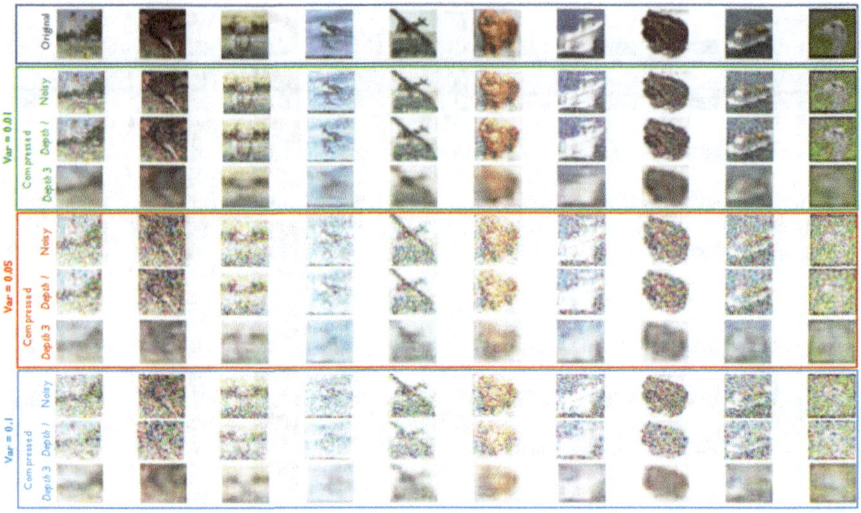

Fig. 15 Samples of the *CIFAR-10* D_{Original}, D_{Noisy}, and $D_{\text{Compressed Noisy}}$ for the *Uniform* noise

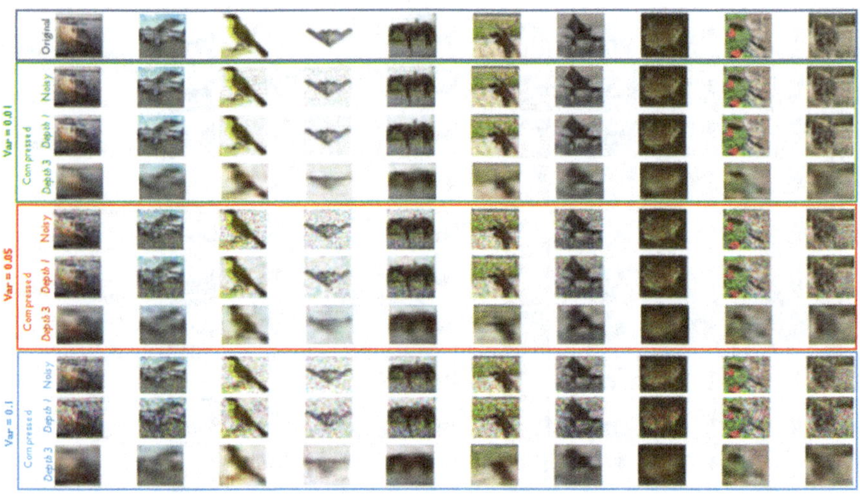

Fig. 16 Samples of the *CIFAR-10* D_{Original}, D_{Noisy}, and $D_{\text{Compressed Noisy}}$ for the *Speckle* noise

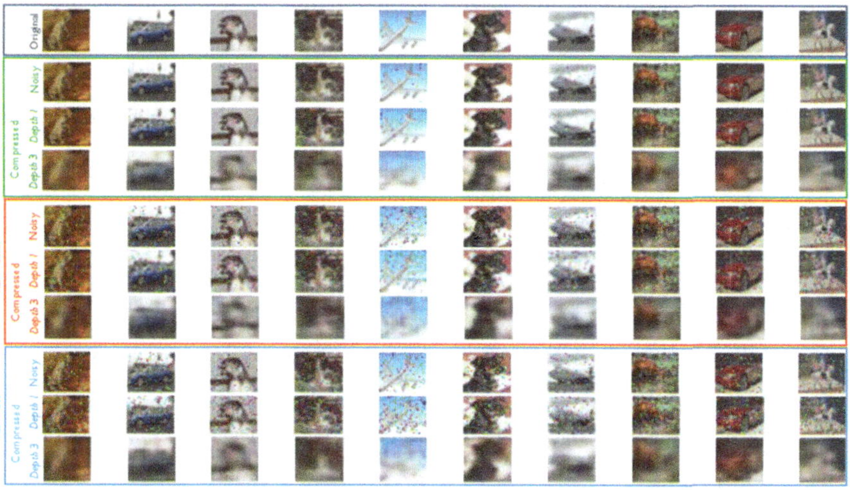

Fig. 17 Samples of the *CIFAR-10* D_{Original}, D_{Noisy}, and $D_{\text{Compressed Noisy}}$ for the *Salt & Pepper* noise

References

1. Anwar, S., Khan, S., & Barnes, N. (2018). A deep journey into super-resolution: A survey. arXiv:1904.07523.
2. Dejean-Servières, M., Desnos, K., Abdelouahab, K., Hamidouche, W., Morin, L., & Pelcat, M. (2018). Study of the impact of standard image compression techniques on performance of image classification with a Convolutional Neural Network. *HAL Archives*.
3. Chen, Z., Lin, W., Wang, S., Xu, L., & Li, L. (2017). Image quality assessment guided deep neural networks training. arXiv:1708.03880v1.
4. Al-Shaykh, O. K., & Mersereau, R. M. (1998). Lossy compression of noisy images. *IEEE Transactions on Image Processing, 7*(12), 1641–1652.
5. Valenzise, G., Purica, A., Hulusic, V., & Cagnazzo, M. (2018). Quality assessment of deep-learning-based image compression. In *IEEE 20th International Workshop on Multimedia Signal Processing*.
6. He, K., Zhang, X., Ren, S., & Sun, J. (2016). Deep residual learning for image recognition. In *IEEE Conference on Computer Vision and Pattern Recognition*.
7. LeCun, Y., Boser, B. E., Denker, J. S., Henderson, D., Howard, R. E., Hubbard, W. E., & Jackel, L. D. (1990). Handwritten digit recognition with a back-propagation network. In *Advances in neural information processing systems* (pp. 396–404). San Francisco: Morgan Kaufmann Publishers Inc.
8. Prabhakar, G., Kailath, B., Natarajan, S., & Kumar, R. (2017). Obstacle detection and classification using deep learning for tracking in high-speed autonomous driving. In *IEEE Region 10 Symposium (TENSYMP)*.
9. Baldi, P. (2012). Autoencoders, unsupervised learning, and deep architectures. *Workshop on Unsupervised and Transfer Learning*.
10. Le Cun, Y., & Bengio, Y. (1995). Convolutional networks for images, speech, and time series. In M. A. Arbib (Ed.), *The handbook of brain theory and neural networks*. Cambridge, MA: MIT Press.
11. http://yann.lecun.com/exdb/mnist

12. https://github.com/zalandoresearch/fashion-mnist
13. https://www.cs.toronto.edu/kriz/cifar.html
14. https://www.kaggle.com/elcaiseri/mnist-simple-cnn-keras-accuracy-0-99-top-1
15. https://www.kaggle.com/fuzzywizard/fashion-mnist-cnn-keras-accuracy-93
16. https://appliedmachinelearning.blog/2018/03/24/achieving-90-accuracy-in-objectrecognition-task-on-cifar-10-dataset-with-keras-convolutional-neural-networks/
17. Hitawala, S., Li, Y., Wang, X., & Dongyang, Y. (2018). Image super-resolution using VDSR-ResNeXt and SRCGAN. arXiv:1810.05731.
18. Shen, Z., Toh, K. C., & Yun, S. (2011). An accelerated proximal gradient algorithm for frame based image restoration via the balanced approach. *SIAM Journal on Imaging Sciences, 4*(2), 573–596.
19. Cai, J.-F., Chan, R. H., Shen, L., & Shen, Z. (2009). Simultaneously inpainting in image and transformed domains. *Numerische Mathematik, 112*(4), 509–533.
20. Cai, J.-F., Chan, R. H., & Shen, Z. (2008). A framelet-based image inpainting algorithm. *Applied and Computational Harmonic Analysis, 24*(2), 131–149.
21. Chan, R. H., Riemenschneider, S. D., Shen, L., & Shen, Z. (2004). Tight frame: An efficient way for high-resolution image reconstruction. *Applied and Computational Harmonic Analysis, 17*(1), 91–115.
22. Cai, J.-F., Ji, H., Shen, Z., & Ye, G.-B. (2014). Data-driven tight frame construction and image denoising. *Applied and Computational Harmonic Analysis, 37*(1), 89–105.

Recognition of Handwritten Nandinagari Palm Leaf Manuscript Text

Prathima Guruprasad and Guruprasad K. S. Rao

1 Introduction

The human mind outperforms machines in reading, seeing, and acquiring knowledge. Research across several decades evolved various techniques in areas such as machine wisdom and pattern recognition. The first challenge of knowledge preservation was convincing individuals and institutions having ancient manuscripts of the importance of treatment and storage in digitized form. The next challenge is to convert these innumerable volumes to textual form. This will aid the common person to find data with a handy search engine easily. The third challenge is the interpretation and reuse.

In any language, writing is the symbolic and graphical representation. Exclusively for language, every written manuscript developed a script. Due to various socio-cultural factors, script changes as time goes by. A modified script can serve as an alternative to the older script for convenience. Some of its characters may stay intact, while some may change. From Nandinagari, we see that there is an addition of Shirorekha in Devanagari to group characters to words for ease of identification apart from a change in some part of the vocabulary. This is the earlier version of Devanagari scripts. Nandinagari script was widely used in India. Nandinagari scanned handwritten documents pose more challenges when they need to be converted into textual form and interpreted. The lack of knowledge and other external constraints hinder this, although an ocean of wisdom lies there in the scanned form. This work helps to overcome the challenge of Nandinagari text

P. Guruprasad (✉)
Department of Information Science and Engineering, RVITM, Bengaluru, India

G. K. S. Rao
Communications, Media and Technology, Mindtree Limited, A Larsen and Toubro Group Company, Bengaluru, India

A. Deshpande et al. (eds.), *Computational Intelligence Methods for Super-Resolution in Image Processing Applications*, https://doi.org/10.1007/978-3-030-67921-7_9

identification by attempting to propose a model system architecture for processing Nandinagari images and conversion to text.

1.1 Background

Nandinagari is a Brahmi-based script derived from Nagari script. It was used in abundance in the Indian mainland for producing Sanskrit manuscripts from seventh till nineteenth century AD. Many manuscripts and metal/stone inscriptions in Sanskrit in south Maharashtra, Karnataka, and Andhra Pradesh are all in Nandinagari. Sri Ananda Thirtha (Madhwacharya), a saint of the thirteenth century who founded the Dwaita school of Vedanta, has hundreds of manuscripts written in Nandinagari on the palm leaves. The concept of Tatvavada propounded by this spiritual guru is very practical, relevant, and followed by many followers, including ISKCON throughout the world. Numerous catalogs of Sanskrit manuscripts covering over 47 subfields of knowledge, from Anthology to Yoga, including Sri Vaishnavism and Saivism, are written in Nandinagari [1]. Very few scholars are now available to read and interpret these rare scripts and hence the need for research. It is our collective responsibility to preserve in proper format and pass on without any loss to the next generation.

The two types of Nagari scripts are Nandinagari and Devanagari. Nandinagari was used to write the Sanskrit language. Golden era script is during the Vijayanagara period in which most of the Sanskrit copper plate inscriptions were written using the Nandinagari script [2]. The Nandinagari manuscript is scanned and kept in digital libraries. The scholars of Sanskrit language and literature cannot remain ignorant of this script though Nandinagari script is no longer in style, neither for printing nor for writing. Learning the Nandinagari script is mandatory for the students of epigraphy and paleography. Nandinagari is useful in reading and studying Madhwa Vaishnava, Virashaiva, and Jain Nagari script with less effort [3].

1.2 Meaning and Relevance of Nandinagari Script Studies

The meaning of the term Nandi is sacred or auspicious. Manuscript writing conveys wise thoughts and enlightens many lives across generations.

This investigation is the first attempt to recognize handwritten Nandinagari text, which is only available in manuscript form. This is not available in any paper document and printed form. Because of the nonexistence of the Nandinagari database, our work will aid in database creation and standardization. It will also help to reduce manual effort since, currently, one document takes a year to identify and interpret. If we automate this process, it is very useful to society, and the time of interpretation could be saved. Nandinagari scripts contain key knowledge, which includes Vedas, medicine, philosophy, management, science, astronomy, religion, politics, astrology, arts, and culture. These are preserved in the manuscript libraries.

1.3 Image Enhancement

Numerous techniques can explore digitized documents, the content of the scripts, and contribute to cataloging accumulated knowledge since the advent of writing. Nevertheless, the main challenge arises because these documents can be partially damaged by coloration issues, faded ink, mold, and burning, among other problems. Even a minor change in illumination while elaborating the document can complicate text recognition and data extraction [20–22]. This manuscript studies ways to retrieve/reconstruct Nandinagari scripts and identify its characters in such cases where the text is not unmistakably visible owing to a lack of contrast or pigmentation. There exist several image processing methods and operations to investigate these historical pieces and compare them to existing knowledge [23].

Commonly used image processing techniques are contrast enhancement, binarization, edge detectors, filters, and morphological operators. In general, combining thresholding and adaptive binarization results in a simple but vastly advantageous way to identify/retain/restore the maximum amount of written information while lessening the undesirable pigmentation simultaneously. The image processing outcomes can be inspected and weighed up with the aid of image quality metrics [24].

Image enhancement helps many applications to improve the quality of images, which is paramount for human interpretation due to their sepia-tone while including special kinds of visualizations and investigations. These techniques can rely on the following:

1. Self-similarity
2. Dictionary learning
3. Deep learning

Self-similarity-based procedures combine low-resolution and high-resolution patches to ameliorate the quality of low-resolution (or distorted) images.

Dictionary learning is a branch of signal processing and machine learning that finds a frame called a dictionary in which some training data admits sparse representation.

Under the term artificial intelligence comes deep learning where the human brain is imitated to process data in order to recognize speech, detect objects, make decisions, and translate languages. This learning could be supervised, semi-supervised, or unsupervised.

In this work, the palm leaf manuscript image is enhanced using an adaptive histogram equalization method using tuned parameters.

1.4 Super-Resolution (SR) Imaging

To increase the resolution of an image super-resolution imaging techniques are used. The diffraction limit of systems in transcended in the optical super resolution. Whereas, the resolution of digital imaging sensor is enhanced in geometrical super resolution.

It has been a challenging task to reconstruct a high-resolution image from a low-resolution image. The challenge gets harder when the input is a single low-resolution image.

Super resolution (SR) is an estimation of high resolution (HR) from a single low-resolution (LR) image. In a nutshell, the predicted high resolution is the SR, the ground truth is the HR and single image input is the LR. LR images are usually down sampled HR image with noise added along with blurring in machine and deep learning applications.

2 Features of Nandinagari Script

(a) The basic Nandinagari character set has 52 characters, of which 15 are vowels and 37 are consonants.
(b) Two or more consonants unite to form conjunct characters.
(c) In Nandinagari, Shirorekha or headline for words is not used.
(d) The Nandinagari character set has many similar-shaped characters, and this makes Nandinagari character recognition system a challenging problem.

2.1 The Snippet of Nandinagari Script

The Nandinagari palm-leaf manuscript is shown in Fig. 1.

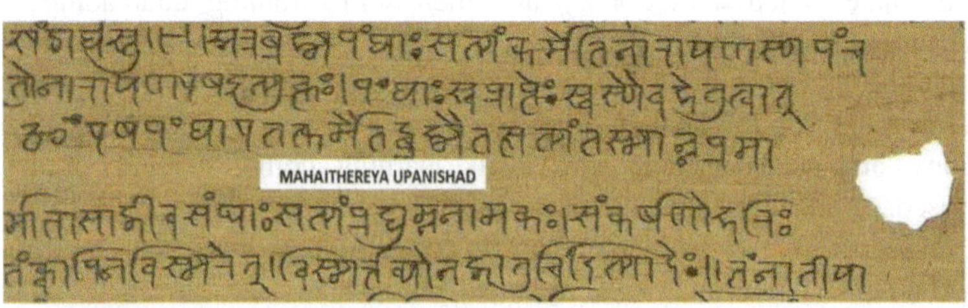

Fig. 1 Nandinagari handwritten palm-leaf manuscript

Fig. 2 Nandinagari character set – vowels and consonants

2.2 The Character Set of Nandinagari Script

The Nandinagari character set is shown in Fig. 2.

3 Problem Statement

Automatic identification of Nandinagari characters obtained from handwritten images using machine-learning techniques. We propose and implement invariant feature extraction and classification techniques for recognition.

4 Database Creation

The Spiritual texts in palm-leaf manuscript form are obtained from prominent Sanathan literature sources such as "Tantra Saara Sangraha," courtesy Acharya Madhwa, twelfth century AD, Shree Pejawar Adhokshaja Mutt, Udupi. Based on the manuscripts, two reference character sets are prepared to interpret the literature by discussing it with scholars. The variations in handwriting are created by providing it to ten users of different age groups to get different writing styles. The type of pen and ink color is also varied to get variations in thickness and color.

Where x_j and $c_{i,j}$, respectively, denote the jth component of the descriptor x considered of its corresponding visual word c_i, and component of v is obtained as a sum over all the image descriptors.

7.2 *Identification Phase – Second Phase*

A query image of an unknown character is taken into an account, SIFT key points are extracted, and represented in 128 descriptor format.

Based on the generated visual codebook from the learning phase, the query vector is generated using VLAD vectorization technique.

Retrieve the top N results of the images which matches the query image after comparing the query vector with the visual vocabulary.

7.3 *Proposed Architecture*

The proposed architecture is shown in Fig. 4, consisting of the following steps.

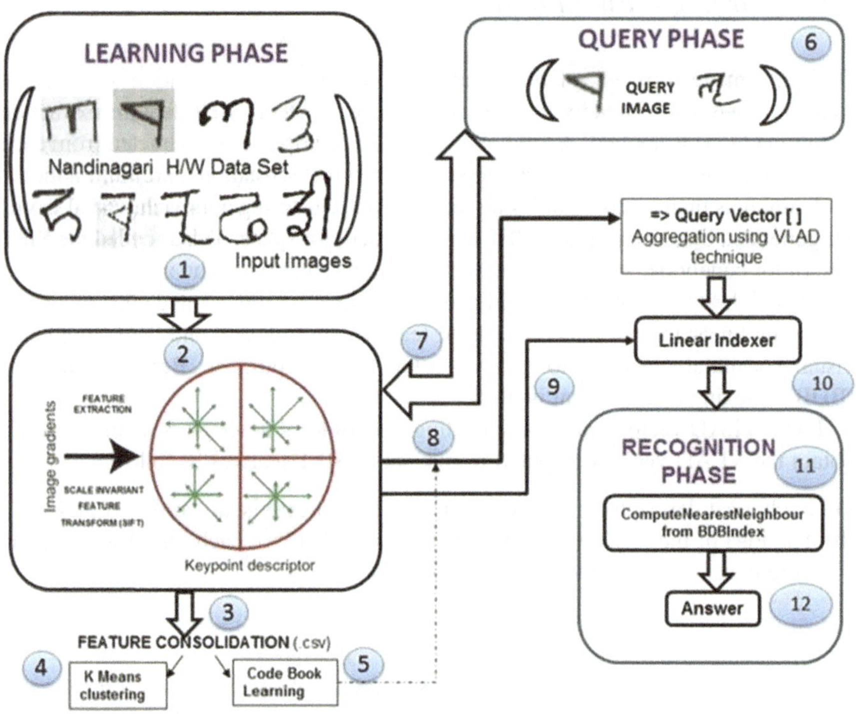

Fig. 4 Proposed architectural framework for handwritten Nandinagari character recognition

1. The handwritten character set of 52 characters is provided to two writers for varied styles and details of scanned images for handwriting style 1 consisted of different sizes 256 × 256, 384 × 384, 512 × 512, and 640 × 640 and different orientation angles 0°, 45°, 90°, 135°, and 180°. This constituted a 1049 characters font 1 database. In the second style, we have different categories of images such as normal images of sizes 128 × 128 and 512 × 512 and convolved images, thinned images, blurred images, images with noise, images rotated by 15°, and images translated by 5 pixels. All these images have various sizes including 92 × 92, 112 × 112, and 128 × 128. This formed 578 images in the dataset making a total of 1627 handwritten Nandinagari character images.
2. SIFT feature extraction: SIFT features of images (.jpg or .png files) were extracted and written text files. For each image, this first key point of each image is identified and 128 feature descriptors are generated for each of these key points. A typical view of query and base image with SIFT point is shown in Fig. 5.
3. Figure 6 shows a sample of the characters with the SIFT feature match points for the letter "Ba."
4. Generating super-resolution images: The size of the input is always reduced by the convolution operation. Thus deconvolution is used in order to get the output image size which is 4 times the input size. Larger numbers of scanned HR image

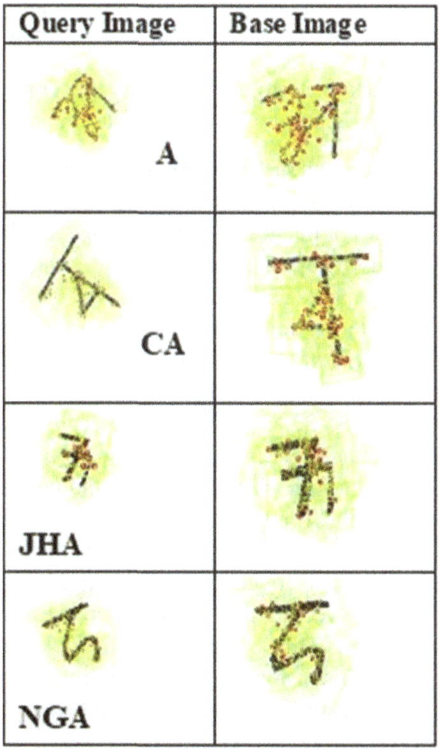

Fig. 5 SIFT interest points for query images and base images

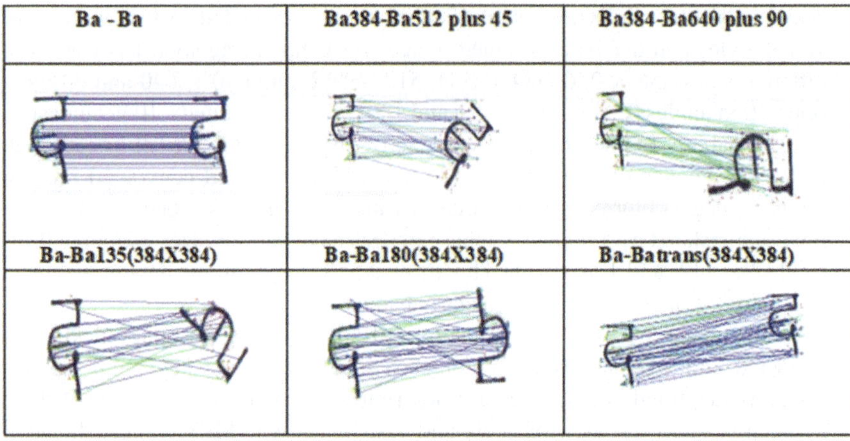

Fig. 6 SIFT interest points matching for the Nandinagari characters

are taken by the training data step. These are further downsized by a factor of 4, hence making them our LR images. To generate a high-resolution image, the low-resolution image is trained. The main objective is to reduce the mean squared error (MSE) between the ground truth image and the pixels of the generated image. The model is able to generate good-quality HR images if the error nears zero.

A sample programmatically generated high-resolution and its low-resolution image counterparts for a character image is shown in Table 1.

5. Thus, total of 1627 input images are fed to a module to extract feature. The feature files are written in a directory with .csv extension representing different styles and types of images. In the next stage, all the feature files are merged into a single file and processed as input. Figure 7 shows the feature analysis.
6. Codebook generation: This step generates the SIFT codebook using K means clustering approach [10–19]. The codebook file stores only cluster centers and the number of cluster centers depends on the number of clusters formed based on the value of K. To analyze the performance, we have generated the codebook with 16, 24, 32, and 52 cluster centroids. These are called visual words. These internally get mapped to all key points present in that particular cluster. Table 2 tabulates the clustering time and codebook size for different values of K. Even though clustering time is slightly high, this is acceptable for a large robust feature set because the codebook is generated only once and the same is used for further steps. The codebook is the compact representation of SIFT feature descriptors.
7. Image vectorization: In this step, we aggregate a set of SIFT features into a VLAD vector. This reads the codebook of K centroids that was learned using 128-dimensional SIFT features. From this VLAD vector [6, 7] is generated by aggregating SIFT features mapped to the nearest cluster centroid.

Table 1 Various images for generating super-resolution images

First set – 256×256	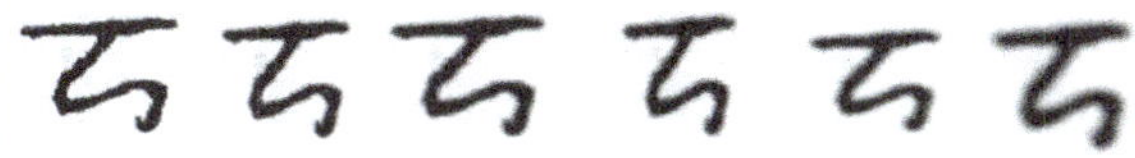
Second set – 128×128	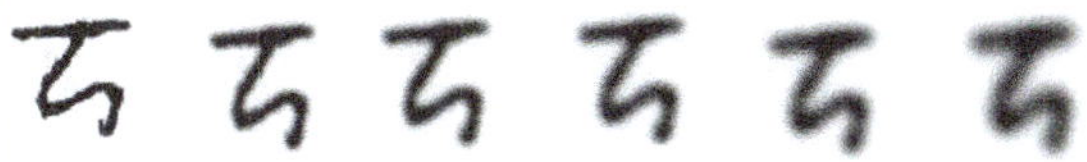
Third set – 64×64	

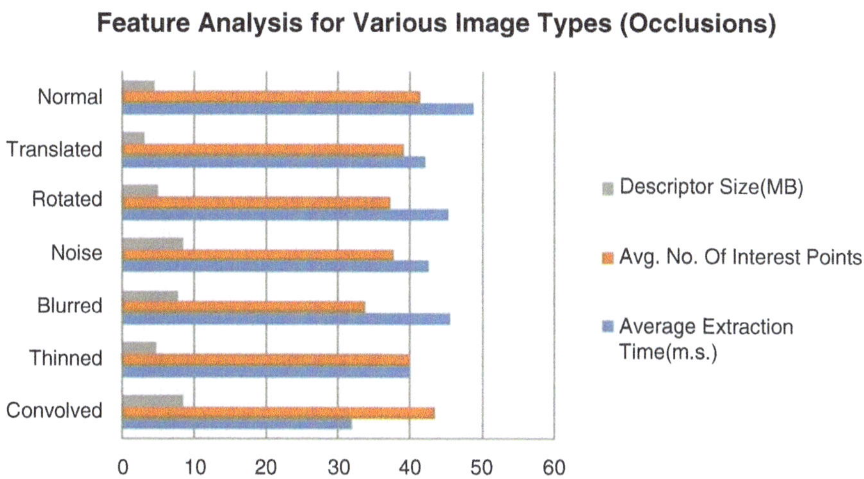

Fig. 7 Average number of interest points, average execution time, and feature descriptor size for type 2 images

Table 2 Codebook size and clustering time for different number of clusters

Sl. no.	No. of clusters	Code book size (KB)	Clustering time (m.s.)
1	16	42	69,777
2	24	62	97,743
3	32	83	118,895
4	52	134	178,367

8. Indexing and retrieval of query images: This is used to generate an index for all database images and we use BDB (Berkeley database) library for indexing purposes. This is based on the persistent storage of the linear indexing approach and assigns the index for the images stored in the image folder. The index size varies from 25.8 to 84 MB for different values of K and so does the indexing time as shown in Table 3.

Table 3 Index size and indexing time different number of clusters

Sl. no.	No. of clusters	Index size (MB)	Indexing time (m.s.)
1	16	25.8	67,455
2	24	39.3	68,961
3	32	51.75	69,552
4	52	84	76,179

Table 4 Mean average precision and retrieval time of a different number of clusters

Sl. no.	No. of clusters	mAP	Retrieval time (m.s.)
1	16	0.9504	39.25
2	24	0.9874	47.5
3	**32**	**0.9876**	**56.625**
4	52	0.9862	74.25

The retrieval of similar images is obtained by comparing the query vector against the vectorized images using the VLAD aggregator [8, 9]. Thus we compute the answer using the index of the images folder and the query image vector with the nearest neighbor method. Here top N similar images are retrieved from the visual vocabulary. From the image set, a representative of 8 query samples as shown in Table 4 is taken for analyzing the performance of our framework.

The retrieval performance is measured by precision at 10, precision at 20, precision at 30, precision at 40, and average precision (AP). Then mean average precision (mAP) and retrieval time for all query images are computed as shown in Table 4.

From Table 4, we conclude that mAP is maximum for a specific number of clusters where the recognition accuracy is also more. Thus for the proposed system on handwritten Nandinagari character recognition, the optimal clusters to be chosen are 32 as shown in Fig. 8. Our approach, which supports the compact representation of a large set of features, is compared with traditional classification algorithms and performance is analyzed. VLAD vectorization gives a better result as compared to other approaches.

The processing is done with an Intel(R) Core(TM) i5–5200 U CPU @ 2.20 GHz, 12 GB RAM, 64-bit operating system (Windows 8.1), ×64-based processor having Java 1.8 installed, and the test is done in multithread with Java client.

8 Conclusion and Future Scope

This is the first attempt to develop an identification system for Nandinagari handwritten scripts. Sufficient work is done in each module as handwritten Nandinagari characters pose specific challenges to recognition systems. These scripts have vast knowledge covering different areas such as politics, management, science, arts, philosophy, culture, and religion. Immeasurable information could be passed on to the next generation if we preserve its proper format. A lot of labor and human effort could be saved if we can automate this process, and it is very useful to society.

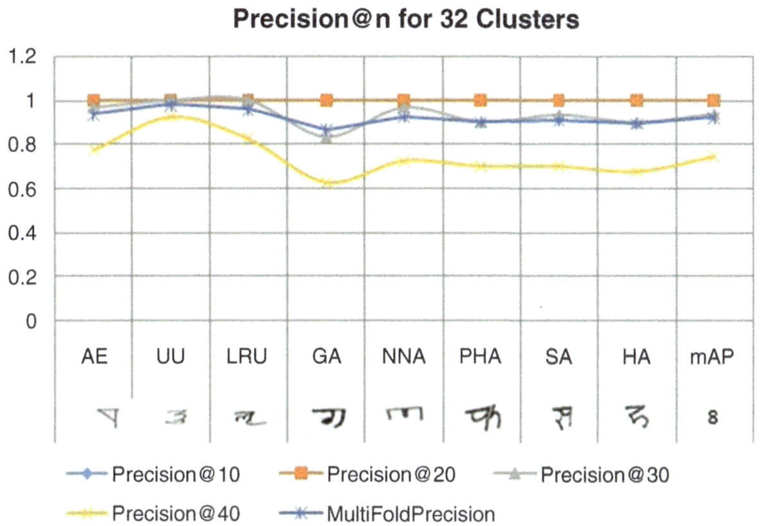

Fig. 8 Precision at 10, 20, 30, and 40 for 32 clusters

The identification of handwritten Nandinagari character is done by extracting the invariant features using the SIFT method and for recognition, the VLAD vectorization concept is used. The research work will be continued further for the recognition of characters and to improve the performance of the same. The scope of this work could be further improved by possible automation of segmentation, which needs the system to be first trained by a scholar knowledgeable of the Nandinagari manuscript.

We could also increase the Nandinagari character vocabulary and expedite the process of image retrieval by storing the indexed features in a search engine. The other way is to use deep learning concepts to find the accuracy of recognition by training and testing large image sets. Many types of deep convolutional layers have to be designed, and characters need to be trained. This approach, although tedious from a computation and resource perspective, does have a lot of promise for large-scale recognition since every pixel will give its input to the feature. With today's availability of high-performance systems, this is achievable. In deep learning, we could deploy super-resolution imaging (SR) as a class of technology to further enhance the resolution of the imaging system.

References

1. Rath, S. (2009). Nandinagari manuscripts: Distinctive features, geographical and chronological range. In *14th World Sanskrit Conference*, Kyoto University, Japan.
2. Mukhopadhyaya, S. (2005). *Paleographical importance of Nandinagari*. New Delhi: SUN.
3. Visalakshi, P. (2003). *Nandinagari script* (1st ed.). Thiruvananthapuram: DLA Publication.

4. Lowe, D. G. (2004). Distinctive image features from scale-invariant key points. *International Journal of Computer Vision, 60*(2), 91–110.
5. Mortensen, E. N., Deng, H., & Shapiro, L. (2005). A SIFT descriptor with global context. In *2005 IEEE computer society conference on computer vision and pattern recognition* (Vol. 1, pp. 184–190). Los Alamitos: IEEE.
6. Jegou, H., Douze, M., Schmid, C., & Perez, P. (2013). Aggregating local descriptors into a compact image representation. *IEEE Transactions on Pattern Analysis and Machine Intelligence, Institute of Electrical and Electronics Engineers, 34*(9), 1704–1716.
7. Delhumeau, J., Gosselin, P. H., Jegou, H., & Perez, P. (2013). *Revisiting the VLAD image representation*. Barcelona: ACM Multimedia.
8. Arandjelovic, R., & Zisserman, A. (2013). All about VLAD. In *IEEE Conference on Computer Vision and Pattern Recognition*, pp. 1578–1585.
9. Picard, D., & Gosselin, P.-H. (2011). Improving image similarity with vectors of locally aggregated tensors. In *IEEE International Conference on Image Processing*, Sept 2011, Brussels, Belgium, pp. 669–672.
10. Fischler, M. A., & Bolles, R. C. (1981). Random sample consensus: A paradigm for model fitting with applications to image analysis and automated cartography. *Journal on Communications of the ACM*, ACM Press, New York, *24*(6), 381–395.
11. Antonopoulos, P., Nikolaidis, N., & Pitas, I. (2007). Hierarchical face clustering using SIFT image features. In *IEEE Symposium on Computational Intelligence in Image and Signal Processing*, April 2007.
12. Gaur, A., & Yadav, S. (2015). Handwritten Hindi character recognition using K-means clustering and SVM. In *2015 4th international symposium on emerging trends and technologies in libraries and information services*. IEEE. Noida, India
13. Alsabti, K., Ranka, S., & Singh, V. (1997, January 1). An efficient K-means clustering algorithm. *Electrical Engineering and Computer Science*, 43.
14. Shi, Y., Xu, F., & Ge, F.-X. (2014). SIFT-type descriptors for sparse-representation based classification. In *10th international conference on natural computation*. Piscataway: IEEE.
15. Guruprasad, P., & Majumdar, J. (2016). Multimodal Recognition Framework: An accurate and powerful Nandinagari handwritten character recognition model. *Procedia Computer Science*, Elsevier BV, *89*, 836–844.
16. Guruprasad, P., & Majumdar, J. (2017). Machine learning of handwritten Nandinagari characters using VLAD vectors. *ICTACT Journal on Image and Video Processing, 8*(2), 1633–1638.
17. Guruprasad, P., & Majumdar, J. (2017). Optimal clustering technique for handwritten Nandinagari character recognition. *International Journal of Computer Applications Technology and Research, 6*(5), 213–223, ISSN: 2319-8656 (Online).
18. Guruprasad, P., & Majumdar, J. (2017). Handwritten Nandinagari image retrieval system based on machine learning approach using bag of visual words. *International Journal of Current Engineering and Scientific Research (IJCESR), 4*(4), 163–168, ISSN (Print): 2393-8374, (Online): 2394-0697.
19. Guruprasad, P., & Guruprasad. (2020). An accurate and robust handwritten Nandinagari recognition system. *ICTACT Journal on Image and Video Processing, 10*(3), 2119–2124.
20. Hu, A., & Razmjooy, N. (2020). Brain tumor diagnosis based on metaheuristics and deep learning. *International Journal of Imaging Systems and Technology*, 1–13. https://doi.org/10.1002/ima.22495
21. Xu, Z., et al. (2020). Computer-aided diagnosis of skin cancer based on soft computing techniques. *Open Medicine, 15*(1), 860–871.
22. Liu, Q., et al. (2020). Computer-aided breast cancer diagnosis based on image segmentation and interval analysis. *Automatika, 61*(3), 496–506.
23. Razmjooy, N., Estrela, V. V., & Loschi, H. J. (2020). Entropy-based breast cancer detection in digital mammograms using world cup optimization algorithm. *International Journal of Swarm Intelligence Research (IJSIR), 11*(3), 1–8.
24. Estrela, V. V., et al. (2019). Why software-defined radio (SDR) matters in healthcare? *Medical Technologies Journal, 3*(3), 421–429.

Deep Image Prior and Structural Variation-Based Super-Resolution Network for Fluorescein Fundus Angiography Images

R. Velumani, S. Bama, and M. Victor Jose

1 Introduction

Advancements in medical imaging have accelerated the clinical protocols concerned with diagnosis, treatment, and prognosis of chorioretinal disorders. Examples of such diseases are diabetic retinopathy, glaucoma, retinal vein occlusion, exudative degeneration, and sclerosis. Fluorescein fundus angiography (FFA) introduced by Novotny and Alvis [1] in 1961 is one of the widely used imaging methodologies today in quantifying blood flow in the retinal region and studying retinal vasculature. A recent investigation [2] has shown that FFA is highly effective in identifying lesions and microaneurysms compared to the modalities such as swept-source optical coherence tomography angiography (SS-OCTA) and ultra-widefield FFA.

However, FFA's low spatial resolution hinders the analysis of microscopic features in the retinal vasculature. Enhancing FFA images' resolution is essential to examine the photoreceptors and fine retinal capillaries in the diagnosis and therapeutic assessment of several ocular disorders. A recent study by Okada [3] and co-authors advocates the need for high-resolution FFA. It proposes a commercial nonadaptive imaging system in capturing fine retinal capillaries at a high resolution.

High-resolution FFA clinical images call for expensive hardware, impractical in low-profile clinical settings, especially in developing countries. However, machine learning-based image super-resolution (SR) approaches seem to be potential

R. Velumani (✉)
IEEE, New York, NY, USA

S. Bama
Kalasalingam Academy of Research and Education, Krishnankovil, Srivilliputtur, Tamilnadu, India

M. V. Jose
Noorul Islam Centre for Higher Education, Kanyakumari, Tamilnadu, India

A. Deshpande et al. (eds.), *Computational Intelligence Methods for Super-Resolution in Image Processing Applications*, https://doi.org/10.1007/978-3-030-67921-7_10

solutions for enhancing FFA images' resolution. Fundus image SR problems can be modeled twofold: single image SR (SISR) and multiple image SR (SISR). This classification depends on the number of low-resolution images involved in the construction of high-resolution imageries. Compared to MISR, SISR is highly challenging as a high-resolution image with an enhanced diagnostic value obtained from a single FFA image captured from a subject.

SISR approaches can be of one of these three types: interpolation, reconstruction, and learning-based methods. With the advent of deep learning, variants of convolutional neural networks (CNN) have successfully deployed SISR problems. The super-resolution CNN (SRCNN) proposed by Dong et al. [4] is one of the benchmarks widely adopted by researchers. This network performs a sequence of three nonlinear transformations to map low-resolution to high-resolution images. However, this model and its variants learn SR from training images, ignoring domain expertise represented as image priors.

Yang et al. [5] have shown that high-resolution images can be constructed from their incomplete representations, exploiting statistical natural image priors. Wang et al. [6] have established that a sparse prior can be represented as a unified sparse coding network (SCN), realized by integrating sparse coding and deep learning, exhibiting best results for super-resolution of natural images. Further, SCNs can be cascaded to form a deep network for super-resolving pictures at arbitrary and large scaling factors.

It is well known that image SR problems characterized by non-unique solutions can be regularized, defining solution constraints consistent with image priors. Ulyanov et al. [7] have shown that rather than training deep networks with voluminous inputs, the best results occur by designing these networks to resonate with the underlying data structure. The authors propose such a network as a deep image prior (DIP), a generative network initialized with random weights. An image reconstruction problem can be modeled as a conditional image generation problem, ensuring a maximum likelihood of reconstructing the image employing the degraded one. The DIP regularizes the image reconstruction solution by optimizing the random weights to generate an image closely matching the degraded one.

Total variation (TV) is a typical regularization process employed for reducing spurious and excessive details in image denoising and reconstruction problems. However, since TV assumes that every signal is piecewise smooth, it suffers from loss of contrast, fine details and spatial resolution, and new artifacts. Hence, the conventional TV minimization model is adapted to specific applications in recent years. A recent study [8] on prior models and regularization techniques exclusively for SISR presents these approaches' comprehensive classifications. This work strongly advocates the usage of flexible priors, featuring image-dependent selection of priors and regularization approaches for SISR.

Inspired by the DIP and TV approaches for SR and the flexibilities of implementation, this text integrates adaptive variants of DIP and structural total variation (STV) to construct a SISR framework called adaptive deep image prior-structural total variation (ADIP-STV). The contributions of the authors in realizing this model are as follows.

1. A state-of-the-art framework encompassing adaptive DIP and STV is conceptualized, which has not been attempted so far.
2. The significance of adaptive, image-driven computations is reinforced in SR that can be extended to other inverse problems.

This model demonstrates the best-quality measures for the super-resolved images compared to the state-of-the-art methods for a standard dataset.

This chapter is organized as follows. Section 2 reviews the related work in the context of this research, encompassing the earliest to the most recent literature. In Sect. 3, we describe the dataset and methods underlying the proposed system. In Sect. 4, the authors propose the architecture of the proposed SR model. The authors present the experimental results, analyses, and comparisons with the state-of-the-art methods in Sect. 5. Section 6 ends this chapter with future research directions.

2 Related Work

This section reviews the existing works on the image prior knowledge that depends on SR and SISR approaches for fundus images.

2.1 Deep Image Prior-Based Super Resolution

SISR is an inverse problem that constructs a high-resolution image from a low-resolution image. Being an ill-posed problem, the approximation solution for this problem is enhanced by regularization, ensuring that the answer is consistent with the image prior. Generally, model or learning-based approaches yield image priors. In model-based approaches, image priors intrinsic to the available data are results from sparse coding [9], dictionary learning [10], simultaneous sparse coding [11], etc. Deep neural network (DNN) architectures have been successfully deployed in image restoration problems with the advent of deep learning. These networks perform an end-to-end mapping of the degraded images to the reconstructed images using convolutional operators such as the SRCNN [4] modeled with three layers.

However, several investigations have revealed that substantially increasing the networks' depth expands the solution space for super-resolution problems. Hence, very deep learning networks have gained complete research attention resulting in several super-resolution models such as the deep recursive residual network (DRRN) [12], enhanced deep SR (EDSR) [13] network, and deeply recursive convolutional network (DRCN) [14]. Figure 1 reveals the schematic of the SISR framework.

In a SISR problem, the mathematical relationship between a given low-resolution image $\boldsymbol{Y}$ and the high-resolution image $\boldsymbol{X}$ to be reconstructed is modeled as in (1), where $\boldsymbol{D}$, $\boldsymbol{S}$, and $\boldsymbol{n}$ are the downsampling operator, blurring operator, and noise, respectively. The super-resolved image $\boldsymbol{X}^*$, which closely matches $\boldsymbol{X}$, results from

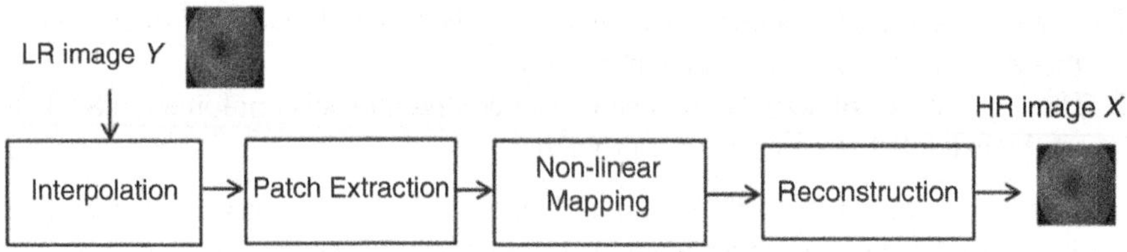

Fig. 1 SISR framework

minimizing the above model as in (2). It is seen that finding a solution to this problem is analogous to a Bayesian posterior estimation *P{X|Y}* using

$$\boldsymbol{Y} = \boldsymbol{DSX} + \boldsymbol{n}, \text{and} \tag{1}$$

$$\boldsymbol{X}* = \min{}_X \left\{ \|\boldsymbol{DSX} - \boldsymbol{Y}\|^2 + \lambda\rho(\boldsymbol{X}) \right\}. \tag{2}$$

In (2), the first term penalizing the difference between the reconstructed image $\boldsymbol{X}$ and the low-resolution image $\boldsymbol{Y}$ is called the fidelity term. The second is the regularization term, which stabilizes the solution. The parameter λ balances the trade-off between the above terms.

The idea of employing a DIP as a regularizer in inverse problems appeared in [7]. It fits the learning parameters of the network to regularize the solution for super-resolution problems implicitly. This article's authors have testified DIP's effectiveness in inverse problems such as denoising, inpainting, restoration of multiple images, and super-resolution. However, the solutions to these problems are comparatively less accurate than unsupervised approaches, as pointed out by Mataev et al. [15]. These authors augmented an explicit regularization with DIP to improve the quality of the reconstructed images. Recent work in this context is reported in [16] that combines total variation (TV) [17] regularization with DIP in image deblurring and denoising problems, achieving significant performance improvements. Similarly, Stein's unbiased risk estimator (SURE) [18], which minimizes the bias-variance trade-off in image construction algorithms, exhibits the best performance in image denoising when combined with DIP.

Regularization by denoising (RED) [19] is an adaptive and effective regularization scheme. RED employs a denoising engine within image deblurring and super-resolution problems. The deep RED approach from [15] merges DIP and RED in image denoising, super-resolution, and deblurring problems.

2.2 Fundus Image Super-Resolution

A detailed analysis of different interpolation and learning-based approaches for super-resolving retinal fundus images comes from [20]. This article investigates the significance of both deterministic and stochastic regularization approaches for modeling the fundus image super-resolution problem using priors. A super-resolution pipeline [21] for FFA images tested with several SISR algorithms demonstrates the best performance with random forests (RFs). A generative adversarial network (GAN), which is built on saliency maps, defining the importance of image pixels, has been employed in the super-resolution of fundus images up to a scale of 16 by Mahapatra et al. [22]. The authors extended this work to construct a multi-stage super-resolution model employing progressive generative adversarial networks (P-GANs) [23]. In this model, the subsequent stages with a triplet loss function enhance the super-resolved image's perceptual quality.

The model in [24] performs super-resolution selectively on arbitrary image zones of retinal fundus images carrying significant diagnostic information. The candidate region for super-resolution stems from a support vector machine (SVM) classifier trained with image features such as contrast sensitivity, energy, and Eigen features. A full CNN for enhancing the resolution of retinal images captured with scanning laser ophthalmoscope (SLO) arises in [25]. This network employs the Adam algorithm, which is invariant to the rescaling of gradients under super-resolution. Further, this algorithm also iteratively updates the network weights, enhancing the quality of the reconstructed images. In [26], an EDSR based on super-resolution residual network (SRResNet) architecture has been employed in the super-resolution of fundus images at the scaling factors 2, 4, and 8.

2.3 Total Variation (TV) Regularization in Inverse Problems

Interpolation is an integral process in deep super-resolution networks, regardless of the network architecture. The widely used bicubic, bilinear, and nearest-neighbor interpolation schemes suffer from high computational cost, the introduction of artifacts in reconstructed images, and loss of high-frequency components scaling factor increases. TV methods capable of preserving the edges are employed in the regularization of the solutions in super-resolution problems. Variational problems with image fidelity and regularization terms can help to model inverse image processing problems. A low-rank total variation (LRTV) [27] based approach for magnetic resonance (MR) image super-resolution demonstrates the best performance metrics compared to interpolation-based methodologies. The SR scheme from [28] obtains an initial estimate of the high-resolution image with Tikhonov regularization. It also employs the modified total variation (MTV) as a regularization term for denoising positron emission tomography (PET) images from animals.

A GAN-based image synthesis approach proposed in [29] improves spatial smoothness in the retinal images generated from tubular structures of blood vessels, incorporating TV in the image generation process. In recent work, strong image priors for preserving the continuity of edges are realized by variational models based on curvature [30]. A variation regularization model [31] for denoising and reconstructing medical images employs a neural network as a regularization functional and trains it as an adversarial model. Similarly, a framework for solving sparse data problem in computed tomography (CT) [32] images also defines a regularizer as a neural network. The total deep variation (TDV) [33] regularizer, which represents its energy with a residual hierarchical network, provides exemplary results for SISR with a few trainable parameters.

From the above review, one perceives the following regarding the existing research in fundus image SISR.

1. DIPs merged with explicit regularization terms demonstrate potential performance gains in image reconstruction problems.
2. Deep learning frameworks for super-resolution of fundus images are gaining research attention recently.
3. Though TV regularizers are employed in inverse problems in medical imaging such as denoising, synthesis, and reconstruction, significant works on the super-resolution of fundus images are not in evidence.

Many designs stemmed from: (i) the SISR frameworks' requirement for FFA images in ophthalmological examinations and (ii) the prospective to realize those integrating DIP and TV regularization. This work recommends a unified SISR model using DIP and structure adaptive TV.

3 Materials and Methods

This section refers to the dataset used in our experiment and the approaches employed in constructing our SISR model with mathematical foundations.

3.1 Dataset

This research utilized the public dataset [34] with 70 FFA images acquired from diabetic patients in Isfahan University of Medical Sciences. The images are available in two classes as normal and abnormal, with 30 and 40 images each. The Abnormal category contains images of four levels of diabetic retinopathy: mild nonproliferative diabetic retinopathy (NPDR), moderate NPDR, severe NPDR, and proliferative diabetic retinopathy (PDR).

3.2 Deep Image Prior (DIP)

Replacing $\boldsymbol{DS}$ by a single degradation matrix $\boldsymbol{H}$, (1) is rewritten as in (3). The high-resolution image $\boldsymbol{X}^*$ is obtained from the low-resolution image $\boldsymbol{Y}$ by solving (4) where $\boldsymbol{X}$ is the desired high-resolution image.

$$\boldsymbol{Y} = \boldsymbol{HX} + \boldsymbol{n}, \text{and} \tag{3}$$

$$\boldsymbol{X}^* = \min_X \frac{\left[\left\{||\boldsymbol{HX} - \boldsymbol{Y}||^2 + \rho(\boldsymbol{X})\right\}\right]}{2}. \tag{4}$$

According to [7], an image prior is a CNN f parameterized by random variables θ on the input $\boldsymbol{Y}$. The super-resolved image $\boldsymbol{X}^*$ becomes the output of f expressed as $\boldsymbol{X}^* = f(z)$, where z is the random vector.

DIP strives to find the solution to the fidelity term in (4), solving the minimization problem in (5):

$$\min_\theta \frac{\left\{||\boldsymbol{H}f_\theta(z) - \boldsymbol{Y}||^2\right\}}{2}. \tag{5}$$

Now, the image reconstruction problem in (4) can be expressed, including the image prior to Eq. (6):

$$\boldsymbol{X}^* = \frac{\left\{||\boldsymbol{H}f_\theta(z) - \boldsymbol{Y}||^2 + \rho(\boldsymbol{X})\right\}}{2}. \tag{6}$$

3.3 Structural Total Variation (STV) Functional

In image restoration, determining the regularization parameter is significant, as it controls the fidelity smoothness trade-off. The idea of using a regularization functional rather than a constant regularization parameter comes from [35]. This work expresses the regularization functional as a function of the restored image, which is iteratively obtained simultaneously from the degraded image. Similarly, a structural total variation (STV) [36] functional based on the structural prior of the ground truth image demonstrates adequate regularization in linear inverse problems compared to the conventional TV. Particularly, this functional has proven to be very effective in medical image reconstruction.

Given an image prior p, the structure regularization functional expressed as an integration of the spatially dependent image gradient is given in Eq. (7), where $\boldsymbol{u}$ is the reconstructed image. Now, $\boldsymbol{u}$ can be reconstructed by minimizing this functional as in (8) where $\boldsymbol{D}$ is the discrepancy term, and o is the linear operator. Here term $\boldsymbol{D}$ is characterized by the inherent noise in the image, which is the L_2 norm for the Gaussian noise. Similarly, $\boldsymbol{K}$ refers to the image formation model in the specific problem domain. Then, it follows that

$$J(\boldsymbol{u}) = \int_{\Omega} j_p(\boldsymbol{x}, \nabla \boldsymbol{u}(\boldsymbol{x})\ \mathrm{dx}, \text{and} \tag{7}$$

$$\min{}_u \int_{\Omega} j_p(\boldsymbol{x}, \nabla \boldsymbol{u}(\boldsymbol{x}))\ dx + (\lambda \boldsymbol{D} o K)(\boldsymbol{u}). \tag{8}$$

When the prior p is the ground truth image, the point-wise function of the image gradient of p on any arbitrary point z in the domain Ω is given (9). Now, the structure regularization functional can be expressed as in (10).

$$j_p(\boldsymbol{x}, z) = \frac{1}{|\nabla p(\boldsymbol{x})|}|z|, \text{and} \tag{9}$$

$$J(\boldsymbol{u}) = \int_{\Omega} \frac{1}{|\nabla p(\boldsymbol{x})|}|\nabla \boldsymbol{u}(\boldsymbol{x})|dx. \tag{10}$$

The regularization term in (6) is replaced by the structure variation functional, and the reconstructed image is given by

$$\boldsymbol{X}^* = \min{}_{\theta} \frac{\left\{||\boldsymbol{H}\boldsymbol{f}_{\theta}(z) - \boldsymbol{Y}||^2 + \lambda J(\boldsymbol{u})\right\}}{2}. \tag{11}$$

Augmented Lagrangian methods have been successfully applied in several constrained optimization problems, such as the one in (11). The alternating direction method of multiplier (ADMM) [38], a variant of these methods, solves the above problem, dividing it into unconstrained problems.

4 Proposed Super-Resolution Framework

This section introduces the recommended super-resolution framework comprising the deep image prior and structural-functional with illustrations and mathematical representations.

4.1 Deep Prior by Residual Learning

The previous section described (i) the DIP, and the structural regularization functional, (ii) the underlying mathematical foundations of the proposed super-resolution framework, and (iii) expressed the super-resolved image in terms of the DIP and the structural–functional regularizer. This section presents the proposed super-resolution framework realized as an ADIP-STV model based on a deep residual network.

In (11), the first term is the DIP that is a trained CNN parameterized by $\boldsymbol{\theta}$. In the context of image super-resolution, this term is optimized as in Eq. (12). Initially, the CNN is constructed with random values for variables $\boldsymbol{\theta}$, which are iteratively optimized to ensure that the CNN output matches $\boldsymbol{X}$. Here, $\boldsymbol{\theta}^*$ and $\boldsymbol{X}^*$refer to the respective optimized values in each iteration.

$$\boldsymbol{\theta}^* = \min{}_{\theta} \frac{\left\{ ||\boldsymbol{Y} - \boldsymbol{H} f_{\theta}(z)||^2 \right\}}{2} \tag{12a}$$

$$\text{subject to} \qquad \boldsymbol{X}^* = f_{\theta^*}(z). \tag{12b}$$

From (12a and 12b), we understand that the CNN must be trained such that the DIP matches the high-resolution equivalent of $\boldsymbol{Y}$, the original low-resolution image. The proposed CNN stacks the residual learning units in a residual learning network manner in line with this. The schematic of the residual learning block is given in Fig. 2. Though the rectified linear unit (ReLU) activation is commonly followed in residual learning units, we have replaced it with the parameterized rectified linear unit (PReLU) activation function capable of learning the slope parameter itself.

In a residual network, the mapping from the input $\boldsymbol{x}$ to the output $\boldsymbol{y}$ is realized by initially mapping $\boldsymbol{x}$ to the residual $\boldsymbol{F(x)} = \boldsymbol{y}$-$\boldsymbol{x}$. Here, the output $\boldsymbol{y}$ results from adding the input $\boldsymbol{x}$ with $\boldsymbol{F(x)}$. Stacking these units yields very deep super-resolution (VDSR) [37] networks with a 20-layer depth. This design leads to high-resolution images.

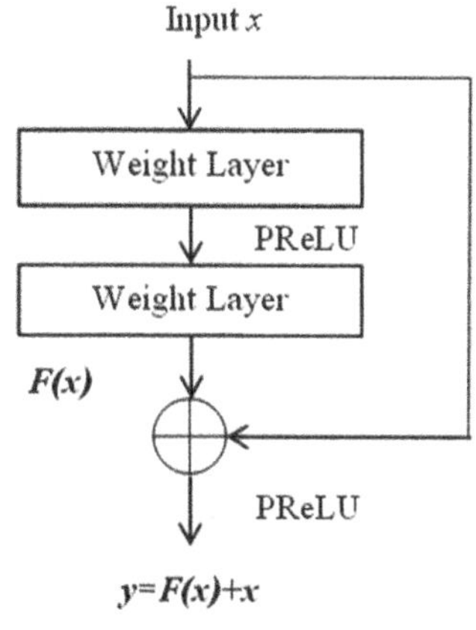

Fig. 2 Residual learning unit

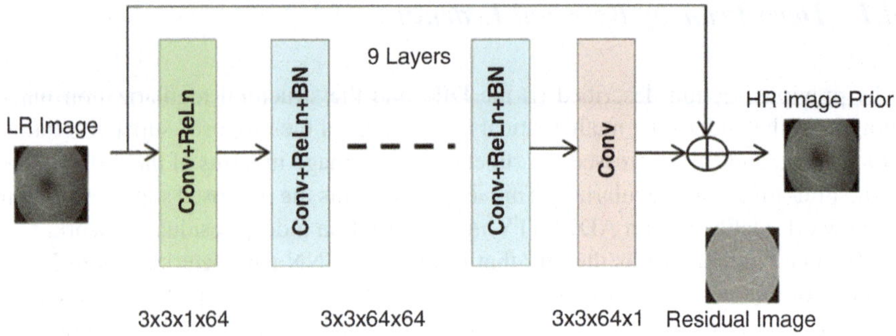

Fig. 3 Deep residual image prior

This text exploits residual learning to determine the image prior. The schematic of the proposed 17-layer residual learning network appears in Fig. 3.

The first network layer convolves the low-resolution image utilizing 64 3×3 convolutional filters. Then comes the PReLU activation. This layer generates 64 feature maps, each of the size of the original image. Next, there are nine layers, each performing convolution, batch normalization (BN), and PReLU activation. The feature maps are convolved with 64 convolution filters of the size $3 \times 3 \times 64$ to generate 64 feature maps. Finally, these feature maps are convolved with a single kernel of size $3 \times 3 \times 64$ to construct the residual image, which is summed up with the low-resolution image to generate the high-resolution image prior.

4.2 Image Super-Resolution with Adaptive DIP-STV

As the reconstructed image $\boldsymbol{X}^*$results from regularizing the prior with the structural functional, one expresses $\boldsymbol{X}^*$ as in (13), replacing $J(\boldsymbol{u})$ in (11) and resulting in

$$\boldsymbol{X}^* = \min{}_\theta \frac{\left\{||\boldsymbol{H}f_\theta(z) - \boldsymbol{Y}||^2\right\}}{2} + \lambda \int_\Omega \frac{1}{|\nabla p(\boldsymbol{x})|} |\nabla \boldsymbol{u}(\boldsymbol{x})| dx. \tag{13}$$

The super-resolved image $\boldsymbol{X}^*$comes from applying the ADMM on (13), introducing an augmented Lagrangian in (14), where L is the Lagrange multiplier and ρ is the free parameter.

$$X^* = \min_\theta \frac{\left\{||Hf_\theta(z) - Y||^2\right\}}{2} + \lambda \int_\Omega \frac{1}{|\nabla p(x)|} |\nabla u(x)| dx$$
$$+ \frac{\rho}{2} ||f_\theta(z) - |\nabla u(x)| + L||^2. \tag{14}$$

The above equation has three unknowns $f(z)$, $\nabla u(x)$, and L, which can be obtained by iteratively solving the Eqs. (15, 16, and 17).

$$f_\theta^{k+1}(z) = \min_\theta \frac{\left\{||Hf_\theta^k(z) - Y||^2\right\}}{2} + \frac{\rho}{2} ||f_\theta^k(z) - \nabla u^k(x) + L^k||^2, \tag{15}$$

$$\nabla u^{k+1}(x) = \prod_\Omega \left[f_\theta^{k+1}(z) + L^{k+1}\right], \text{and} \tag{16}$$

$$L^{k+1} = L^k + f_\theta^{k+1}(z) - \nabla u^{k+1}(x). \tag{17}$$

where ρ is evaluated in each iteration as in Eq. (18).

$$\rho^{k+1} = \begin{cases} \tau^{\text{incr}} \rho^k \text{if} ||r^k||_2 \succ \mu ||s^k||_2 \\ \tau^{-\text{decr}} \rho^k \text{if} ||s^k||_2 \succ \mu ||r^k||_2 \\ \rho^k \text{otherwise} \end{cases} \tag{18}$$

where $r^k = f_\theta^k(z) - \nabla u(x)$ and $s^k = \rho^k(\nabla u^k(x) - \nabla u^{k-1}(x))$ are primal and dual residuals, respectively. Here, the constants' typical values are 2 for τ and τ', and 10 for μ as defined in [38]. It is seen that larger values of ρ tend to reduce prime residuals while smaller amounts reduce the dual residuals. This self-adaptive ADMM can be explored to construct the super-resolved image X^* by merging the DIP and the structural–functional regularizer adaptively.

5 Experimental Results and Discussions

The dataset consists of 30 normal and 40 abnormal low-resolution images of dimension 576 × 720. The images are super-resolved by two scaling factors 2 and 4 to generate high-resolution images with dimensions 1152 × 1440 and 2304 × 2880, respectively. The proposed ADIP-STV is implemented in Matlab 2019b software and tested in an i7-7700K processor with 16GB DDR4 RAM and NVIDIA GeForce GTX1060 3GB Graphics card. The images' perceptual quality is evaluated with the peak signal-to-noise ratio (PSNR) and structural similarity index measure (SSIM) metrics.

The present text authors have evaluated their model and benchmarked it with the developed dataset, and present the objective performance metrics for comparison.

Figure 4a shows a pair of normal and abnormal images for the super-resolved images scaled by 4, corresponding to the PSNR and SSIM values in Fig. 4b.

The super-resolved FFA images' quality is evaluated with the PSNR and SSIM metrics evaluated with Eqs. (18) and (19), respectively.

$$\mathrm{PSNR} = 10\log_{10}\left(\frac{R^2}{\mathrm{MSE}}\right), \text{ and} \tag{18}$$

$$\mathrm{MSE} = \frac{1}{\mathrm{MN}}\sum_{m,n}[\boldsymbol{I}_1(m,n) - \boldsymbol{I}_2(m,n)]^2.$$

$$\mathrm{SSIM}(\boldsymbol{x},\boldsymbol{y}) = l^{\alpha}(\boldsymbol{x},\boldsymbol{y})\, c^{\beta}(\boldsymbol{x},\boldsymbol{y})\, s^{\gamma}(\boldsymbol{x},\boldsymbol{y}), \tag{19}$$

where

$$\mathrm{l}(\boldsymbol{x},\boldsymbol{y}) = \frac{2\mu_x\mu_y + C_1}{\mu_x^2 + \mu_y^2 + C_1},$$

$$c(\boldsymbol{x},\boldsymbol{y}) = \frac{2\sigma_x\sigma_y + C_2}{\sigma_x^2 + \sigma_y^2 + C_2}, \text{ and}$$

$$s(\boldsymbol{x},\boldsymbol{y}) = \frac{\sigma_{xy} + C_3}{\sigma_x\sigma_y + C_3},$$

With μ_x, μ_y, σ_x, σ_y, and σ_{xy} representing the local means, standard deviations, and cross-covariance for images $\boldsymbol{x}$ and $\boldsymbol{y}$ if $\alpha = \beta = \gamma = 1$.

It is evident from Fig. 4 that the best performance metrics are obtained for the ADIP-STV super-resolution model. Further, the objective metrics for state-of-the-

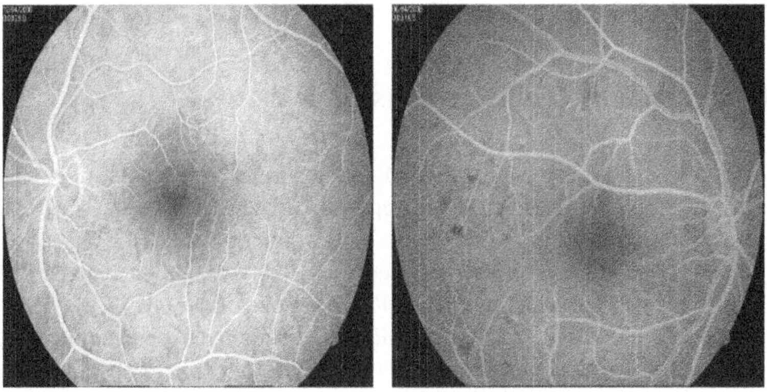

Normal FFA Abnormal FFA

Fig. 4 (**a**) Low-resolution FFA. (**b**) Super-resolved images

Method	Reconstructed HR Image with PSNR/SSIM	
	Normal	Abnormal
ADIP-STV (Proposed)	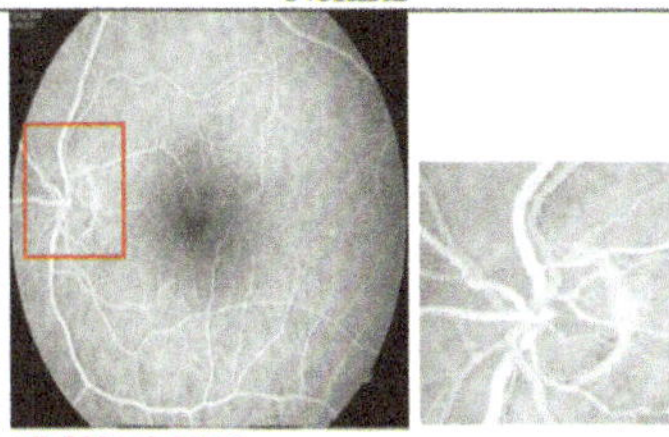 47.5901/0.9946	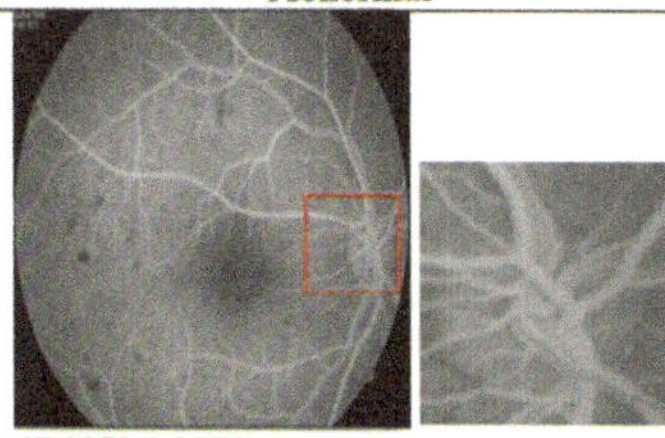 47.1352/0.9874
VDSR [12] (2016)	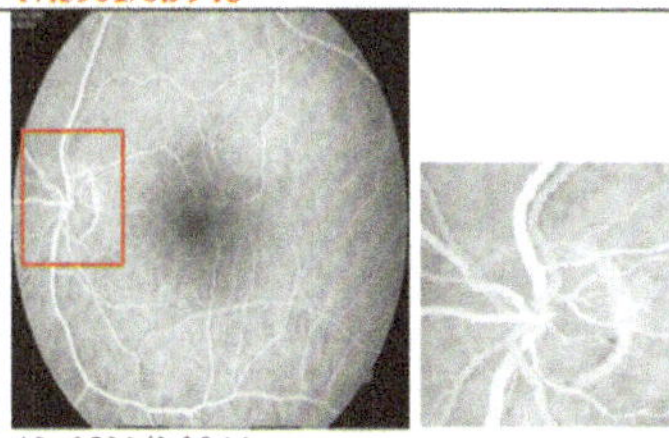 43. 0531/0.9844	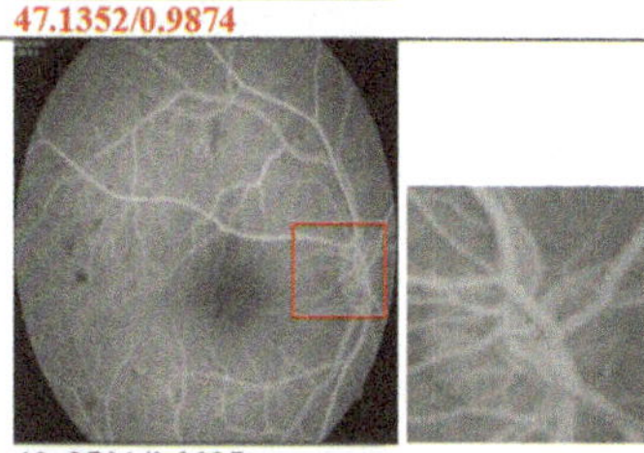 43. 2731/0.9827
GAN[23] (2019)	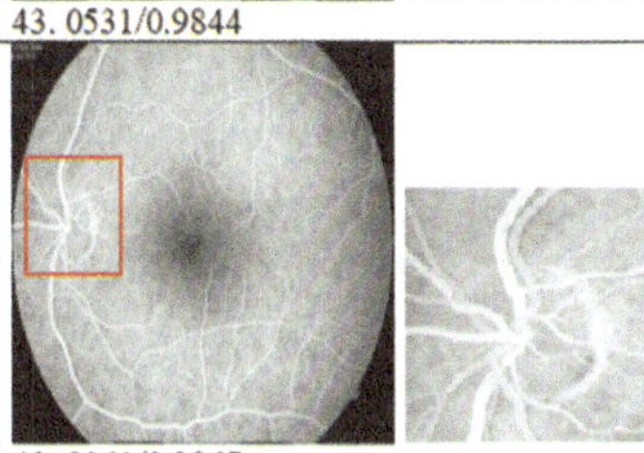 43. 8061/0.9867	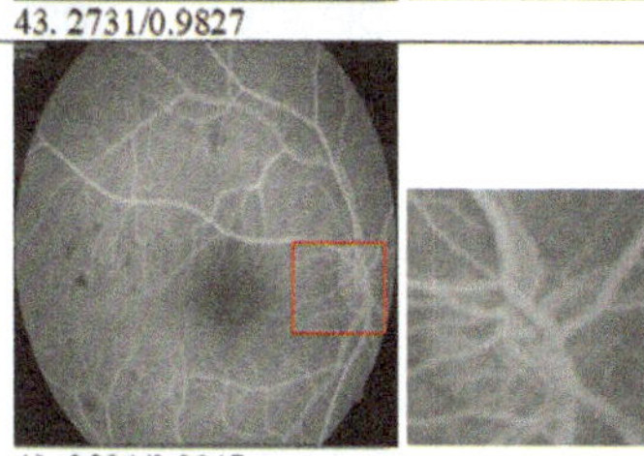 43. 2594/0.9847
EDSR [13] (2017)	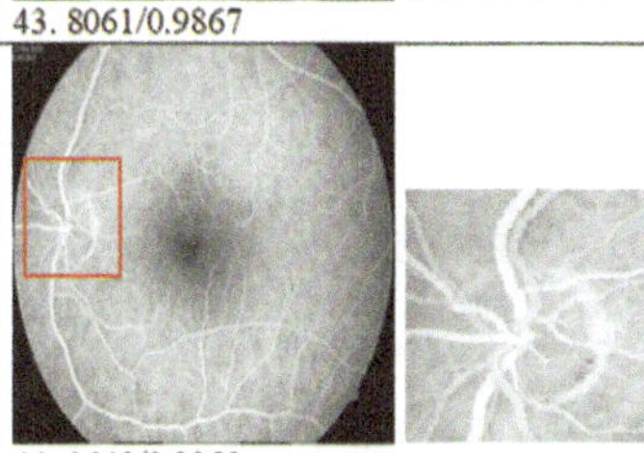 44. 1640/0.9958	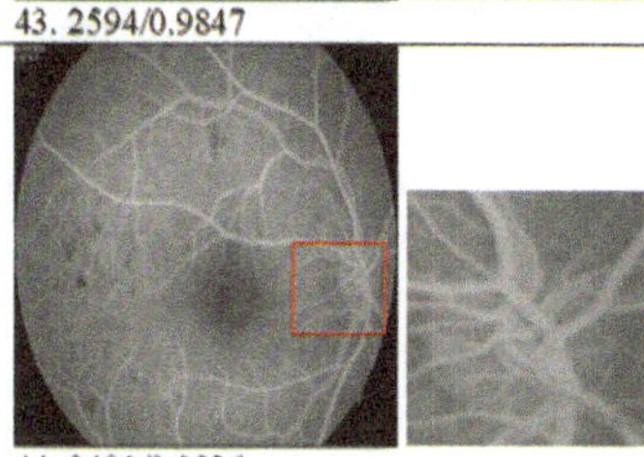 44. 0491/0.9926
SRF[21] (2018)	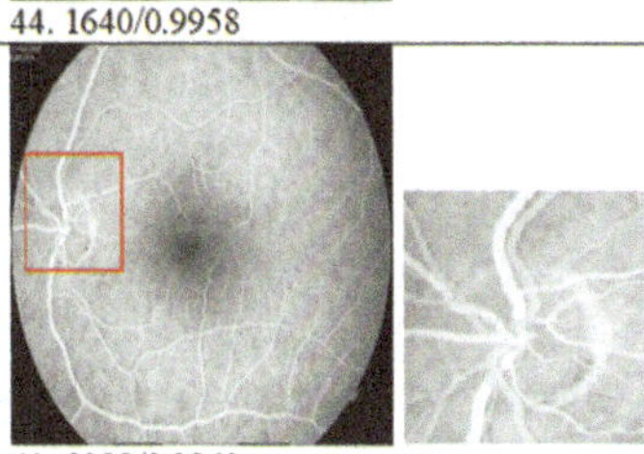 41. 8382/0.9862	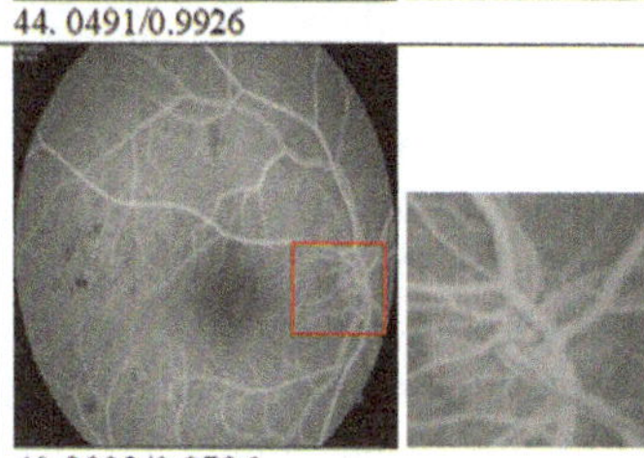 40.9982/0.9726

Fig. 4 (continued)

Table 1 Performance metrics comparison

Method	Scale factor = 2		Scale factor = 4	
	PSNR	SSIM	PSNR	SSIM
Proposed	47.4919	0.9893	47.4238	0.9941
VDSR	45.8756	0.9838	43.9044	0.9767
GAN	43.7421	0.9847	41.5258	0.9647
EDSR	41.8470	0.9693	43.5519	0.9847
SRF	41.1354	0.9561	44.0901	0.9765

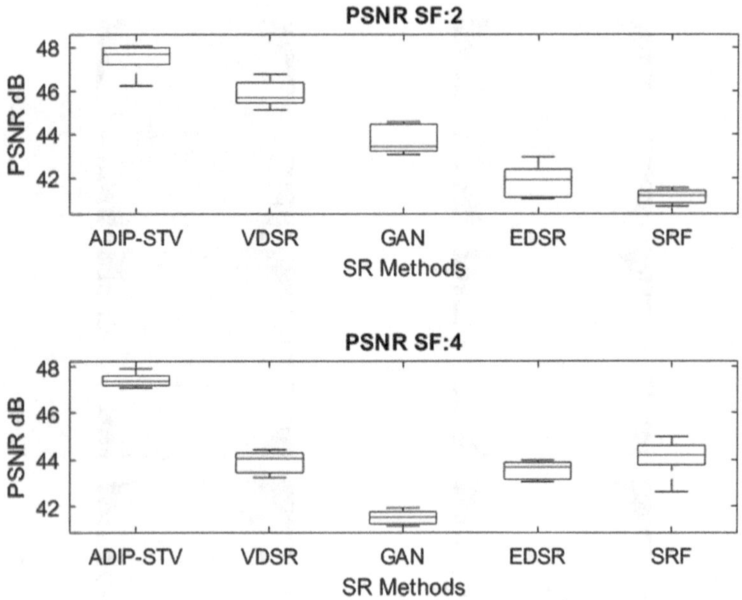

Fig. 5 Kruskal–Wallis test – PSNR

art, and benchmark approaches have been evaluated, as shown in Table 1 for the dataset. One observes that the PSNR and SSIM values are equally superior for the proposed model for both scaling factors.

A high PSNR value metric indicates good perceptual image quality, and a high SSIM signifies the structural intactness of the reconstructed image. The performance metrics show that the residual learning-based image prior and the structural regularization functional capture the image's essential features on super-resolution.

Further, we perform the Kruskal–Wallis test with the null hypothesis that the proposed ADIP-STV and other models considered for comparison are equivalent. This test is performed for the PSNR and SSIM metrics for the scale factors 2 and 4. The box plots generated in this statistical analysis appear in Figs. 5 and 6.

The p-values for the PSNR and SSIM metrics are given in Table 2 to evaluate the methods' statistical significance at a 5% significance level. It is seen that these values

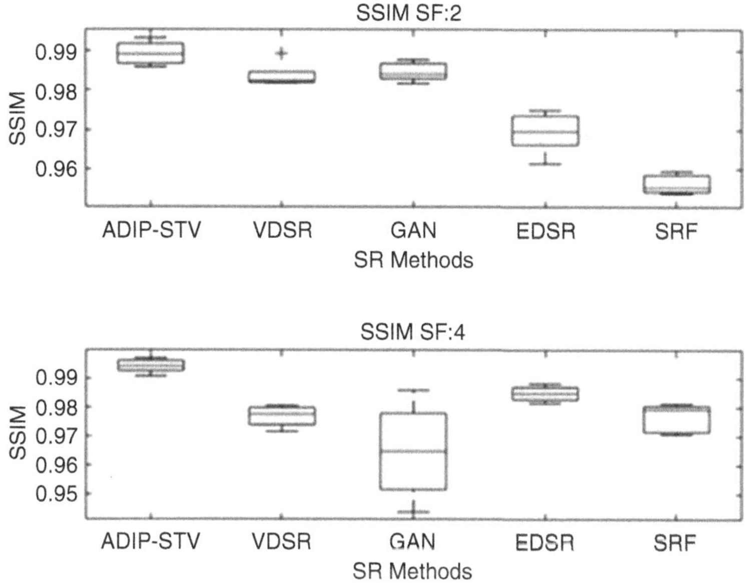

Fig. 6 Kruskal–Wallis test – SSIM

Table 2 Statistical significance

Metrics	Scale factor = 2	Scale factor = 4
PSNR	2.1026e-04	5.8141e-04
SSIM	3.8857e-04	5.1079e-04

Table 3 Computational time comparison

Method	Computational time (ms)	
	Scale factor = 2	Scale factor = 4
Proposed	13.73	20.14
VDSR	23.51	27.46
GAN	25.29	29.41
EDSR	28.65	33.05
SRF	36.32	38.01

are very low for the metrics for both the scaling factors. This indicates that the null hypothesis assuming that all the super-resolution methods are equivalent is rejected. Further, it is also seen that the mean values of the metrics are very stable for the proposed ADIP-STV model, which demonstrates the reliability of our model under different scaling factors.

The computational times for all the models employing the target dataset have been evaluated as in Table 3. It is seen that the computational times are the lowest for our proposed approach for both the scaling factors. Further, it is seen that these values are small compared to the very deep SR (VDSR) and EDSR, both of which are residual networks. Earlier, we have also shown that the proposed model's performance metrics are better than these models. From this, we understand that efficient, super-resolution models can be constructed with residual priors rather than very deep networks. Further, we have also demonstrated the potential of STV in capturing the structural elements in the super-resolved images.

In the conventional deep residual super-resolution networks, the perceptual quality of the super-resolved images is limited by the networks' depth. The proposed model learns the DIP adaptively from the image gradients characteristic. Further, these priors are merged with adaptive structural functionals for regularization. We understand that the proposed model is purely adaptive compared to the conventional models in which the design parameters are static. We have shown that the deep prior realized as a residual network is optimized with learning parameters. In our model, the DIP is parameterized only by the PReLU slope parameter, which is learned by the residual network. Further, the free parameter ρ associated with the ADMM in optimizing the solution is also iteratively updated.

Ascribed to its high adaptiveness, the proposed model exhibits superior performance toward image quality and computational time, the highly desirable features for a medical image super-resolution system.

6 Conclusion

This chapter recommends a novel system for FFA images' super-resolution based on image prior and total variation. The framework consists of an integral model harnessing the potential of deep residual networks with structural total regularization. This system's performance surpasses the recent benchmarked super-resolution models regarding visual and objective quality metrics. The architectural model relies on strong mathematical foundations and design considerations. It also uses adaptive computations to determine the image prior and perform the regularization. Statistical evaluations of the performance metrics and their interpretations reinforce the efficacy of the proposed model. Further, the proposed model also features very low computational time. Existing ophthalmological protocols can be strengthened with the proposed model in scarce-resource settings, enhancing the accuracy of FFA image examinations.

Though the present model is tested with normal and abnormal images, it has not been evaluated from the perspective of specific disorders. FFA images reveal a diverse range of abnormalities, such as microaneurysms, exudates, and dilated capillaries. Rather than super resolving the FFA images as a whole, upscaling specific regions of abnormalities can further improve the quality of diagnosis. The proposed model can be adapted to these disorders' distinct structural components in the FFA images. This requires the design of DIPs and regularizers specific to the particular conditions, which is a prospective issue for further research.

References

1. Novotny, H. R., & Alvis, D. L. (1961). A method of photographing fluorescence in circulating blood in the human retina. *Circulation, 24*(1), 82–86.
2. La Mantia, A., Kurt, R. A., Mejor, S., Egan, C. A., Tufail, A., Keane, P. A., & Sim, D. A. (2019). Comparing fundus fluorescein angiography and swept-source optical coherence tomography angiography in the evaluation of diabetic macular perfusion. *Retina, 39*(5), 926–937.
3. Okada, M., Heeren, T. F., Mulholland, P. J., Maloca, P. M., Cilkova, M., Rocco, V., & Tufail, A. (2019). High-resolution in vivo fundus angiography using a nonadaptive optics imaging system. *Translational Vision Science & Technology, 8*(3), 54–54.
4. Dong, C., Loy, C. C., He, K., & Tang, X. (2014, Sept). Learning a deep convolutional network for image super-resolution. In *European conference on computer vision* (p. 184199). Cham: Springer.
5. Yang, C. Y., Ma, C., & Yang, M. H. (2014, Sept). Single-image super-resolution: A benchmark. In *European conference on computer vision* (pp. 372–386). Cham: Springer.
6. Wang, Z., Liu, D., Yang, J., Han, W., & Huang, T. (2015). Deep networks for image super-resolution with sparse prior. In *Proceedings of the IEEE international conference on computer vision* (pp. 370–378).
7. Ulyanov, D., Vedaldi, A., & Lempitsky, V. (2018). Deep image prior. In *Proceedings of the IEEE conference on computer vision and pattern recognition* (pp. 9446–9454).
8. Pandey, G., & Ghanekar, U. (2020). Classification of priors and regularization techniques appurtenant to single image super-resolution. *The Visual Computer*, 36, 1291–1304.
9. Egiazarian, K., & Katkovnik, V. (2015, Aug). Single image super-resolution via BM3D sparse coding. In *2015 23rd European signal processing conference (EUSIPCO)* (p. 28492853). IEEE.
10. Mairal, J., Bach, F., Ponce, J., & Sapiro, G. (2009, June). Online dictionary learning for sparse coding. In *Proceedings of the 26th annual international conference on machine learning* (pp. 689–696).
11. Mairal, J., Bach, F., Ponce, J., Sapiro, G., & Zisserman, A. (2009, Sept). Non-local sparse models for image restoration. In *2009 IEEE 12th international conference on computer vision* (pp. 2272–2279). IEEE.
12. Tai, Y., Yang, J., & Liu, X. (2017). Image super-resolution via deep recursive residual network. In *Proceedings of the IEEE conference on computer vision and pattern recognition* (pp. 3147–3155).
13. Lim, B., Son, S., Kim, H., Nah, S., & Mu Lee, K. (2017). Enhanced deep residual networks for single image super-resolution. In *Proceedings of the IEEE conference on computer vision and pattern recognition workshops* (pp. 136–144).
14. Kim, J., Kwon Lee, J., & Mu Lee, K. (2016). Deeply-recursive convolutional network for image super-resolution. In *Proceedings of the IEEE conference on computer vision and pattern recognition* (pp. 1637–1645).
15. Mataev, G., Milanfar, P., & Elad, M. (2019). DeepRED: Deep image prior powered by RED. In *Proceedings of the IEEE international conference on computer vision workshops* (p. 00).
16. Liu, J., Sun, Y., Xu, X., & Kamilov, U. S. (2019, May). Image restoration using total variation regularized deep image prior. In *ICASSP 2019–2019 IEEE international conference on acoustics, speech and signal processing (ICASSP)* (pp. 7715–7719). IEEE.
17. Rudin, L. I., Osher, S. J., & Fatemi, E. (1992). Nonlinear total variation based noise removal algorithms. *Physica D: Nonlinear Phenomena, 60*, 259–268.
18. Metzler, C. A., Mousavi, A., Heckel, R., & Baraniuk, R. G. (2018). Unsupervised learning with Stein's unbiased risk estimator. arXiv preprint arXiv:1805.10531.
19. Romano, Y., Elad, M., & Milanfar, P. (2017). The little engine that could: Regularization by denoising (RED). *SIAM Journal on Imaging Sciences, 10*(4), 1804–1844.

20. Thapa, D., Raahemifar, K., Bobier, W. R., & Lakshminarayanan, V. (2014). Comparison of super-resolution algorithms applied to retinal images. *Journal of Biomedical Optics, 19*(5), 056002.
21. Jiang, Z., Yu, Z., Feng, S., Huang, Z., Peng, Y., Guo, J., et al. (2018). A super-resolution method-based pipeline for fundus fluorescein angiography imaging. *Biomedical Engineering Online, 17*(1), 125.
22. Mahapatra, D., Bozorgtabar, B., Hewavitharanage, S., & Garnavi, R. (2017, Sept). Image super resolution using generative adversarial networks and local saliency maps for retinal image analysis. In *International conference on medical image computing and computer-assisted intervention* (pp. 382–390). Cham: Springer.
23. Mahapatra, D., Bozorgtabar, B., & Garnavi, R. (2019). Image super-resolution using progressive generative adversarial networks for medical image analysis. *Computerized Medical Imaging and Graphics, 71*, 30–39.
24. Das, V., Dandapat, S., & Bora, P. K. (2019). A novel diagnostic information based framework for super-resolution of retinal fundus images. *Computerized Medical Imaging and Graphics, 72*, 22–33.
25. Chen, Z., Wang, X., & Deng, Y. (2019, Dec). A super-resolution method of retinal image based on laser scanning ophthalmoscope. In *AOPC 2019: AI in optics and photonics* (Vol. 11342, p. 1134206). International Society for Optics and Photonics.
26. Gulati, T., Sengupta, S., & Lakshminarayanan, V. (2020, Feb). Application of an enhanced deep super-resolution network in retinal image analysis. In *Ophthalmic technologies XXX* (Vol. 11218, p. 112181K). International Society for Optics and Photonics.
27. Shi, F., Cheng, J., Wang, L., Yap, P. T., & Shen, D. (2015). LRTV: MR image super-resolution with low-rank and total variation regularizations. *IEEE Transactions on Medical Imaging, 34* (12), 2459–2466.
28. Mejia, J., Mederos, B., Ortega, L., Gordillo, N., & Avelar, L. (2017). Small animal PET image super-resolution using Tikhonov and modified total variation regularisation. *The Imaging Science Journal, 65*(3), 162–170.
29. Zhao, H., Li, H., Maurer-Stroh, S., & Cheng, L. (2018). Synthesizing retinal and neuronal images with generative adversarial nets. *Medical Image Analysis, 49*, 14–26.
30. Chambolle, A., & Pock, T. (2019). Total roto-translational variation. *NumerischeMathematik, 142*(3), 611–666.
31. Lunz, S., Öktem, O., & Schönlieb, C. B. (2018). Adversarial regularizers in inverse problems. In *Advances in neural information processing systems* (pp. 8507–8516).
32. Li, H., Schwab, J., Antholzer, S., & Haltmeier, M. (2020). NETT: Solving inverse problems with deep neural networks. *Inverse Problems, 36*, 065005.
33. Kobler, E., Effland, A., Kunisch, K., & Pock, T. (2020). Total deep variation for linear inverse problems. arXiv preprint arXiv:2001.05005.
34. Alipour, S. H. M., Rabbani, H., & Akhlaghi, M. (2014). A new combined method based on curvelet transform and morphological operators for automatic detection of foveal avascular zone. *Signal, Image and Video Processing, 8*(2), 205–222.
35. Kang, M. G., & Katsaggelos, A. K. (1995). General choice of the regularization functional in regularized image restoration. *IEEE Transactions on Image Processing, 4*(5), 594–602.
36. Hintermüller, M., Holler, M., & Papafitsoros, K. (2018). A function space framework for structural total variation regularization with applications in inverse problems. *Inverse Problems, 34*(6), 064002.
37. Kim, J., Kwon Lee, J., & Mu Lee, K. (2016). Accurate image super-resolution using very deep convolutional networks. In *Proceedings of the IEEE conference on computer vision and pattern recognition* (pp. 1646–1654).
38. Boyd, S., Parikh, N., Chu, E., Peleato, B., & Eckstein, J. (2011). Distributed optimization and statistical learning via the alternating direction method of multipliers. *Foundations and Trends in Machine Learning, 3*(1), 1–122.

Lightweight Spatial Geometric Models Assisting Shape Description and Retrieval

S. Priyanka and M. S. Sudhakar

1 Introduction

The ability to define and distinguish various objects from their shape evidence has led to their comprehensive utilization in diverse fields, namely, object recognition, detection, and retrieval. Its employment in characterization highly brings down the feature size necessary in content-based image retrieval (CBIR). The quest for determining efficient shape descriptors and matching metrics remains a challenging task in the current period. These things happen because most of the documented descriptors claiming invariance suit image segmentation and retrieval highly [1]. These descriptors' initial intention is to yield a higher retrieval rate by compromising their dual computational complexity. This demands the realization of lightweight shape-based extraction schemes assisting image matching and retrieval.

Usually, these characterization schemes classify areas as a region or a contour [2] to further part it into global and local approaches based on the information localization. Global strategies build descriptors by considering the entire image data while local ones deal with image segments for shape characterization. There is a description of the existing descriptors' chronological ordering with their technical virtues and shortcomings below.

The widely popular shape context (SC) [3] resembles a global descriptor that capitulates the spatial relationship between contour points using correlation. Then, a

S. Priyanka
Department of Electronics and Communication Engineering, Sreenivasa Institute of Technology and Management Studies, Chittoor, India
e-mail: spriyanka915@gmail.com

M. S. Sudhakar (✉)
School of Electronics Engineering, VIT Vellore, Vellore, India
e-mail: sudhakar.ms@vit.ac.in

A. Deshpande et al. (eds.), *Computational Intelligence Methods for Super-Resolution in Image Processing Applications*, https://doi.org/10.1007/978-3-030-67921-7_11

sixty-dimensional histogram packs the resulting features that represent the spatial correlation of the contour points. Herein, the Euclidean metric quantifies the shape similarity followed by dynamic programming to support matching and retrieval. The articulation invariant shape feature of Ling et al. [4] blended the inner-distance captured across multidimensional scales. Global approaches [5, 6] acquire boundary points that are then localized into triangular area representation (TAR) across different scales and along with the polarity variations indicating their direction [14]. The triangulation scheme localizes the curvature of the contour points. The attached polarity identifies their concavity or convexity. The Hough transform statistics neighborhood (HTSn) scheme from Souza et al. [7] characterizes object shapes using the Hough transform statistics. The ability of HTSn to produce a fixed feature vector length for varying image sizes in the Hough space is the core merit of this scheme. The approach of [8] decomposed the shape information using piecewise approximation into conic-section coefficients. Upon extracting the coefficients, only the projective conic invariant pairs were considered for shape description. The above global descriptors cogitate the entire shape information to formulate feature vectors, whereas shape is highly confined.

Moment-based characterizer such as the Tchebichef [9] links the lower order invariants to build the shape vector. Similarly, Hu et al. [1] acquired features of the common base triangle area (CBTA) to formulate the descriptor, and dynamic programming was involved in matching and retrieval. Circle views [10] signatures were extracted using Fourier transform from shape boundaries to express feature vectors. Feature points from shape contours were extracted by the invariant multi-scale descriptor [11] using dynamic programming. The shape feature of Kaothanthong et al. [12, 13] utilized line segment distribution and their localized intersection patterns for description. The discussion above shows that dynamic programming plays a vital role in determining their performance. The addendum further heightens the matching and retrieval process's complexity, thereby limiting its employment for real-time applications.

This chapter discusses and demonstrates a couple of lightweight feature descriptors with warranted retrieval performance to minimize computational costs. This strategy is helpful, especially in areas related to image processing and computer vision. Algorithmic hybridization is accomplished in the rendered models by cross-linking trigonometry concepts with point-wise pixel processing to provide precise and straightforward shape descriptors that deliver a higher retrieval rate. The consequential descriptors enforce retrieval accuracy by rigorously imposing localization on shape characterization. Also, it offers to enhance the retrieval accuracy by incorporating a computationally intelligent model to the feature classification process.

The remainder of this chapter is compartmentalized as follows: Section 2 lists the generic shape retrieval framework with two geometric models dealing with feature extraction from binary shapes. Efficacy analysis of the presented descriptors on diverse shape datasets is elaborated in Sect. 3. Section 4 discusses super-resolution for spatial geometric models assisting shape description and retrieval. In conclusion, Section 5 outlines the descriptors' highlights with the scope for further improvement.

2 Generic Shape Retrieval Framework

A generic retrieval framework represents a search and retrieval mechanism that mainly comprises four modules. They are the query (input) module, followed by feature extraction, matching, and display. Generally, module 1 is where the user inputs the query image to the search mechanism. The features are then extracted by module 2 and then matched with the feature database to determine the relevant matches' given input. Upon matching, images that closely resemble the query are displayed in ranking order. The processes mentioned above are shown in Fig. 1.

In compliance with the above scheme, two geometrical models blended with diverse classification techniques aiding retrieval are discussed in this chapter. An elaborate discussion of the same arises under the relevant header.

2.1 *Tetrakis Square Tiling-Based Triangulated Feature Descriptor*

Feature extraction and characterization play a vital role in determining the efficacy of several retrieval systems. Subsequently, the classification mechanism for rendering relevant matches for the given query during the retrieval phase determines the similarity among the acquired features. This denotes that characterization schemes offering effective performance with minimal computational load remain a challenging problem for researchers. Thus, this sub-section introduces a lightweight shape descriptor rendering optimal retrieval performance with reduced computation cost labeled as triangulated feature descriptor (TFD) [26]. Initially, TFD operates on the

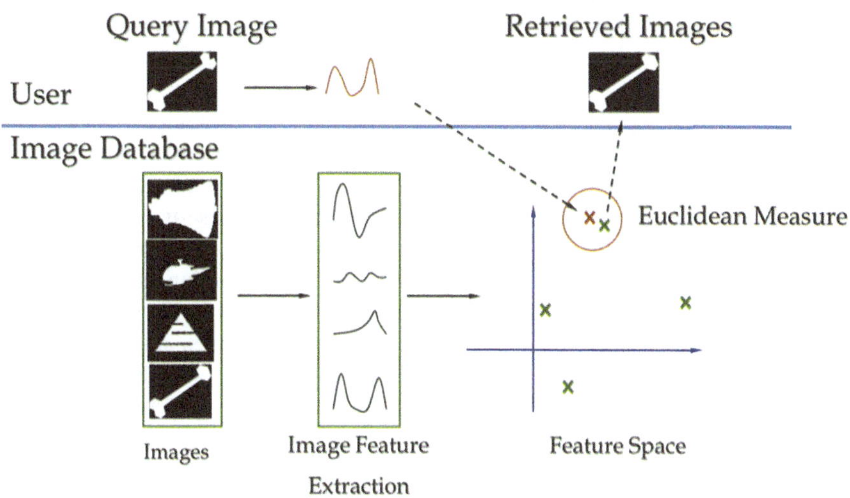

Fig. 1 The generic shape retrieval mechanism

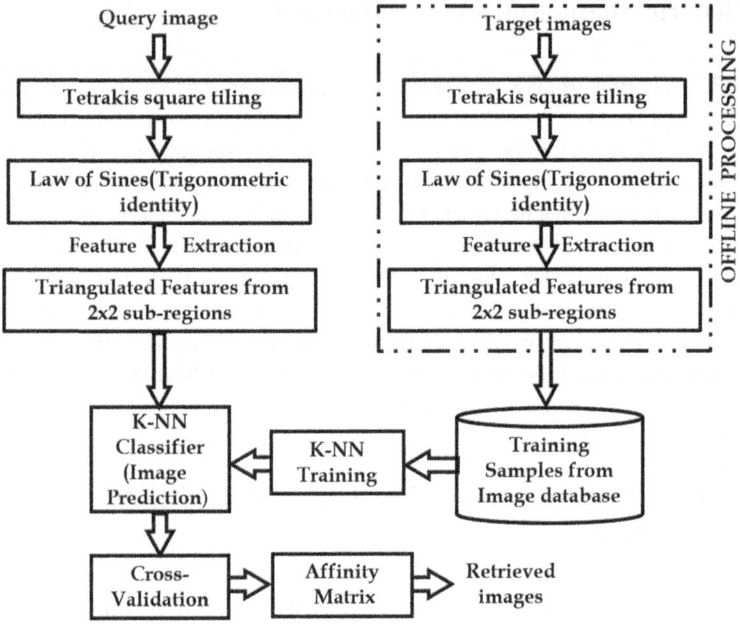

Fig. 2 Overall TFD extraction and retrieval process

given query to extract the features that are then accumulated into shape histograms. Likewise, the images in the shape dataset are processed by TFD to formulate the feature database. The accumulated features are then spatially categorized by the K-nearest neighbor (K-NN) algorithm. During the retrieval stage, the query features are mapped with the feature dataset to determine the relatively closer matches and deliver them across the output window. The TFD realization process appears in Fig. 2.

At the onset, input shapes are tiled into squared 2×2 sub-regions using the Tetrakis tiling scheme. Further, they are subdivided into 4-right angled triangles. The side intensity differences of each right-angled triangle, along with the corresponding angles, are subsequently processed using the law of sines. Similarly, the features from the remaining triangles are attained. The features of these triangles are then merged by the maximum operator to yield highly acute features representing that sub-region. These features are then binarized and blended using the logical OR operation to produce highly localized features. This process is performed on the remaining image to render an octal image representing the given shape finally. As the triangulation process accomplishes feature localization, the resulting feature is termed as aTFD. The entire process of extracting and formulating TFD histograms is presented in Fig. 3.

TFD features are organized as follows: vertical (S_v), horizontal (S_H), and diagonal (S_D) of each right-angled triangle. The relationship between these features and the matrix defining the shape $I(i,j)$becomes

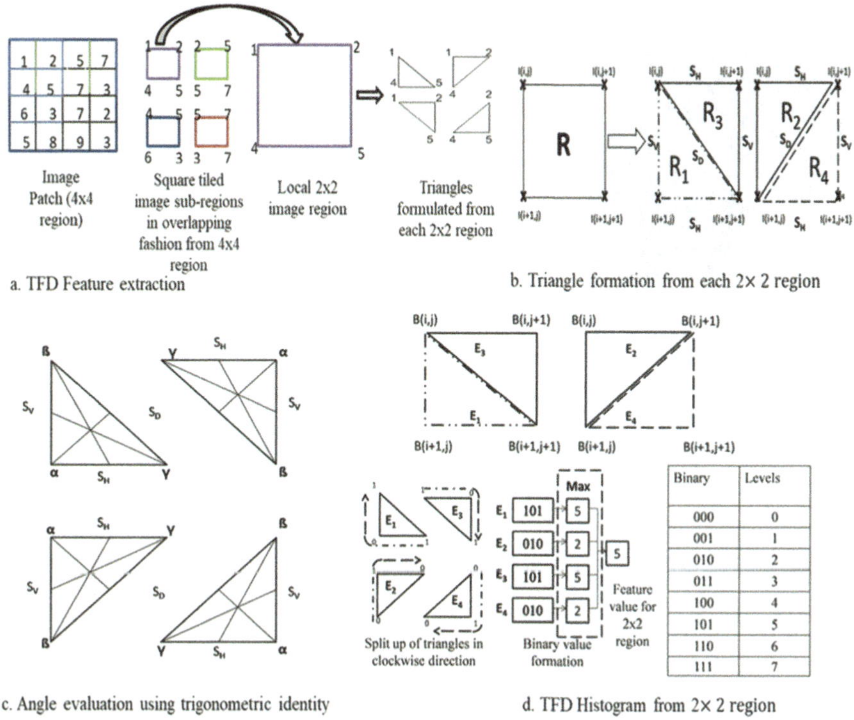

Binary	Levels
000	0
001	1
010	2
011	3
100	4
101	5
110	6
111	7

Fig. 3 TFD realization process

$$[S_V S_H S_D] = [\boldsymbol{I}(i,j)\boldsymbol{I}(i+1,j+1)\boldsymbol{I}(i+1,j)\boldsymbol{I}(i+1,j)\boldsymbol{I}(i,j)\boldsymbol{I}(i+1,\\ ,j+1)\boldsymbol{I}(i+1,j+1)\boldsymbol{I}(i+1,j)\boldsymbol{I}(ij)][10-1]. \tag{1}$$

The resulting triangular features are merged by the maximum operation to yield the side response (R_i) upon applying the law of sines according to

$$R_i = \text{argmax}\left\{\frac{S_V}{\sin 45}, \frac{S_H}{\sin 45}, \frac{S_D}{\sin 90}\right\}, \tag{2}$$

where R_i attained from each right-angled triangle is then binarized, and logically ORed becomes

$$R_i = \text{Dec2bin}(\text{R}_1)|\text{Dec2bin}(\text{R}_2)|\text{Dec2bin}(\text{R}_3) \mid \text{Dec2bin}(\text{R}_4). \tag{3}$$

The resulting binary features $B(i,j)$are reorganized into triangles and chained, as shown in Fig. 3d. The attained chain values are then mapped with the corresponding octal values to produce the ternary side feature E_1, E_2, E_3, and E_4 as

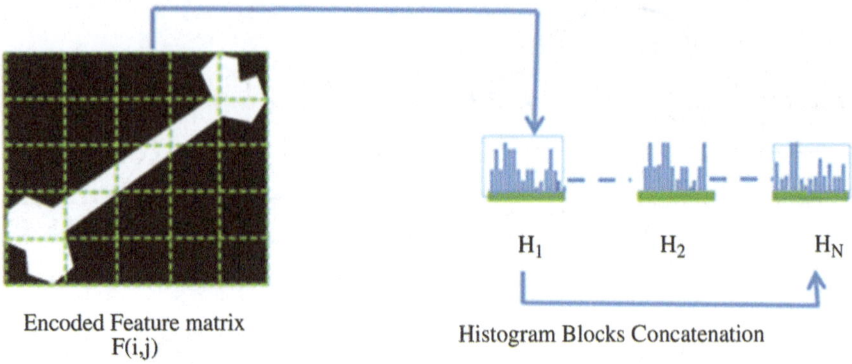

Fig. 4 Realization of TFD histograms

$$E_1 = [B(i,j), B(i, j+1), B(i+1, j+1)], \quad (4)$$

$$E_2 = [B(i,j), B(i, j+1), B(i+1, j)], \quad (5)$$

$$E_3 = [B(i,j), B(\mathrm{i}+1, j), B(i+1, j+1)], \text{ and} \quad (6)$$

$$E_4 = [B(i+1, j), B(i, j+1), B(i+1, j+1)], \quad (7)$$

The maximum operator again combines these edges according to

$$F(i,j) = \max(E_N), N = 1, 2, 3, 4. \quad (8)$$

The final feature $F(i,j)$ replaces each pixel in the given input shape, and this operation is repeated on the entire image. This arrangement finally results in a strongly linked feature map that is then subdivided into non-overlapping blocks $F = \{F_1, F_2, F_3, \ldots, F_n\}$ of size 10 × 10 from which the local histograms $H = \{H_1, H_2, H_3, \ldots, H_n\}$are transformed into a global shape descriptor as illustrated in Fig. 4.

Mathematically, histogram concatenation becomes

$$\mathrm{TFDH} = \bigcup_{i=1}^{n} H_i, \quad i = 1, 2, \ldots n, \quad (9)$$

where n = MN is the image size

For a 256 × 256 image, the resulting descriptor has a length of 256 × 8 = 2048 upon acquiring TFD features followed by a quantization into eight bins. The realized TFD is a mere representation of the given shape attained by merging the captured local features to achieve a global histogram. The yielded histogram blends both

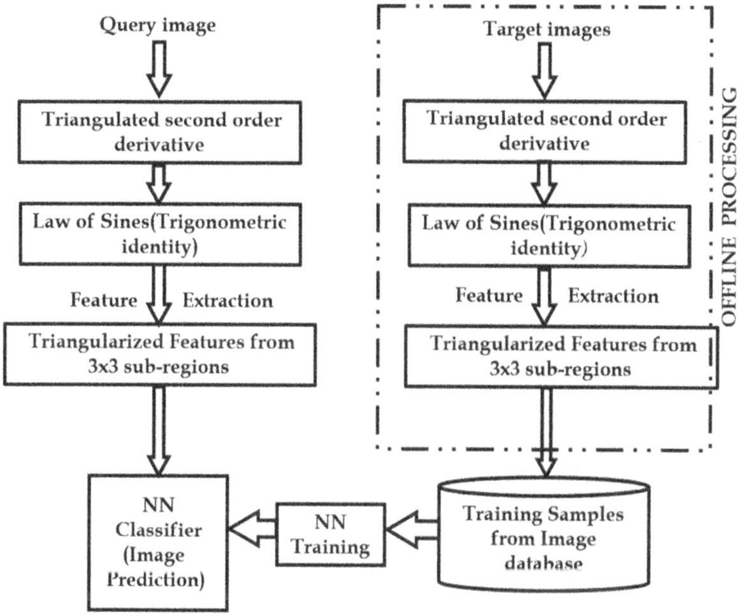

Fig. 5 TSOSD formulation process

coarse and fine features by the accumulation of spatial resolution across different neighborhoods.

The next descriptor is obtained by slightly modifying the TFD as follows:

(i) The calculation of the gradients for feature characterization is further extended for precision using the second-order derivatives.
(ii) The other variation corresponds to the sub-region size used for the localization of shape features.

The determination of features relying on triangular geometrical shape remains the same with the resulting operator known as triangulated second-order shape derivative (TSOSD) [27]. Similar to the TFD, TSOSD decomposes the shapes into square sub-regions followed by triangulation. Subsequently, the triangle sides' derivatives interact locally using the law of sines to yield the angle-based feature map. Using TFD, the attained feature maps are fabricated into shape histograms representing the given image. An overview of the TSOSD formulation process is presented in Fig. 5.

At the onset, each shape is processed by the Tetrakis square tiling scheme to decompose them into square sub-regions. Later, evaluating the second-order derivatives from their sides provides localized features from each triangulated sub-region. The maximum among these features replaces the edges of the squared sub-regions, thereby representing a feature. The attained features resulting from the four triangles are binary transformed. These triangles logically interact using the OR operation, and their concatenation results in an octal value. Next, this process is repeated on the

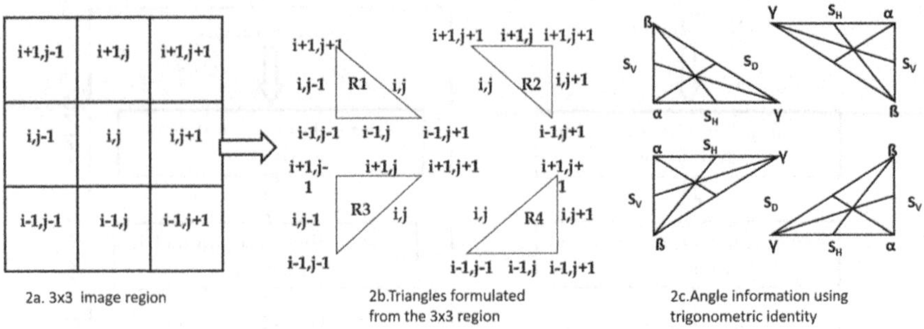

Fig. 6 Tessellation process involved in realizing TSOSD

entire image, and the corresponding feature map is attained. These feature maps are then fabricated into shape histograms representing the TSOSD feature. As the TSOSD is only an extension of TFD, the variation in terms of the sub-region size and the mathematical operations are discussed below. Initially, each 3 × 3 sub-region is decomposed into four triangles, as depicted in Fig. 6.

The squared sub-region R is mathematically decomposed as follows

$$\mathbf{R} = (R_1, R_2, R_3, R_4).\text{QUOTE} \tag{10}$$

Similar to the TFD process, at the beginning of feature localization, the sides of each right-angled triangles interact locally using the absolute difference operation to yield the side responses in accordance with the direction, namely, vertical $Sv(S_{V1}, S_{V2})$, horizontal $(S_{H1}, S_{H2}\, S_H)$, and diagonal (S_{D1}, S_{D2}) and is attained using Eqs. (11) to (17).

$$S_{V1} = \text{abs}(\boldsymbol{I}(i+1, j-1) - \boldsymbol{I}(i, j-1) - \boldsymbol{I}(i-1, j-1)), \tag{11}$$

$$S_{H1} = \text{abs}(I(i-1, j-1) - I(i-1, j) - I(i-1, j+1)), \tag{12}$$

$$S_{D1} = \text{abs}(I(i+1, j-1) - I(i, j) + I(i-1, j+1)), \tag{13}$$

$$S_{V2} = \text{abs}(I(i+1, j+1) - I(i, j+1) - I(i-1, j+1)), \tag{14}$$

$$S_{H2} = \text{abs}(I(i+1, j-1) - I(i+1, j) - I(i+1, j+1)), \text{and} \tag{15}$$

$$S_{D2} = \text{abs}(I(i+1, j+1) - I(i, j) - I(i-1, j-1)), \tag{16}$$

In the next step, the second-order derivative of S_{V1} is obtained by differentiation of Eq. (11) to produce

$$\frac{\partial S_{V1}}{\partial \boldsymbol{I}} = \text{abs}(\boldsymbol{I}(i+2,j) - \boldsymbol{I}(i+1,j) - \boldsymbol{I}(i+1,j) + \boldsymbol{I}(i,j-1) - \boldsymbol{I}(i,j) + \boldsymbol{I}(i-1,j-1)). \tag{17}$$

Equation (17) treats the displacements beyond 1-pixel. By image geometry, these displacements should be constrained within a unity distance. Hence, the spatial displacement of -1 is performed in (17) to yield

$$\frac{\partial S_{V1}}{\partial \boldsymbol{I}} = \text{abs}(\boldsymbol{I}(i+1,j-1) - \boldsymbol{I}(i,j+2) - \boldsymbol{I}(i,j-1) + \boldsymbol{I}(i-1,j-2) - \boldsymbol{I}(i-1,j-1) + \boldsymbol{I}(i+2,j+2)). \tag{18}$$

The unit subtrahend of the spatial coordinates introduces displacements that are more than unit distance. Hence, neglecting displacements greater than unity in the previous expression leads to

$$\frac{\partial S_{V1}}{\partial \boldsymbol{I}} = \text{abs}(\boldsymbol{I}(i+1,j-1) - \boldsymbol{I}(i,j-1) - \boldsymbol{I}(i-1,j-1)). \tag{19}$$

This indicates the invariance of TSOSD toward translation. Further, differentiating Eq. (19) and following the above constraints which merely represent S_{V1}, result in

$$\frac{\partial^2 S_{V1}}{\partial \boldsymbol{I}^2} = \text{abs}(\boldsymbol{I}(i+1,j-1) - \boldsymbol{I}(i,j-1) - \boldsymbol{I}(i-1,j-1)). \tag{20}$$

Triangles exhibit congruency, as shown in Eqs. (19) and (20). They depict similar characteristics since pixels are placed at a unit distance. Likewise, the second-order derivatives on the other sides are attained using the following equations:

$$\frac{\partial^2 S_{V1}}{\partial I^2} = \text{abs}(I(i+1,j-1) - I(i,j-1) - I(i-1,j-1)), \tag{21}$$

$$\frac{\partial^2 S_{H1}}{\partial I^2} = \text{abs}(I(i-1,j-1) - I(i-1,j) - I(i-1,j+1)), \tag{22}$$

$$\frac{\partial^2 S_{D1}}{\partial I^2} = \text{abs}(I(i+1,j-1) - I(i,j) + I(i-1,j-1) - I(i-1,j+1)), \tag{23}$$

Fig. 9 Images from the TARI-1000 dataset

Fig. 10 Images from the KIMIA-99 dataset

Kimia-99 Dataset

Several shape matching and object recognition applications employ this dataset for efficacy analysis. Although two variants, namely Kimia25, Kimia216, are available, this dataset covers almost all images in the datasets mentioned previously and hence not utilized here. The 99 images in Kimia-99 are divided into nine classes, with each containing 11 images. Figure 10 depicts sample images contained in the Kimia-99 dataset.

Table 1 Parameters adopted for NN training

Sl. no	Parameters	Values
1	Architecture	Feedforward MLP
2	No. of input neurons	2048 for MPEG-7,TARI-1000,KIMIA-99
3	No. of output neurons	70 for MPEG-7,9 for Kimia-99, 50 for TARI-1000.
4	No. of hidden layers	15
5	No. of hidden layer neurons	3
6	Learning algorithm	Back propagation algorithm (BPA)
7	Activation function	Sigmoid
8	Initial weights	0–1
9	Initial bias	1
10	Learning rate	0.2
11	No. of epochs	87
12	No. of iterations	100
13	Type of error	Mean square error(MSE)
14	Tolerance	0.001

Performance Assessment on TFD

The presented descriptors' performance is assessed using the widespread bull's eye retrieval (BER) rate from the images obtained across the following datasets. *BER quantifies the ratio of the total number of correct matches to the maximum number of correct matches.*

The performance of the above descriptors is assessed via a retrieval framework and the relative metrics. In the dual implementations related to TFD and TSOSD for image classification, the following classifiers along with their relative configuration are as follows:

- In TFD's case, the K-NN classifier has the following settings:
  1. Number of observations N is mapped to the dataset size.
  2. Number of neighbors is fixed as 2.
  3. Euclidean Distance with $\boldsymbol{X}$ loaded with feature values (based on the size of the image) and $\boldsymbol{Y}$ mapped with labels depending on the size of the database.

  Performance Assessment on TSOSD
- In the TSOSD case, the neural network with the architectural settings detailed in Table 1 is used.

  The relevant metrics outlining the performance of the NN classifier and its definition are listed below.

1. Confusion matrix outlines the classifier performance by representing the predicted classifications and defined classifications along with the columns and rows, respectively. Likewise, the other metrics are as follows:

(a) True positive (TP): Correctly classified test set samples, given positive sample
(b) False positive (FP): Accepted test set samples that should have been rejected
(c) False negative (FN): Rejected test set samples that should have been accepted
(d) True negative (TN): Correctly classified test set samples, given a negative sample

These metrics assist in calculating the below measures that aid in analyzing the classifier performance.

2. *Accuracy*: It is the fraction of the total number of classifications that were correct given by

$$\text{Accuracy} = (\text{TP} + \text{TN})/(\text{TN} + \text{FP} + \text{FN} + \text{TN}). \tag{29}$$

3. *False-positive rate/false acceptance rate (*FAR*):* It is the fraction of the classifications in which an unacceptable sample is invalidated using

$$\text{False Acceptance Rate} = \text{FAR} = \text{FP}/(\text{FP} + \text{TN}). \tag{30}$$

4. *False-negative rate/false rejection rate* (FRR)*:* It is the proportion of the classifications in which the valid sample is misclassified or rejected according to

$$\text{False Rejection Rate} = \text{FRR} = \text{FN}/(\text{TP} + \text{FN}). \tag{31}$$

5. *True-negative rate/true rejection rate* (TRR)*:* It is the fraction of the classifications, where an invalid sample is exactly classified as invalid following

$$\text{True Rejection Rate} = \text{TRR} = \text{TN}/(\text{TN} + \text{FP}). \tag{32}$$

The involvement of NN in the TSOSD retrieval framework is mainly attributed to its computationally intelligent behavior and hence, used by several information retrieval systems. NNs enhance the matching process by learning from the referential data, thereby escalating TSOSD's retrieval accuracy and hence blended into the retrieval model. The frameworks relating to both descriptors are implemented in MATLAB 7.0 running on a1.6 GHz dual-core CPU operating on Windows 7 platform. A comparative analysis of both the descriptors with their predecessors using the datasets mentioned before is discussed.

BER Analysis for the MPEG7 Dataset

This is done by initially constructing the TFD histogram and then formulating the affinity matrix. Later, the sum of the correct matching shapes with the 20 shapes belonging to each class and dividing this sum by the affinity matrix's dimensions to attain BER. Also, to build the predictive model, k-NN is five-fold cross-validated. Competitive analysis of TFD, and TSOSD with proposed schemes are shown in Table 2.

Table 2 Relative retrieval analysis of diverse descriptors on the MPEG-7 dataset

Method	Retrieval rate (%)	Average retrieval time (ms)
Visual part [5]	76.65	NA
Shape context [12]	76.51	NA
Distance sets [6]	78.38	NA
MCC+DP [20]	84.93	2.30×10^4
SC+DP [12]	86.80	2.17×10^4
IDSC +DP [17]	85.40	1.81×10^4
CDTW + SC [21]	86.73	2.77×10^4
HPM [16]	86.35	1.40×10^2
TAR+DP [15]	87.13	2.77×10^4
Shape tree [15]	87.70	NA
Angular and binary angular pattern [18]	81.20	NA
Shape vocabulary [3]	90.41	2.24×10^2
CBTA [19]	89.45	NA
CBTA+SC [19]	93.65	2.2×10^2
TFD	**95.35**	1.34×10^4
TSOSD	**98.75**	1.64×10^2

Table 2 demonstrates the strength of both TFD and TSOSD feature formulation schemes in terms of the retrieval rate and time. The intensified local interactions in both these schemes have highly escalated the retrieval rate as shown in Table 2. Moreover, the ease in achieving the feature vectors for both these schemes lessened the realization period, as recorded in Table 2. Despite containing complex shape variations in the MPEG-7 dataset, both descriptors can offer higher retrieval accuracy. These descriptors are deemed fit for real-time shape retrieval schemes owing to their faster realization. Also, NNs learning ability from the feature data has brought down the retrieval time compared with its peers with substantial retrieval improvement, a characteristic owed to its computationally intelligent behavior.

TARI-1000 Dataset

Upon inferring the images in TARI-1000, it can be stated that this dataset contains shapes with more articulation changes than the MPEG7. Hence, attaining higher retrieval rates is a slightly more significant challenge than the results obtained in several other datasets. The achieved BERs for TFD and TSOSD are relatively stated in Table 3.

The stated BER values of both these schemes confirm the consistency rendered by them in both the datasets. IDSC tops Table 3 for this dataset, but the consistent retrieval performance more significant than 90% of the TFD, TSOSD in both these datasets is not witnessed for the same.

Kimia-99 Dataset

As this dataset contains 99 images, 60% of them were selected for training, resulting in 63 images with seven belonging to each class. The remaining 40% of images formed the testing that resulted in 36 images with four from each class. Retrieval

Table 3 Retrieval rates observed on the TARI-1000 dataset

Method	Retrieval rate(%)
SC(DP) [12]	94.17
IDSC(DP) [12]	95.33
ASC [22]	95.44
IDSC+LP [23]	99.35
SC+LP [23]	97.79
IDSC+LCDP [24]	99.7
ASC+LCDP [24]	99.79
SC+IDSC+ Co-Transduction [24]	99.995
Angular and Binary angular pattern [18]	93.02
TSOSD	**95.22**
TFD	**93.45**

Table 4 Retrieval performance observed on the Kimia-99 dataset

Method	Top 1	Top 2	Top 3	Top 4	Retrieval rate (%)
L2(Baseline) [12]	25	15	12	10	38.75
SC [11]	20	10	23	5	36.25
MDS+SC [12]	36	26	17	15	58.75
IDSC [12]	40	34	35	27	85
CBTA [19]	40	38	32	23	81.87
TFD	**39**	**38**	**38**	**32**	**85.57**
TSOSD	**40**	**40**	**39**	**36**	**94.35**

precision is the number of top closest matches from the correct object as top1 to top 4. The retrieval rates achieved by TFD and TSOSD are relatively stated in Table 4.

Table 4 shows that TSOSD excels in terms of the retrieval rate. Additionally, this quality is attributed to the highly localized interactive nature of the descriptor and, together with NN, has achieved an improved accuracy of 94.35%. Thereby, this proves that local–global approaches have profoundly influenced the presented descriptor's retrieval performances across every retrieval phase.

The scheme's classification performance is further analyzed using the above-described metrics, namely *Accuracy*, *FAR*, *TRR*, and *FRR* on the MPEG-7 and TARI-1000 datasets. The related results appear in Table 5, indicating the very low *FAR*, high *TRR* with significantly high recognition accuracy.

These outcomes demonstrate the outstanding performance of the system in rejecting the false samples and in classifying. Moreover, with the tan sigmoid activation function's deployment, the NN converges at a faster rate compared to the log sigmoid to achieve satisfactory performance goals. This analysis proves that when TSOSD merged with NN is precise in matching images as reflected in the recognition accuracy.

Table 5 Experimental results on the MPEG-7 and TARI-1000 datasets using the set NN

Activation function in the hidden layer	Number of epochs used during training phase		Recognition accuracy (%)		FAR (%)		TRR (%)		FRR (%)	
	MPEG-7	TARI-1000	MPEG-7	TARI-1000	MPEG-7	TARI-1000	MPEG-7	TARI-1000	MPEG-7	TARI-1000
Log sigmoid	10	8	98.85	98.92	0.58	0.53	99.42	99.47	0.60	0.92
Log sigmoid	09	7	99.07	99.22	0.48	0.41	99.51	99.59	0.25	0.17
Log sigmoid	10	6	98.92	99.43	0.51	0.35	99.48	99.65	0.33	0.97
Log sigmoid	08	6	98.89	99.39	0.57	0.32	99.42	99.68	0.50	0.45
Log sigmoid	07	6	99.2	99.37	0.44	0.34	99.55	99.66	0.5	0.71
Log sigmoid	06	6	99.14	99.48	0.43	0.28	99.56	99.27	0.52	0.68
Tan sigmoid	17	9	98.7	98.87	0.67	0.60	99.32	99.40	0.25	0.13
Tan sigmoid	11	7	98.98	99.19	0.51	0.43	99.48	99.75	0.65	0.27
Tan sigmoid	09	7	99.17	99.39	0.40	0.32	99.59	99.68	0.52	0.45
Tan sigmoid	09	7	98.82	99.13	0.60	0.37	99.39	99.36	0.22	0.23
Tan sigmoid	07	6	98.96	99.34	0.51	0.35	99.48	99.65	0.58	0.97
Tan sigmoid	07	6	99.15	99.06	0.40	0.21	99.59	99.79	0.36	0.64

4 Extension to Super-Resolution (SR) Retrieval Applications

In many under-sampled imageries, the corresponding image quality and spatial resolution can be improved by inserting a point spread function (PSF) to model the problems plaguing image acquisition (noise) from several sources. Information about the signal-to-noise ratio (SNR) can also enhance image quality [32].

Such under sampling may lead to aliasing artifacts and reduced image utility. Designing imaging systems for specific applications entails navigating a sophisticated space in terms of conflicting requirements and involves balancing factors such as optical resolution, the field of view, aliasing, SNR, frame rate, and size, in addition to power [28]. Similar reflections are involved in the design of microscopy systems [29]. Using multiple frames with sub-pixel interframe motion allows one to obtain a denser sampling of $f(x, y)$ than may be possible with a single image. The resulting samples may or may not meet the Nyquist criterion. The image sampling will be non-uniformly distributed unless there is a careful interframe motion control [28].

With the extensive application of surveillance and diagnostic systems, imagery investigation technology has a paramount role, and image searches become valuable. However, in real investigations, due to distance and equipment limitations, most relevant details in images are of low quality. The SR technology can utilize information from databases to render high-definition image versions, which can effectively enhance analysis and restore image features with more detail. This technology is crucial to improve the clarity of images, increase recognition accuracy, and increase the number of successfully treated and solved cases.

A better image gives rise to more precise shape semantics models and posterior image knowledge. Better algorithms can improve the robustness of both single and multiple modalities of SR algorithms and help reference other images with increased retrieval accuracy. Hence, SR can extend the shape model descriptor, which leads to better information about image semantics [31].

CBIR investigates the perceptible image contents from different imaging modalities, e.g., color, shape, and texture related to representing the image features. CBIR research looks toward ameliorating other exploration procedures, explaining, organizing, and indexing the dense databases. Some schemes rely on color distribution entropy (CDE), color level co-occurrence matrix (CLCM), and high-quality edge detection. CDE considers the correlativity of an image's spatial color distribution. That is, it effectively tells the spatial color information of images. CLCM takes into account the texture features of a picture whose base is from the old procedure gray level co-occurrence matrix (GLCM), which only makes the gray level images. Still, in CLCM, it takes colored texture images. It is a colored alternative to old texture, recognizing GLCM. High-quality edge detectors facilitate the detection of the image boundaries to extract shape features bearing in mind matters like translation, changes in scale, and rotation. Higher and faster retrieval outcomes can result from carefully taking several and primitive image descriptors [30]. It is evident from the above discussions that spatially localized feature extraction highly assists applications dealing with the retrieval of SR images.

Accordingly, to encounter SR image retrieval from the relevant dataset initially, image registration is performed. To supplement retrieval, [32] offers an SR curve to superimpose two affine regions and subsequently extract the high-frequency features. A further extension to agricultural field analysis [33] led to the detection of land boundaries from satellite images using a contour detection network. Later, an evaluation metric [34] for the extracted boundaries is calculated and is then matched with the reference boundaries to analyze the matching performance. An enhancement to the aforesaid methods [35] offered a novel surface SR scheme to detect the repetitive patterns in 3D images. Initially using these SR registration techniques, the retrieval scheme [36] offered 96.4% accuracy by constructing three low-level descriptors. The method [37] replaced low-level features with several distance metrics for retrieval and improved retrieval accuracy when tested on a generic image collection. Further, [38] relied upon the shape information extracted using Fourier descriptors, moment-based features, hierarchical centroids, and histogram of oriented gradients. These features were then fused using discriminant correlation analysis and rendered 90% retrieval accuracy when compared with its competitors.

The above discussion enunciates the superiority of shape for SR image retrieval. Consequently, the presented descriptors in this chapter highly consider the local information for building the shape descriptors. Therefore, extending the same for retrieval applications handling SR images very much favors improved accuracy.

5 Conclusion

This chapter discusses two novel, simple, and effective shape descriptors labeled TFD and TSOSD for shape retrieval. These descriptors aim at exploiting the spatial relationship existing between local points in neighborhoods. These descriptors resulted from merging simple geometrical concepts with image processing techniques. Initially, TFD is constructed by tessellating the local regions into right-angled triangles that were then localized using the law of sines to capture the feature. The attained features were then fabricated into shape histograms using a novel feature representation scheme that successfully improved its distinction ability. BER validations on available datasets establish the excellent retrieval accuracy achieved by TFD.

An extension of the concept of TFD resulted in the formulation of TSOSD. The significant difference between these schemes is the involvement of the second-order derivative operator and the increase in the sub-regions' size, with the remaining formulation process being the same. Similar to the TFD, the analysis of TSOSD was performed on diverse datasets. From the comparison, TSOSD demonstrated excellent retrieval accuracy as compared with its peers. Lightweight computational models involved in realizing TFD and TSOSD make it recommendable for dynamic retrieval without compromising retrieval accuracy. The main highlight of both these contributions is maintaining consistent retrieval rates in diverse datasets, which is not seen for its competitors.

In distinction, its peers delivered excellent results in either of the databases but failed in the others. This statement substantiates that these descriptors can match their counterparts in part, especially in terms of the retrieval rate, by realizing them using simple processing techniques. The inherent localization involved in understanding these descriptors justifies its inclusion, as it becomes evident from the improved retrieval accuracy over its predecessors. This supports their extension to retrieval mechanisms dealing with super-resolution images. Concretely, the computational intelligent NN favors retrieval time within a shorter span without comprising on the TSOSDs retrieval accuracy.

References

1. DellAcqua, F., & Gamba, P. (Sep.1998). Simplified modal analysis and search for reliable shape retrieval. *IEEE Transactions on Circuits and Systems for Video Technology, 8*, 656–666.
2. Rui, Y., Huang, T. S., & Chang, S. F. (Mar.1999). Image retrieval: Current techniques, promising directions, and open issues. *Journal of Visual Communication and Image Representation, 10*, 39–62.
3. Kokare, M., Chatterji, B. N., & Biswas, P. K. (2002). A survey on current content based image retrieval methods. *IETE Journal of Research, 48*, 261–271.
4. SardeyMohini, P., & Kharate, G. K. (2015). A comparative analysis of retrieval techniques in content based image retrieval. *IETE Journal of Research*, arxiv: 1508.06728.
5. Bartolini, P. C., & Patella, M. (2005). WARP: Accurate retrieval of shapes using phase of Fourier descriptors and Time warping distance. *IEEE Transactions on Pattern Analysis and Machine Intelligence, 27*, 142–1476.
6. Mokhtarian, F., Abbasi, S., & Kittler, J. (1997). Efficient and robust retrieval by shape content through curvature scale space. In *Image databases and multi-media search* (Vol. 8, pp. 51–58). Danvers.
7. Rui, Y., She, A. C., & Huang, T. S. (1997). *A modified Fourier descriptor for shape matching in MARS, Series on software engineering and knowledge engineering* (Vol. 8, pp. 165–180). Singapore: World Scientific Publishing.
8. Attalia, E., & Siy, P. (2005). Robust shape similarity retrieval based on contour segmentation polygonal multi resolution and elastic matching. *Pattern Recognition, 38*, 2229–2241.
9. Shu, X., & Wu, X. J. (2011). A novel contour descriptor for 2D shape matching and its application to image retrieval. *Image and Vision Computing, 29*, 286–294.
10. Ling, H., & Okada, K. (2007). An efficient Earth mover's distance algorithm for robust histogram comparison. *IEEE Transactions on Pattern Analysis and Machine Intelligence, 29*, 840–853.
11. Belongie, S., Malik, J., & Puzicha, J. (2002). Shape matching and object recognition using shape contexts. *IEEE Transactions on Pattern Analysis and Machine Intelligence, 24*, 509–522.
12. Ling, H., & Jacobs, D. W. (Feb. 2007). Shape classification using the inner distance. *IEEE Transactions on Pattern Analysis and Machine Intelligence, 29*, 286–299.
13. Xu, C. J., Liu, J. Z., & Tang, X. (2009). 2D shape matching by contour flexibility. *IEEE Transaction on Pattern Analysis and Machine Intelligence, 31*, 180–186.
14. El Rube, I., Alajlan, N., Kamel, M., Ahmed, M., Freeman, G. (2005). *Robust multi-scale triangle-area representation for 2D shapes*. Proc. IEEE Int. Conf. on Image Processing, Genoa, Italy, vol. 1, pp. I–545, Sep 2005.

15. Alajlan, N., El Rube, I., Kamel, M. S., & Freeman, G. (2007). Shape retrieval using triangle-area representation and dynamic space warping. *Pattern Recognition, 40*, 1911–1920.
16. McNeill, G., & Vijayakumar, S. (2006). Hierarchical Procrustes matching for shape retrieval. In *Proc. IEEE Conf. CVPR, New York, USA* (pp. 885–894).
17. Felzenszwalb, P. F., & Schwartz, J. D. (2007). Hierarchical matching of deformable shapes. In *Proc. IEEE CVPR, Minneapolis, MN, USA* (pp. 1–8).
18. Hu, R., Jia, W., Ling, H., Zhao, Y., & Gui, J. (2014). Angular pattern and binary angular pattern for shape retrieval. *IEEE Transactions on Image Processing, 23*(3), 1118–1127.
19. Hu, Dameng, Weiguo Huang, Jianyu Yang, Li Shang, & Zhongkui Zhu. (2015). Shape matching and object recognition using common base triangle area. *Computer Vision, IET, 9* (5), 769–778.
20. Yang, J., Wang, H., Yuan, J., Li, Y., & Liu, J. (2016). Invariant multi-scale descriptor for shape representation, matching and retrieval. *Computer Vision and Image Understanding, 145*, 43–58.
21. Bai, X., Wang, B., Yao, C., Liu, W., & Tu, Z. (2012). Co-transduction for shape retrieval. *IEEE Trans. on Image Processing, 21*(5), 2747–2757.
22. Premachandran, V., & Kakarala, R. (2013). Perceptually motivated shape context which uses shape interiors. *Pattern Recognition, 46*(8), 2092–2102.
23. Ling, H., Yang, X., & Latecki, L. J. (Sep. 2010). Balancing deformability and discriminability for shape matching. In *Proc. Eur. Conf. Computer Vision* (pp. 411–424). Berlin/Heidelberg: Springer.
24. Bai, Xiang, Cong Rao, & Xing gang Wang. (2014). Shape vocabulary: A robust and efficient shape representation for shape matching. *IEEE Transactions on Image Processing, 23*, 3935–3949.
25. Adamek, T., & O'Connor, N. E. (2004). A multi-scale representation method for non-rigid shapes with a single closed contour. *IEEE Transactions on Circuits and Systems for Video Technology, 14*(5), 742–743.
26. Priyanka, S., & Sudhakar, M. S. (2018). Tetrakis square tiling-based triangulated feature descriptor aiding shape retrieval. *Elsevier Journal of Digital Signal Processing, 79*, 125–135.
27. Priyanka, S., & Sudhakar, M. S. (2018). Geometrically modelled derivative feature descriptor aiding supervised shape retrieval. *Springer Journal of Applied Intelligence, 45*, 4960.
28. Bhusal, S., Bhattarai, U., & Karkee, M. (2019). Improving Pest Bird detection in a Vineyard Environment using super-resolution and deep learning. *IFAC-PapersOnLine, 52*, 18–23.
29. Zhao, Z., Xin, B., Li, L., & Huang, Z. (2017). High-power homogeneous illumination for super-resolution localization microscopy with large field-of-view. *Optics Express, 25*(12), 13382–13395.
30. Kaur, S., & Jindal, S. (2018). Content based image retrieval scheme using color, texture and shape features along edge detection.
31. Peng, Y., Jian, H., & Sun, J. (2013). A research of face super-resolution technology in low-quality environment.
32. Xia, M., & Liu, B. (2003). *"Super-resolution curve" and image registration*. 2003 IEEE international conference on acoustics, speech, and signal processing, 2003. Proceedings (ICASSP '03), 3, III-509.
33. Masoud, K. M., Persello, C., & Tolpekin, V. A. (2020). Delineation of agricultural field boundaries from Sentinel-2 images using a novel super-resolution contour detector based on fully convolutional networks. *Remote Sensing, 12*, 59.
34. Dey, E. K., & Awrangjeb, M. (2020). A robust performance evaluation metric for extracted building boundaries from remote sensing data. *IEEE Journal of Selected Topics in Applied Earth Observations and Remote Sensing, 13*, 4030–4043.
35. Hamdi-Cherif, A., Digne, J., & Chaine, R. (2018). Super-resolution of point set surfaces using local similarities. *Computer Graphics Forum, 37*, 60–70.
36. El-Henawy, I. M., & Ahmed, K. (2016). Content-based image retrieval using multiresolution analysis of shape-based classified images. ArXiv, abs/1610.02509.

37. Mistry, Y., Ingole, D. T., & Ingole, M. D. (2018). Content based image retrieval using hybrid features and various distance metric. *Journal of Electrical Systems and Information Technology, 5*, 874–888.
38. Abro, M., Talpur, S., Soomro, N. Q., & Brohi, N. A. (2019). Shape based image retrieval using fused features. IoT 2018.

Dual-Tree Complex Wavelet Transform and Deep CNN-Based Super-Resolution for Video Inpainting with Application to Object Removal and Error Concealment

Gajanan Tudavekar, Sanjay R. Patil, and Santosh S. Saraf

1 Introduction

Inpainting is a method of completing an image with missing pixels by filling in the missing regions with visually plausible information for the surrounding area [1 4]. Image inpainting can be used for image-editing tasks, such as damaged image restoration, scratch treatment, and object removal. Tremendous work has been done in image inpainting to the extent of using tools such as "content-aware fill" which are available for image inpainting. Despite tremendous work in image inpainting, it is still challenging to apply these techniques to inpaint videos due to the added time dimension. The relative lack of work in video inpainting can be attributed to high computational complexity. Nevertheless, video inpainting has numerous applications, for example, in remote sensing [5], augmented reality [6–8], diminished reality [6–8], and navigation [9–12] to name a few. The video inpainting algorithms are classified into patch- and object-based.

The patch-based methods' rationale relies on the notion of copying and pasting a small video patch into the missing area. These patch-based methods provide a feasible way of restoring texture and structure in a video initially to treat texture

G. Tudavekar (✉)
Department of Electronics & Communication Engineering, Angadi Institute of Technology and Management, Belagavi, Karnataka, India

KLS Gogte Institute of Technology, Affiliated to Visvesvaraya Technological University, Belagavi, Karnataka, India

S. R. Patil
Department of Electronics & Telecommunication Engineering, Marathwada Institute of Technology, Aurangabad, Maharashtra, India

S. S. Saraf
Department of Electronics & Communication Engineering, KLS Gogte Institute of Technology, Belagavi, Karnataka, India

A. Deshpande et al. (eds.), *Computational Intelligence Methods for Super-Resolution in Image Processing Applications*, https://doi.org/10.1007/978-3-030-67921-7_12

synthesis [13–15], and later these were extended to image inpainting [16, 17]. The patch-based approaches copy and paste the patches into the missing region in a greedy manner, leading to a lack of global coherence. Patwardhan et al. [18] extended the patch-based method to video inpainting. Nevertheless, these methods assume static [18] or constrained [19] camera motion. To maintain global coherence, the author Wexler et al. [20] optimized global energy minimization for 3D patches by performing patch search and reconstruction in succession. This method was further improved by Newson et al. [21] by introducing spatio-temporal extension to patch match to speed up patch matching. Huang et al. [22] unified the spatial patch-based optimization and the flow-field estimation to realize spatial and temporal inpainting [11]. Although these methods are identified as state of the art, their significant drawbacks are high computational complexity.

The object-based methods segment the video into foreground and background objects. Then it independently reconstructs each part and then merges the results to get an inpainted video. The author proposed homography-based algorithms based on graph cut [23]. The object-based methods' major drawback is that the unknown regions are to be copied from the known region only. Hence the object-based approaches are generally susceptible to swift appearance changes like scale variation.

A hierarchical approach is used for video inpainting [24, 25] to tackle the issue of high-computational time. In this approach, a low-resolution (LR) image is obtained and is inpainted using an exemplar-based tactics. Then the low-resolution inpainted image is converted into the high-resolution (HR) image using a single image super-resolution (SISR).

Super-resolution (SR) is a process of obtaining a high-resolution image from one or multiple low-resolution images [26–33]. In either case, the challenge is of determining the missing high-frequency details in the input images. Zhang et al. [34] proposed edge-guided interpolation methods to preserve edge sharpness. A simplified linear minimum mean square error estimation-based algorithm was used to reduce the computational time. An edge-directed interpolation method based on Markov random field (MRF) was proposed by Li et al. [35]. This method produces smoothness along the edges and sharpness across the edges. Iterated back protection algorithm has been proposed by Deshpande et al. [27] to super-resolve iris polar images. The drawback of the interpolation-based super-resolution approach is that it introduces jaggy artifacts near the edges.

Reconstruction-based super-resolution methods can perform image processing in tasks like image enhancement, deconvolution, and image denoising. Irani et al. [36] proposed an iterative back-projection algorithm in which the error between the observed and estimated LR image is used iteratively to estimate the HR image. Joshi et al. [37] modeled the HR image as a Markov random field (MRF) and estimated model parameters for the most zoomed observation. The super-resolved image was estimated using the maximum a posteriori (MAP) estimate. Qiang et al. proposed a super-resolution algorithm-based spatially weighted total variation (TV) [38]. The algorithm was able to lower the artifacts and preserve edge information. A method to suppress blocking artifacts was proposed by Ren et al. [39]. Gaussian process regression (GPR) and TV-based methods were proposed by Deshpande et al. [31] and other authors [40–43]. This method uses a diamond search algorithm to compute the motion vectors. The GRP kernel uses the covariance function to obtain the HR images. The reconstruction-based methods suffer from discontinuity and loss of pixels near the edges.

The learning-based SR methods have a high magnification factor [33]. Tai et al. [44] proposed an SR method that combines edge-directed SR and learning-based SR. This method uses a single image for super-resolution. A single image super-resolution method was recommended by Zhang et al. [45]. The author utilizes redundancy within the same and over different scales in the LR image to obtain the HR image. Another method to super-resolve image using an external database was proposed by He et al. [46]. In this method, GPR is used to predict pixels by its neighbors. Xie et al. [47] proposed a scheme in which squared exponential covariance detects patches' similarity. The images super-resolved using this method had fewer artifacts. Dang et al. [48] utilized the input image and downsampled the input images versions to extract a set of training points by using the Min-Max algorithm. Tangent spaces were estimated from the extracted samples and were clustered into a group of manifold neighborhoods. The HR image is reconstructed using these tangent spaces. The learning-based methods require an image database and hence are computationally expensive.

This paper introduces a novel video inpainting framework that combines the dual-tree complex wavelet transform (DTCWT) and deep convolutional neural network (CNN)-based super-resolution. The process of down-sampling, inpainting, and subsequently enhancing the video using a deep CNN-based super-resolution technique reduces the video inpainting time. It improves the quality of the inpainted video. Also, autoregression avoids the discontinuities that arise due to variation in brightness or contrast of patches.

The framework from [25] builds upon the super-resolution-based video inpainting, which is sourced on exemplar-based inpainting and super-resolution. The proposed method improves on the state-of-the-art super-resolution-based inpainting methods by using deep CCN-based super-resolution to increase the quality of the inpainted LR image and reduce the execution time.

Section 2 of this work lays the foundations of the dual-tree complex wavelet transform (DTCWT), applying deep convolutional neural networks (CNN) relying on single image super-resolution (SISR) aka called DTCWT-DCNN. Section 3 displays the experimental results. Conclusions appear in Sect. 4.

2 Proposed Method

The dual-tree complex wavelet transform (DTCWT) employing deep convolutional neural networks (DCNN), called DTCWT-DCNN, for single image super-resolution (SISR) is the combination of two sequential operations, as shown in Fig. 1. The first operation fills in the missing regions of low-resolution (LR) image using Criminsi-based inpainting algorithm [49]. The second operation super-resolves the inpainted LR image to get a high-resolution (HR) image. The block diagram appears below.

Fig. 1 Proposed block diagram for video inpainting

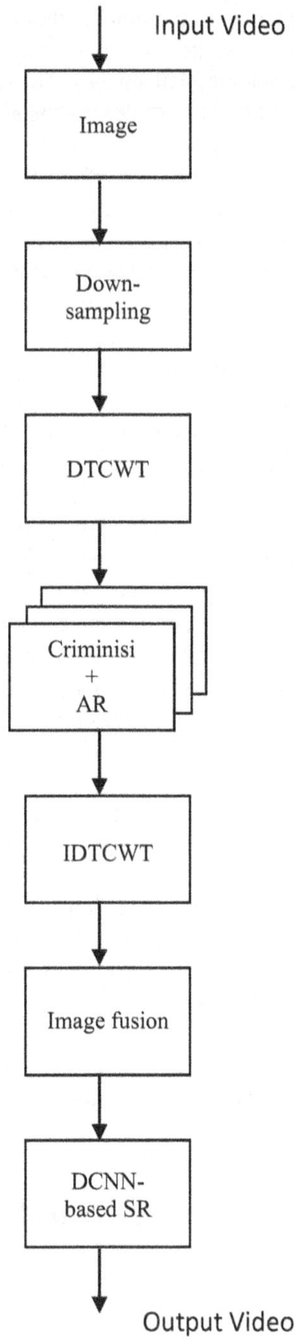

2.1 Frame Extraction and Downsampling

The frames are extracted from a video. Then, these frames are downsampled by a factor of two to get the LR images. LR image inpainting is not dependent on noise [50], and hence computational time is low compared to inpainting an HR image. After getting the LR image, subtract the current image from the previous image. If the residual is zero or less than the pixels' threshold value, the inpainting process is skipped as it indicates that the current frame is the same as the previous frame. This step avoids the inpainting of redundant frames. If the residual is non-zero, the inpainting process is carried out, as discussed in the following sections.

2.2 Dual-Tree Complex Wavelet Transform (DTCWT)

The author Kingsbury [51] suggested that dual-tree complex wavelet transform can overcome the drawbacks of traditional wavelet transforms. The complex wavelet transform (CWT) is the complex-valued continuation of traditional discrete wavelet transform (DWT). CWT uses complex value filtering that breaks up the signal into real and imaginary parts in the transform domain. The phase and the amplitude are determined by using the real and imaginary coefficients. DTCWT has separated sub-bands for positive and negative orientations.

The DTCWT decomposes the image into 16 sub-bands; 4 bands contain the average image contents, and 12 sub-bands contain high-frequency information. All sub-bands comprise the edge information aligned at several angles. The sub-bands in the DTCWT are aligned at an angle of $\pm 15°$, $\pm 45°$, $\pm 75°$. Therefore we obtain the aligned sub-bands A1–A12. The rest of the sub-bands contains average information [52] and can be dropped.

2.3 Inpainting Method

Inpainting algorithm Criminisi's is applied to four sub-bands of the image formed after applying DTDWT. The inpainting approach is as follows:

(i) For each pixel k, on boundary $\partial\Omega$.
(ii) Determine the priority

$$P(k) = C(k)D(k), \tag{1}$$

with $C(k)$ is a confidence metric that measures the information content in the neighborhood of a pixel with a $D(k)$ data term. In this,

$$C(k) = \frac{\sum_{q\epsilon(\psi_k(R-\Omega))} C(q)}{|\,|\psi_k|\,|}, \tag{2}$$

where, $||\psi_k||$ – area of the patch ψ_k of the pixel, Ω – target region of image R. This confidence term processes a high value near the initial mask border and decays near the center of $\mathbf{\Omega}$. The algorithm initially inpaints the pixels with the most reliable neighbors. The initial conditions of confidence term $C(q)$ are calculated as follows:

$$C(q) = \begin{cases} 0, & q \in R - \Omega \\ 1, & q \in \Omega \end{cases}.$$

The confidence term is updated as below after the patch has been filled with the pixels $C(q) = C(k)$. The term $D(k)$ takes care of structures in ψ_k. In this method, a structure containing tensors [38] calculates the data term

$$D(k) = \left\|\mathbf{W_p}\,\overrightarrow{\mathbf{n_k}}\right\|, \tag{3}$$

where, $\boldsymbol{W_p} = \sum_{q\epsilon(\psi_k \cap (I-\Omega))} \boldsymbol{w}\overrightarrow{\nabla \boldsymbol{I}_k}\,\overrightarrow{\nabla \boldsymbol{I}_k}^T$. The color gradient $\overrightarrow{\nabla \boldsymbol{I}_k}$ at k, $\boldsymbol{n_k}$ is a unit vector orthogonal to boundary $\partial\Omega$ at k, and w is a normalized 2D Gaussian function centered at p.

(iii) Once computing the priorities, the patch $\psi_{k'}$ about the pixel k' with the highest priority is selected for inpainting. As the pixels on the region's boundary $\partial\Omega$ get more priority to be inpainted, the selected patch $\psi_{k'}$ will invariably be made up of the known as well as unknown pixels. Hence, we need to use contents or similar patches called exemplars. The patch $\psi_{k'}$ is compared with a patch $\psi_{l'}$ around every pixel l in the entire image to find an exemplar. The larger the image, the more is the time required to search an exemplar. Nevertheless, similar patches may be found in the nearby region. Instead of searching the whole image, the search for candidate patches is restricted to a large-sized search window $\boldsymbol{W}_{k'}$ of size 90×90 (set empirically) around the patch that is to be filled up. This step reduces the number of computations required. The correlation between patches is computed by employing the sum of squared differences (SSD). The patch $\psi_{l'}$ which gives the minimum SSD is considered as an exemplar E_k. Due to variation in intensity or contrast, the SSD calculated for $E_{k'}$ will be high. In such cases, patches will be visible in the inpainted image. To achieve better inpainting, estimate the neighborhood pixels using autoregression (AR) model.

(iv) Estimate the missing pixels in the $\psi_{k'}$ with the help of the exemplar $E_{k'}$ and AR parameters. The methods employed by Zenzo et al. [53] and Perez et al. [54] are used to copy the pixels from the known area in an image into the unknown area in the same or another image. When a patch is inpainted, it is considered as a part of a known region. The subsequent iterations are carried on the updated unknown area, and the unknown area narrows with every iteration. The inpainting is said to be completed when there are no more missing pixels to be inpainted.

(v) Repeat the above steps for all target pixels in each sub-band at the current level until all the missing pixels are filled.

2.4 *Inverse Dual-Tree Complex Wavelet Transform (IDTCWT)*

The wavelet images are inpainted by patch size having $7 \times 7, 9 \times 9$, and 11×11 and also 11×11 rotated by 1800 to allow the filling order to robustify the proposed method. By this, four inpainted LR images are obtained. Then inverse DTCWT is applied to reconstruct the original image.

2.5 *Image Fusion*

The obtained LR images are combined to get one LR image using the Fusion technique based on variance calculated in the discrete cosine transform (DCT) domain [55]. The DCT coefficients are calculated for image blocks of size 8×8 pixels, and the Fuzzy logic-based histogram equalization [56] is used to enhance the image. This method preserves image brightness and improves the local contrast of the original image.

2.6 *Super-Resolution*

Super-resolution is a technique to reconstruct the high-resolution image using low-resolution images. The high-frequency content lost during the image acquisition process must be recovered in the super-resolution techniques.

The super-resolution algorithm's primary concern is to reproduce high-resolution images and reconstruct high-quality images from blurred, noisy, and corrupted images. Super-resolution algorithms can be classified as interpolation-based, example-based, and reconstruction-based algorithms. The interpolation-based methods are fast, but the accuracy is less. The example-based SR methods are computationally quick, and the quality of the super-resolved image is better than that of interpolation-based methods. The reconstruction-based SR methods use prior information to reduce the solution space to generate fine details of the image. The performance of these methods deteriorates when the scale factor grows. Another drawback of these methods is that they are time consuming. Over the last few years, deep learning (DL)-based SISR methods have exhibited superior accuracy and reduced execution time. The traditional deep models require enormous computational resources, therefore unsuitable for devices like mobiles and tablets.

In DL-based SISR, the choice of the activation function has a significant effect on deep networks' performance. Currently, ReLU is the most popular activation function. The author [57] proposed a Swish Function that matches or outperforms ReLU on deep networks. The Swish function's advantage is that apart from being unbounded above and bounded below, it is also non-monotonic and smooth. The non-monotonicity Swish increases expressivity and improves gradient flow, and this property provides robustness to different initializations and learning rates. The proposed method incurs a lower computational expense by using deep CNN with residual net (ResNet), Skip connection, and Network in Network. The proposed SISR CCN model is inspired by Yamanaka et al. [58], consisting of a feature extraction layer and reconstruction layer.

Feature extraction network: Traditional DL-based methods use up sampled images as input for the feature extraction network. Such techniques require a total of 25–35 layers and suffer from computational complexity. To overcome this, we use the original image as an input for the feature extraction network. Here, seven sets of 3 × 3 CNN, bias, and SWISH are cascaded. The output of the unit is passed to the next unit and the reconstruction network. In SISR, local features are more important than the global features, and hence we extract only the local feature, thereby reducing the computational time.

Image reconstruction network: Traditional transposed convolutional layer is used for image detail reconstruction and to obtain better results, these layers are stacked deeply, leading to heavy computation. To reduce the computation time, 1 × 1 CNN structure is used. The 1 × 1 CNN structure reduces the previous layer's dimensions and results in a lesser loss of information. The 1 × 1 CNN has nine times less computation than 3 × 3 layers, and hence the reconstruction network is much lighter than the other deep learning-based methods. The last CNN is represented in Fig. 2 outputs four channels, and each channel represents each corner – pixel of the upsampled image. The DCSCN reshapes the fourth LR image to an HR image, and then finally, it is added to the cubic-splined-up sampled original input image. The model learns from the residual output, which dramatically improves performance even with a shallow network.

3 Results

We use the dataset from [59] for training, validation, and testing. It has six groups of 131 real-world videos having more than 84,000 frames as separate JPEG files. One JPEG for each frame of the input video. We train our model for 150 epochs with a batch size of 64. The atom optimizer is used with $\beta = (0.9, 0.999)$ and a learning rate of 0.001. The performance of the proposed framework is evaluated for two applications: object removal and error concealment.

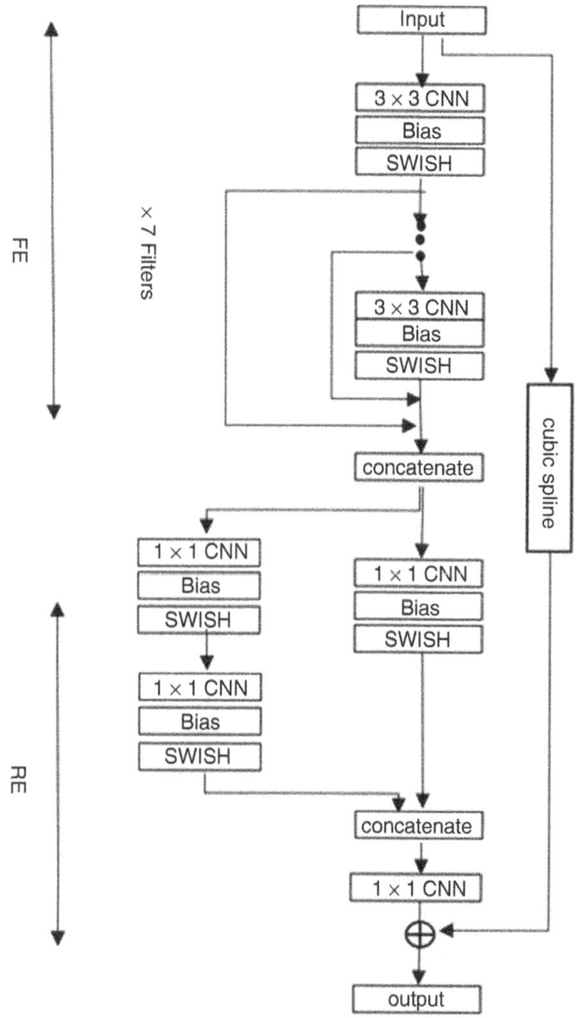

Fig. 2 Proposed CNN-based SISR model. The last CNN outputs the channels of the square of the scale factor. Then it will be reshaped to an HR image. Feature extraction network (FE), reconstruction network (RE)

Object Removal

The performance of the proposed algorithm is tested on 31 video sequences having 7700 frames. In the form of binary maps, the reference image is provided for all these videos to indicate where the changes occur. The input videos used to evaluate the proposed framework's performance appear in Table 1. The samples of the input frame and the output frame arise in Figs. 3 and 4.

For image quality analysis, a new object, as a mask, is inserted into original frames and these frames are inpainted by using the proposed framework. Then the quality of the inpainted frames is analyzed using [60–63] for peak signal-to-noise ratio (PSNR), structural similarity index matrix (SSIM), and visual information fidelity in pixel domain (VIFP). The analysis of the influence of discrete wavelet

Table 1 Input video sequences

Category	Video	Frames	Frames size	Description
General	Akiyo	300	960 × 720	Lena and Akiyo videos are manually degraded by removing a few pixels from each frame
	Akiyo	300	960 × 720	
Bad weather	Blizzard	800	720 × 576	Videos captured in bad weather
	Snowfall	400	360 × 240	
Dynamic background	Fountain02	300	320 × 240	Dynamic background
	Canoe	300	320 × 240	
	Fall	200	360 × 240	
	Overpass	200	320 × 240	

Fig. 3 Input videos

Fig. 4 Output videos

transform (DWT), non-subsampled Shearlet transform (NSST), and DTCWT on the DTCWT-DCNN design appears in Table 2. From the analysis, as shown in Table 2, DCTWT performs well compared to other algorithms. Therefore, DTCWT is used in this framework.

The super-resolved images' quality is evaluated with the *PSNR*, *SSIM*, and *VIF* metrics evaluated with Eqs. (4) and (5), respectively, [10, 11, 60, 61].

Table 2 Influence of various wavelet transform on the DTCWT-DCNN framework

Video	Algorithm	PSNR	SSIM	VIFP	Time (seconds)
Akiyo	DWT	37.02	0.892	0.871	44
	NSST	38.26	0.893	0.885	62
	DTCWT	39.28	0.934	0.898	37
Snowfall	DWT	33.92	0.781	0.802	61
	NSST	36.73	0.882	0.831	77
	DTCWT	38.17	0.913	0.883	49
Blizzard	DWT	32.21	0.772	0.732	47
	NSST	35.81	0.846	0.833	71
	DTCWT	37.25	0.905	0.875	37

$$\mathrm{PSNR} = 10\log_{10}\left(\frac{R^2}{\mathrm{MSE}}\right), \quad \text{and} \tag{4}$$

$$\mathrm{MSE} = \frac{1}{MN}\sum_{m,n}\left[\boldsymbol{I}_1(m,n) - \boldsymbol{I}_2(m,n)\right]^2.$$

$$\mathrm{SSIM}(\boldsymbol{x},\boldsymbol{y}) = l^{\alpha}(\boldsymbol{x},\boldsymbol{y})\, c^{\beta}(\boldsymbol{x},\boldsymbol{y})\, s^{\gamma}(\boldsymbol{x},\boldsymbol{y}), \tag{5}$$

where

$$l(\boldsymbol{x},\boldsymbol{y}) = \frac{2\mu_x\mu_y + C_1}{\mu_x^2 + \mu_y^2 + C_1} \tag{6}$$

$$\boldsymbol{c}(\boldsymbol{x},\boldsymbol{y}) = \frac{2\sigma_x\sigma_y + C_2}{\sigma_x^2 + \sigma_y^2 + C_2}, \quad \text{and} \tag{7}$$

$$s(\boldsymbol{x},\boldsymbol{y}) = \frac{\sigma_{xy} + C_3}{\sigma_x\sigma_y + C_3}. \tag{8}$$

with μ_x, μ_y, σ_x, σ_y, and σ_{xy} representing the local means, standard deviations, and cross-covariance for images $\boldsymbol{x}$ and $\boldsymbol{y}$ if $\alpha = \beta = \gamma = 1$.

The distortion model relies on additive noise model:

$$\boldsymbol{Y} = \boldsymbol{HX} + \boldsymbol{N}, \tag{9}$$

Where $\boldsymbol{Y}$ is the output image, $\boldsymbol{X}$ is the input image distorted by $\boldsymbol{H}$, and $\boldsymbol{N}$ is a stationary additive zero-mean Gaussian noise with variance $\boldsymbol{C}_N = \sigma_N{}^2\boldsymbol{I}$. Let

$\boldsymbol{C}_N = \{C_1, \ldots, C_N\}$ designate the vector containing all blocks from a given sub-band.

Vectors $\boldsymbol{S}_N$, $\boldsymbol{B}_N$, and $\boldsymbol{E}_N$ be defined similarly.

$\boldsymbol{s}_N$ be the maximum likelihood estimate of $\boldsymbol{S}_N$ given $\boldsymbol{C}_N$ and $\boldsymbol{C}_U$.

The amount of information extracted from the reference J becomes

$$J(\boldsymbol{C}_N;\boldsymbol{E}_N|\boldsymbol{S}_N=\boldsymbol{s}_N)=\frac{1}{2}\sum_{i=1}^{N}\log_2\left(\frac{\left|s_i^2\boldsymbol{C}_U+\sigma_N^2\boldsymbol{I}\right|}{\left|\sigma_N^2\boldsymbol{I}\right|}\right), \tag{10}$$

and the amounts of information mined from the test imagery become

$$J(\boldsymbol{C}_N;\boldsymbol{E}_N|\boldsymbol{S}_N=\boldsymbol{s}_N)=\frac{1}{2}\sum_{i=1}^{N}\log_2\left(1+\frac{s_i^2\lambda_k}{\sigma_N^2}\right),\quad \text{and} \tag{11}$$

$$J(\boldsymbol{C}_N;\boldsymbol{F}_N|\boldsymbol{S}_N=\boldsymbol{s}_N)=\frac{1}{2}\sum_{i=1}^{N}\log_2\left(1+\frac{g_i^2s_i^2\lambda_k}{\sigma_v^2+\sigma_N^2}\right). \tag{12}$$

In the previous expressions, $\boldsymbol{E}$ and $\boldsymbol{F}$ denote the output image or video as perceived by the HVS model for the reference and the test images or videos, respectively. Hence, the VIF can be defined as follows:

$$\text{VIF}=\frac{\sum_{j\in\text{sub-bands}} J\left(\boldsymbol{C}_{Nj};\boldsymbol{F}_{Nj}|\boldsymbol{S}_{Nj}=\boldsymbol{s}_{Nj}\right)}{\sum_{j\in\text{sub-bands}} J\left(\boldsymbol{C}_{Nj};\boldsymbol{E}_{Nj}|\boldsymbol{S}_{Nj}=\boldsymbol{s}_{Nj}\right)}. \tag{13}$$

Super-Resolution

SR plays a significant role in reducing the execution time of the framework for inpainting the video sequences. The DTCWT-DCNN methodology turned out to be efficient and faster Single Image Super-Resolution. In traditional methods, upsampled images are used as input to the CCN model, leading to a more significant number of CNN layers and are computationally intense. In this approach, the original image is used as an input to the model so that network can extract features more efficiently. This scheme for removing a feature from an original image helps us achieve a state-of-the-art performance with less computational time.

Error Concealment

The DTCWT-DCNN framework's performance is further analyzed in a context of error concealment [12, 64–66]. In this test, four videos are considered. The state-of-the-art algorithms Newson [21], Janardhana Rao [3], and Yang Li [4] are used as the comparison baseline, as shown in Table 4 of the Appendix. As mentioned in Table 4, three-loss rates are considered, namely, 10%, 25%, and 45% for each video sequence. This loss of pixels is manually done.

As summarized in Table 3, the execution time depends on frame resolution and the number of missing pixels. The execution time is more for more missing pixels. For Lena and Akiyo videos, missing pixels are very less compared to Pets and Blizzard videos. Hence, the time required to inpaint the Lena and Akiyo videos is less compared to other video sequences.

Table 3 Average execution time per frame

Video	Frames	Frames size	Loss of pixels (%)	Time (in seconds)
Akiyo	300	960 × 720	11	39
Snowfall	400	360 × 240	14	49
Blizzard	800	720 × 576	17	44
Lena	300	960 × 720	09	33
Highway	400	320 × 240	15	28
Pets	800	720 × 576	24	56
Office	200	360 × 240	16	29

From Table 4, as the loss of pixels increases, the quality of the inpainted frame decreases. However, compared to state-of-the-art algorithms, the DTCWT-DCNN performs well to increase the loss of pixels. The average execution time required to inpaint the sample video sequences are according to Table 3.

4 Conclusion

In this chapter, we present a two-stage architecture for video inpainting. The DTCWT-DCNN framework reduces the computational time without compromising on the quality of the inpainted video. The basic idea is that LR images' inpainting requires lesser computational load than that for an HR image. Therefore, in the first stage of the DTCWT-DCNN structure, a downsampled version of the input image is inpainted, and in the second stage, the DL-based super-resolution method recovers the image at the original resolution. Experiments and comparisons show that DTCWT-DCNN architecture produces better results than state-of-the-art methods. In future work, we will investigate the possible application of the DTCWT-DCNN arrangement to other restoration tasks such as video deblurring and video dehazing.

It should be pointed out that a more profound investigation on alternative architecture designs can improve the recommended system while treating noise [67–75]. This work did not take into consideration issues like Quality of Service (QoS), Quality of Experience (QoE), and cloud computing [76, 77].

Table 4 Analysis of error concealment application

Loss of pixel (in %)	Video	Quality	Newson [21]	Janardhana Rao [3]	Yang Li [4]	DTCWT-DCNN
10	Akiyo	PSNR (dB)	36.81	37.25	38.23	39.28
		SSIM	0.877	0.883	0.887	0.934
		VIFP	0.842	0.859	0.871	0.898
		Time (seconds)	89	67	53	37
25		PSNR (dB)	36.05	36.62	37.28	38.71
		SSIM	0.803	0.811	0.822	0.845
		VIFP	0.731	0.788	0.797	0.832
		Time (seconds)	91	88	74	54
45		PSNR (dB)	30.59	30.66	31.37	32.36
		SSIM	0.688	0.693	0.712	0.748
		VIFP	0.638	0.671	0.688	0.732
		Time (seconds)	133	117	97	63
10	Snowfall	PSNR (dB)	36.59	36.87	37.14	38.17
		SSIM	0.851	0.876	0.892	0.913
		VIFP	0.826	0.830	0.849	0.883
		Time (seconds)	81	68	56	49
25		PSNR (dB)	34.88	35.05	35.11	35.72
		SSIM	0.753	0.781	0.795	0.832
		VIFP	0.755	0.762	0.769	0.778
		Time (seconds)	115	89	72	59
45		PSNR (dB)	31.73	31.95	32.01	32.81
		SSIM	0.672	0.694	0.701	0.742
		VIFP	0.607	0.613	0.619	0.694
		Time (seconds)	181	152	134	101
10	Blizzard	PSNR (dB)	35.76	35.93	36.21	37.25
		SSIM	0.881	0.886	0.892	0.905
		VIFP	0.839	0.854	0.859	0.875
		Time (seconds)	65	59	48	37
25		PSNR (dB)	33.91	34.16	34.52	35.54
		SSIM	0.804	0.811	0.817	0.845
		VIFP	0.749	0.751	0.759	0.792
		Time (seconds)	115	92	79	58
45		PSNR (dB)	30.57	30.93	31.29	31.97
		SSIM	0.654	0.672	0.681	0.716
		VIFP	0.598	0.601	0.603	0.639
		Time (seconds)	152	113	101	78

References

1. Guillemot, C., & Le Meur, O. (2014). Image inpainting: Overview and recent advances. *IEEE Signal Processing Magazine, 31*(1), 127–144.
2. Bertalmio, M., Sapiro, G., Caselles, V., et al. (2000). Image inpainting. In *Proceedings of the ACM SIGGRAPH Conference on Computer Graphics (SIGGRAPH 2000)*, July 2000, pp. 417–424.
3. Rao, J. B., Chakrapani, Y., & Kumar, S. (2018). Image inpainting method with improved patch priority and patch selection. *IETE Journal of Education, 59*(1), 26–34.
4. Li, Y., Jiang, B., Lu, Y., et al. (2019). Fine-grained adversarial image inpainting with super resolution. In *International Joint Conference on Neural Networks (IJCNN)*, Budapest, Hungary, pp. 1–8.
5. Aroma, R. J., Raimond, K., Razmjooy, N., Estrela, V. V., & Hemanth, J. (2020). Multispectral vs. hyperspectral imaging for unmanned aerial vehicles: Current and prospective state of affairs. In V. V. Estrela, J. Hemanth, O. Saotome, G. Nikolakopoulos, & R. Sabatini (Eds.), *Imaging and sensing for unmanned aircraft systems* (Vol. 2, 7, pp. 133–156). London: IET. https://doi.org/10.1049/PBCE120G_ch7.
6. de Jesus, M. A., Estrela, V. V., Huacasi, W. D., Razmjooy, N., Plaza, P., & Peixoto, A. B. M. (2020). Using transmedia approaches in STEM. In *2020 IEEE Global Engineering Education Conference (EDUCON)*, pp. 1013–1016. https://doi.org/10.1109/EDUCON45650.2020.9125239.
7. Li, J., Gao, W., & Wu, Y. (2019). High-quality 3D reconstruction with depth super-resolution and completion. *IEEE Access, 7*, 19370–19381.
8. Kawai, N., Sato, T., & Yokoya, N. (2014). *From image inpainting to diminished reality*. HCI. Heraklion, Crete, Greece
9. Farfan, W. S., Saotome, O., Estrela, V. V., & Razmjooy, N. (2020). Integrated optical flow for situation awareness, detection and avoidance systems in UAV systems. In V. V. Estrela, J. Hemanth, O. Saotome, G. Nikolakopoulos, & R. Sabatini (Eds.), *Imaging and sensing for unmanned aircraft systems* (Vol. 1, 3, pp. 47–74). London: IET. https://doi.org/10.1049/PBCE120F_ch3.
10. Sidorov, O., & Hardeberg, J. (2019). Deep hyperspectral prior: Single-image denoising, inpainting, super-resolution. In *2019 IEEE/CVF International Conference on Computer Vision Workshop (ICCVW)*, pp. 3844–3851.
11. Zeng, Y., Fu, J., & Chao, H. (2020). Learning joint spatial-temporal transformations for video inpainting. ECCV, arXiv:2007.10247.
12. Razmjooy, N., & Ramezani, M. (2016). Training wavelet neural networks using hybrid particle swarm optimization and gravitational search algorithm for system identification. *International Journal of Mechatronics, Electrical and Computer Technology, 6*(21), 2987–2997.
13. Efros, A., & Leung, T. (1999). Texture synthesis by non-parametric sampling. In *IEEE International Conference on Computer Vision (ICCV)*, pp. 1033–1038.
14. Harrison, P. (2001). A non-hierarchical procedure for re-synthesis of complex texture. In *International Conference in Central Europe on Computer Graphics, Visualization and Computer Vision (WCSG)*.
15. Ashikhmin, M. (2001). Synthesizing natural textures. In *ACM Symposium on Interactive 3D Graphics*, pp. 217–226.
16. Yamauchi, H., Haber, J., & Seidel, H.-P. (2003). Image restoration using multiresolution texture synthesis and image inpainting. In *IEEE Computer Graphics International (CGI)*, pp. 120–125.
17. Rivera, L. A., Estrela, V. V., Carvalho, P. C. P., & Velho, L. (2004). Oriented bounding boxes based on multiresolution contours, Journal of WSCG. In *Proceedings of the 12-th International Conference in Central Europe on Computer Graphics, Visualization and Computer Vision '2004, WSCG 2004*, University of West Bohemia, Campus Bory, Plzen-Bory, Czech Republic, February 2–6, 2004 (Short Papers), pp. 219–212.

18. Patwardhan, K. A., Sapiro, G., & Bertalmio, M. (2005). Video inpainting of occluding and occluded objects. *IEEE International Conference on Image Processing (ICIP), 2*, 69–72.
19. Patwardhan, K. A., Sapiro, G., & Bertalmio, M. (2007). Video inpainting under constrained camera motion. *IEEE Transactions on Image Processing (TIP), 16*(2), 545–553.
20. Wexler, Y., Shechtman, E., & Irani, M. (2007). Space-time completion of video. In *IEEE Transactions on Pattern Analysis and Machine Intelligence (PAMI)*, pp. 463–476.
21. Newson, A., Almansa, A., Fradet, M., et al. (2014). Video inpainting of complex scenes. *Journal on Imaging Sciences, 7*(4), 1993–2019.
22. Huang, J., & Tang, X. (2016). A fast video inpainting algorithm based on state matching. In *9th International Congress on Image and Signal Processing, BioMedical Engineering and Informatics (CISP-BMEI)*, Datong, pp. 114–118.
23. Granados, M., Kim, K. I., Tompkin, J., et al. (2012). Background inpainting for videos with dynamic objects and a free-moving camera. In *Proceedings of the 12th European Conference on Computer Vision (ECCV '12)*, pp. 682–695.
24. Le Meur, O., Ebdelli, M., & Guillemo, C. (2013). Hierarchical super-resolution-based inpainting. *IEEE Transactions on Image Processing, 22*(10), 3779–3790.
25. Tudavekar, G., & Patil, S. R. (2016). Super resolution based video inpainting. In *2016 IEEE International Conference on Computational Intelligence and Computing Research (ICCIC)*, Chennai, pp. 1–3. https://doi.org/10.1109/ICCIC.2016.7919586.
26. Deshpande, A., & Patavardhan, P. (2017). Super resolution of long range captured multiframe iris polar images. *IET Biometrics, 6*(5), 360–368.
27. Deshpande, A., & Patavardhan, P. (2017). Multiframe super-resolution for long range captured iris polar image. *IET Biometrics, 6*(2), 108–116.
28. Deshpande, A., & Patavardhan, p. (2016). Single frame super resolution of non-cooperative iris images. *ICTACT Journal on Image and Video Processing, 7*(2), 1362–1365.
29. Deshpande, A., Patavardhan, P., & Rao, D. H. (2015). Iterated back projection based super-resolution for iris feature extraction, Elsevier. *Procedia Computer Science, 48*, 269–275.
30. Deshpande, A., Patavardhan, P., & Rao, D. H. (2014). Super-resolution for iris feature extraction. In *IEEE International Conference on Computational Intelligence and Computing Research*, pp. 1–4.
31. Deshpande, A., Patavardhan, P., & Rao, D. H. (2016). Gaussian Process Regression based iris polar image super resolution. In *International Conference on Applied and Theoretical Computing and Communication Technology*, pp. 692–696.
32. Deshpande, A., Patavardhan, P., Estrela, V. V., & Razmjooy, N. (2020). Deep learning as an alternative to super-resolution imaging in UAV systems. In V. V. Estrela, J. Hemanth, O. Saotome, G. Nikolakopoulos, & R. Sabatini (Eds.), *Imaging and sensing for unmanned aircraft systems* (Vol. 2, 9, pp. 177–212). London: IET. https://doi.org/10.1049/PBCE120G_ch9.
33. Deshpande, A., Patavardhan, P., & Rao, D. H. (2019). Survey of super resolution techniques. *ICTACT Journal on Image & Video Processing, 9*(3), 1927–1934.
34. Zhang, D., & Wu, X. (2006). An edge-guided image interpolation algorithm via directional filtering and data fusion. *IEEE Transactions on Image Processing, 15*(8), 2226–2238.
35. Li, X., & Nguyen, T. Q. (2008). Markov random field model based edge directed image interpolation. *IEEE Transactions on Image Processing, 7*(7), 1121–1128.
36. Irani, M., & Peleg, S. (1991). Improving resolution by image registration. *CVGIP: Graphical Models and Image Processing, 53*(3), 231–239.
37. Joshi, M. V., Chaudhuri, S., & Panuganti, R. (2005). A learning-based method for image super-resolution from zoomed observations. *IEEE Transactions on Systems, Man, and Cybernetic, Part B: Cybernetics, 35*(3), 441–456.
38. Yuan, Q., Zhang, L., & Shen, H. (2012). Multiframe super-resolution employing a spatially weighted total variation model. *IEEE Transactions on Circuits and Systems for Video Technology, 22*(3), 561–574.
39. Ren, Z., He, C., & Zhang, Q. (2013). Fractional order total variation regularization for image super-resolution. *Signal Processing, 93*(9), 2408–2421.

40. Estrela, V. V., Magalhaes, H. A., & Saotome, O. (2016). Total variation applications in computer vision. In N. K. Kamila (Ed.), *Handbook of research on emerging perspectives in intelligent pattern recognition, analysis, and image processing*. Hershey: IGI Global. https://doi.org/10.4018/978-1-46668654-0.ch002.
41. Mejia, J., Mederos, B., Ortega, L., Gordillo, N., & Avelar, L. (2017). Small animal PET image super-resolution using Tikhonov and modified total variation regularisation. *The Imaging Science Journal, 65*(3), 162–170.
42. Zhang, Y., Tuo, X., Huang, Y., & Yang, J. (2020). A TV forward-looking super-resolution imaging method based on TSVD strategy for scanning radar. *IEEE Transactions on Geoscience and Remote Sensing, 58*, 4517–4528.
43. Villena, S., Vega, M., Babacan, S. D., Molina, R., & Katsaggelos, A. (2013). Bayesian combination of sparse and non-sparse priors in image super resolution. *Digital Signal Processing, 23*, 530–541.
44. Tai, Y.-W., Liu, S., Brown, M. S., & Lin, S. (2010). Super resolution using edge prior and single image detail synthesis. In *Proceedings of IEEE Conference on Computer Vision and Pattern Recognition*, pp. 23–29.
45. Zhang, K. (2013). Single image super-resolution with multi-scale similarity learning. *IEEE Transactions on Neural Network Learning System, 24*(10), 1648–1659.
46. He, H., & Siu, W. C. (2011). Single image super-resolution using Gaussian process regression. In *Proceedings of IEEE Conference Proceedings on Pattern Recognition*, pp. 449–456.
47. Li, J., Qu, Y., Li, C., Xie, Y., Wu, Y., & Fan, J. (2015). Learning local Gaussian process regression for image super resolution. *Neurocomputing, 154*, 284–295.
48. Dang, C., Aghagolzadeh, M., & Radha, H. (2014). Image super-resolution via local self-learning manifold approximation. *IEEE Signal Processing Letters, 21*(10), 1123–1138.
49. Criminisi, A., Perez, P., & Toyama, K. (2004). Region filling and object removal by exemplar-based image inpainting. *IEEE Transactions on Image Processing, 13*(9), 1200–1212.
50. Le Meur, O., Ebdelli, M., & Guillemot, C. (2013). Hierarchical super-resolution-based inpainting. *IEEE Transactions on Image Processing, Institute of Electrical and Electronics Engineers (IEEE), 22*(10), 3779–3790.
51. Kingsbury, N. (2000). A dual-tree complex wavelet transform with improved orthogonality and symmetry properties. In *Proceedings 2000 International Conference on Image Processing* (Cat. No.00CH37101), Vancouver, Sept. 2000, pp. 375–378.
52. Patil, S. R., & Talbar, S. N. (2012). Multiresolution analysis using complex wavelet and curvelet features for CBIR. *International Journal of Computer Applications, 47*(17), 6–10.
53. Di Zenzo, S. (1986). A note on the gradient of a multi-image. *Computer Vision, Graphics, and Image Processing, 33*(1), 116–125.
54. Pérez, P., Gangnet, M., & Blake, A. (2003). Poisson image editing. *ACM Transactions on Graphics, 22*(3), 313–318.
55. Haghighat, M., Aghagolzadeh, A., & Seyedarabi, H. (2011). Multi-focus image fusion for visual sensor networks in DCT domain. *Computers and Electrical Engineering, 37*, 789–797.
56. Magudeeswaran, V., & Ravichandran, C. G. (2013). Fuzzy logic-based histogram equalization for image contrast enhancement. *Mathematical Problems in Engineering, 2013*(5), 891864.
57. Ramachandran, P., Zoph, B., & Le, Q. V. (2018). Searching for activation functions. arXiv preprint arXiv:1710.05941.
58. Yamanaka, J., Kuwashima, S., & Kurita, T. (2017). Fast and accurate image super resolution by deep CNN with skip connection and network in network. In *Proceedings of NIPS* (pp. 217–225). Cham: Springer.
59. http://jacarini.dinf.usherbrooke.ca/dataset2014/. Last accessed in April 2020.
60. Wang, Z., Bovik, A. C., Sheikh, H. R., et al. (2004). Image quality assessment: From error visibility to structural similarity. *IEEE Transactions on Image Processing, 13*(4), 600–612.
61. Sheikh, H. R., Bovik, A. C., & de Veciana, G. (2005). An information fidelity criterion for image quality assessment using natural scene statistics. *IEEE Transactions on Image Processing, 14*, 2117–2128.

62. Lukes, T., Fliegel, K., Klíma, M., et al. (2013). Performance evaluation of image quality metrics with respect to their use for super-resolution enhancement. In *IEEE Fifth International Workshop on Quality of Multimedia Experience (QoMEX)*, Klagenfurt, July 2013, pp. 42–43.
63. Zhou, X., & Bhanu, B. (2008). Evaluating the quality of super-resolved images for face recognition. In *IEEE Computer Society Conference on Computer Vision and Pattern Recognition Workshops*, Anchorage, AK, June 2008, pp. 1–8.
64. Coelho, A. M., Estrela, V. V., Carmo, F. P., & Fernandes, S. R. (2012). Error concealment by means of motion refinement and regularized Bregman divergence. In *Proceedings of the 13th International Conference on Intelligent Data Engineering and Automated Learning*, Natal, Brazil. https://doi.org/10.1007/978-3-64232639-4_78.
65. Coelho, A. M., de Assis, J. T., & Estrela, V. V. (2009). Error concealment by means of clustered blockwise PCA. In *Proceedings of IEEE 2009 Picture Coding Symposium (PCS 2009)*. https://doi.org/10.1109/PCS.2009.5167442.
66. Tudavekar, G., Patil, S., & Saraf, S. (2020). Dual-tree complex wavelet transform and super-resolution based video inpainting application to object removal and error concealment. *CAAI Transactions on Intelligence Technology, 5*(4), 314–319.
67. Razmjooy, N., Estrela, V. V., & Loschi, H. J. (2020). Entropy-based breast cancer detection in digital mammograms using world cup optimization algorithm. *International Journal of Swarm Intelligence Research (IJSIR), 11*(3), 1–8.
68. Zhao, H., Li, H., Maurer-Stroh, S., & Cheng, L. (2018). Synthesizing retinal and neuronal images with generative adversarial nets. *Medical Image Analysis, 49*, 14–26.
69. Yang, G., Ye, X., Slabaugh, G., Keegan, J., Mohiaddin, R., & Firmin, D. (2016). Combined self-learning based single-image super-resolution and dual-tree complex wavelet transform denoising for medical images. *SPIE Medical Imaging, 9784*, 1–7.
70. Chen, B., Cui, J., Xu, Q., Shu, T., & Liu, H. (2019). Coupling denoising algorithm based on discrete wavelet transform and modified median filter for medical image. *Journal of Central South University, 26*, 120–131.
71. Frischer, R., Krejcar, O., Selamat, A., & Kuča, K. (2020). 3D surface profile diagnosis using digital image processing for laboratory use. *Journal of Central South University, 27*, 811–823.
72. Varma, D., Mishra, S., & Meenpal, A. (2020). An adaptive image steganographic scheme using convolutional neural network and dual-tree complex wavelet transform. In *2020 11th International Conference on Computing, Communication and Networking Technologies (ICCCNT)*, pp. 1–7.
73. Si, Y., Zhang, Z., Kong, C., Li, S., Yang, G., & Hu, B. (2020). Looseness condition feature extraction of viscoelastic sandwich structure using dual-tree complex wavelet packet-based deep autoencoder network. *Structural Health Monitoring, 19*, 873–884.
74. Shivagunde, S., & Biswas, M. (2019). Saliency guided image super-resolution using PSO and MLP based interpolation in wavelet domain. In *2019 International Conference on Communication and Electronics Systems (ICCES)*, pp. 613–620.
75. Li, W., Wei, W., & Boni, L. (2020). Sparse representation of image with immune clone algorithm based on harmonic wavelet packet dictionary. *International Journal of Science and Research (IJSR), 9*(3), 519–528.
76. Laghari, A. A., Khan, A., He, H., Estrela, V. V., Razmjooy, N., Hemanth, J., et al. (2020). Quality of experience (QoE) and quality of service (QoS) in UAV systems. In V. V. Estrela, J. Hemanth, O. Saotome, G. Nikolakopoulos, & R. Sabatini (Eds.), *Imaging and sensing for unmanned aircraft systems* (Vol. 2, 10, pp. 213–242). London: IET. https://doi.org/10.1049/PBCE120G_ch10.
77. Estrela, V. V., Hemanth, J., Loschi, H. J., Nascimento, D. A., Iano, Y., & Razmjooy, N. (2020). Computer vision and data storage in UAVs. In V. V. Estrela, J. Hemanth, O. Saotome, G. Nikolakopoulos, & R. Sabatini (Eds.), *Imaging and sensing for unmanned aircraft systems* (Vol. 1, 2, pp. 23–46). London: IET. https://doi.org/10.1049/PBCE120F_ch2.

Super-Resolution Imaging and Intelligent Solution for Classification, Monitoring, and Diagnosis of Alzheimer's Disease

Abhishek Tiwari and Alexey N. Nazarov

1 Introduction

Clinical imaging contains a set of cycles or strategies to make visual depictions of within bits of the body, for instance, to break down, and to treat sickness and wounds. Super-resolution imaging is where significant standard picture is worked from a gathering of testing limited low objective discernments. The continuous advances in significant learning structures have enabled faster and more exact disclosure [12, 13]. Time is an essential factor for clinical end, and early acknowledgment can add an extremely significant time frame to the life of a patient. Astounding comprehension is a required to eliminate an impetus from clinical imaging; yet, human interpretation is limited and slanted to bungles [14, 15]. While in clinical imaging, for instance, MRI is used near PC examination, standard MRI assessment requires significant length of figuring time. To change two MRI analyses, a PC needs to sort a large number of 3D pixels, and the opportunity has arrived exhausting to scale this up and assessment data from a colossal number of patients. Regardless, we can plan neural association to get indications of a comparable illness [16, 17]. The data are dealt with into one completion of the association and a short time later experiences various center points to convey the ideal yield. It is connected with cerebrum degeneration that causes issues, for instance, memory impedance and disarray. In 2015, more than 5,000,000 Americans were lost to this disease [12]. This segment portrays the route toward improving the display of a neural association that describes cut MRI inspects. The key point of convergence of

A. Tiwari (✉)
Shiv Nadar University, Department of Computer Science and Engineering, Greater Noida, Uttar Pradesh, India

A. N. Nazarov
Moscow Institute of Physics and Technology State University (MIPT), Phystech School of Radio Engineering and Computer Technology, Moscow, Russia

A. Deshpande et al. (eds.), *Computational Intelligence Methods for Super-Resolution in Image Processing Applications*, https://doi.org/10.1007/978-3-030-67921-7_13

this of this assessment is the use of various coordinates in the association [18, 19]. The improvement ought to begin from features that are shared between various cuts. Our fixation in this segment is to produce a model classifier with the help of convolutional neural network in deep learning plan that can mastermind the brain pictures with exact strategy known as transfer learning. Ensuring evaluation of various areas, transfer learning starts at now moving subject in the clinical field that we have found [20, 21]. Hence, our task is to describe between two classes with the help of a classifier and evaluate the presentation of classifier.

2 State of the Art/Previous Published Work

After 2013, significant learning took place in clinical science by and large, and has made an enormous impact in the field of assessment. Starting late, a huge segment of the circulated part models relies upon neural association designing. That was where the hugeness of significant learning in Alzheimer disease was revealed and dispersed segments extended rapidly in 2017. Besides clinical administrations, significant learning around there, for instance, object affirmation, object following, and sound game plan got achievement in research work and still work has been going on. The reason for using convolutional neural network in deep picking up designing is that it has a couple of channels/touches to distinguish complex edges and shapes with the help of PC vision. Through various parts, we found that transfer learning with the help of convolutional neural network is in a general sense the point of convergence of authorities, so they can manufacture a model which can predict even more definitely and have least hardship when diverged from past segments. Move learning is connected to planning and associations on a specific dataset as instatement, retraining them on another dataset by changing the networks [22] (Table 1).

3 Proposed Work

As super-goal microscopy has advanced, so too has analysts' ability to dive significantly into the neural framework and the disturbs that total identified with a combination of conditions. Cerebrum disease research has seized on this development to fathom a protein complex that is ordinary in patients with Alzheimer's contamination and with threatening development. The dataset comprises 20 MRI cerebrum pictures. The training, approval, and test dataset are split into 16, 2, and 2 MRI pictures. Through past overview parts, plainly profound learning design is the most broadly utilized strategy for clinical picture examination when contrasted with different strategies. It conquers the disadvantages of recently utilized innovation. The test that this informational collection offers is to construct an exact classifier that can group singular mind pictures into one of three classes: normal (CN) and Alzheimer's disease (AD). The output sets can be utilized in various manners to prepare the classifier. This classifier in profound learning is the convolutional neural

Table 1 Characterization of Alzheimer's MRI utilizing convolutional neural network

Authors	Procedures	Restriction	Dataset	Determination
Amir Ebrahimi-Ghahnavieh et al. (20I9) [12]	2D CNN with LSTM or RNN	Brain shrinkage happens due to both aging and AD. It is very difficult to classify AD from healthy old people only based on MRI scans	AD Neuroimaging Initiative Dataset	LSTM or RNN
Hasan Ucuzal et al. (2019) [22]	To develop an open-source software for DL classification of dementia in MRI scans	Improves performance with more accurate results	Open Access Series of Imaging Studies-2 (OASIS-2)	Performance is evaluated for accuracy, sensitivity, specificity, positive, and negative predictive values
Muazzam Maqsood et al. (2019) [23]	It deals with an effective procedure using figuring out how to group the pictures by tweaking a pre-prepared CNN, AlexNet	The detection of AD stages remains a difficult problem because of the multiclass classification task	Open Access Series of Imaging Studies (OASIS) dataset	The re-trained CNN was then validated using the testing data, giving overall accuracies of 89.6% and 92.8% for binary and multi-class problems, respectively
Xin Hong et al. (2019) [24]	It proposes a predicting model based on LSTM with fully connected layer. An activation layer shows relation between features and next stage of the AD	Improves performance with more accurate results	AD Neuroimaging Initiative Dataset	Compared with existing approaches, this model can carry out future state prediction for AD, rather than classifying the state of a current diagnosis
Shrikant Patro et al. (2019) [25]	They proposed an image processing technique to process MRI of brain from a different plane	Detecting only single image at a time. Neural network and ML can produce more accurate and better results	AD Neuroimaging Initiative Dataset	The amount of enlargement will classify the patient stage: 1, 2, AD, MCI
Weiming Lin et al. (2018) [26]	A DL approach based on CNN is designed to predict MCI to AD conversion	Classifier precisely predicted an MCI. Nonconverters who would not convert to AD within a specific period, but the conversion might still happen	AD Neuroimaging Initiative Dataset	They have developed a framework that only uses MRI data to predict the MCI-to-AD conversion, by applying CNN

(continued)

Table 1 (continued)

Authors	Procedures	Restriction	Dataset	Determination
		half year or even a month later		
Jyoti Islam et al. (2018) [27]	They propose a deep CNN for AD diagnosis using brain MRI data analysis	Improves performance and produces more accurate results	Open Access Series of Imaging Studies (OASIS) dataset	Planned network can be very beneficial for early-stage AD diagnosis.
Biju K.S. et al. (2017) [28]	A modified approach based on the watershed algorithm used for segmenting the hippocampus region	It does not show the stage of the Alzheimer disease and how much brain is affected by the disease	AD Neuroimaging Initiative Dataset	The major requirement was to detect the disease as early as possible. It is cost-effective and requires no expertise
R. Anitha et al. (2016) [29]	A software for detecting the brain abnormalities for detection of AD. It produces 3D representation of the brain MRI slices	It uses a highly sophisticated watershed algorithm to find the diseased area in the scanned image	AD Neuroimaging Initiative Dataset	The outcome of this watershed algorithm of the brain scan is analyzed
Luis Javier Herrera et al. (2013) [30]	The methodology includes wavelet feature extraction, dimensionality reduction, training test and classification using SVMs	Identification of AD and the condition prior to dementia, which is mild cognitive impairment (MCI) with the development of intelligent classifiers	AD Neuroimaging Initiative Dataset	SVM has been used as classification technique and results obtained have shown to be promising

network that contains a few unique channels/bits comprising teachable boundaries which can convolve on a given picture spatially to distinguish features like edges and shapes. These high amounts of channels fundamentally sort out some way to get spatial features from the image reliant on the informed loads, and stacked layers of channels can be used to perceive complex spatial shapes from features. The CNN involves four sections that are convolution, max-pooling, activation work, and totally related layers. Each part has different positions and plays out their own tasks so CNN model can fabricate. Resulting in building this model, the photos are rescaled. We will use learning rate in slant dive. Learning rate is a hyper-limit that controls the sum to change the model on account of the evaluated botch each time the model burdens are invigorated. In case learning rate is little result in a long getting ready measure that could slow down out, however a value too huge may achieve learning a hazardous course of action of burdens too fast or an inconsistent planning measure. The 3D layers are used in order to get more information out of the cerebrum checks. They moreover consolidate an assessment with various associations. Note that their association can perceive two unmistakable classes while holding a high exactness (Fig. 1).

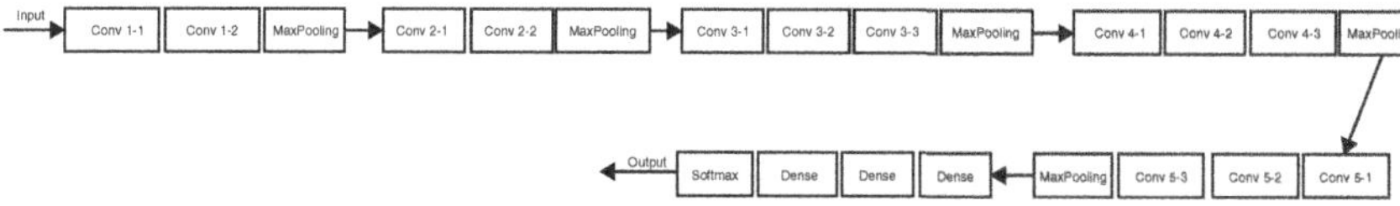

Fig. 1 Architecture of the proposed model

4 Model

The 3D CNN used sequential model, which has max pooling, fully connected layers, convolution layers, and finally, soft max output layer. Since it includes multiple layers, 3D CNNs are hypothesized as the best classification for the Alzheimer's MRI scans dataset. The features stretch over multiple layers and are highly correlated. Pooling operation and 3D convolution are performed spatially over multiple layers instead of only spatially. A suitable architecture for a 3D convolutional model is identified in this section.

Due to the time-consuming experiments of 3d CNNs, it can be initiated to find good architecture with a smaller dataset. Full-sized dataset was then verified for the final architecture. The experiments are developed using a neural network library for python, Keras using a tensor flow backend and Spyder platform. This tool allows for experimentation [23] and performing TL with ImageNet VGG16 architecture. An image is read and processed by image Data Generators. Further experiment was carried on with the size of the batch and the network's depth. Between the quality of the network and the speed at which the network can be trained, there is a trade-off. Speed above quality was selected for the resulting parameters because of limited resources and time constraints. The convolutional network layers have a batch size of 32 with each followed by layer of pooling. Upon completion of the convolution part, there is a flattened layer followed by two fully connected layers which use ReLU activation function. Finally, a fully connected layer using "Softmax" activation function is processed in order to get two classes. By comparing the performance of model to models proposed by other chapters, the recommended model will be evaluated. The accuracy and loss of this model will be compared with the other models. It can be noticed in the confusion matrix that no clinically normal patients (CN) are classified with AD.

4.1 Define Train/Test Data and Classes

The number of training images per category in normal and AD brain images appears in Fig. 2.

The given dataset is divided into training and test data and stored in the directory as shown in Figs. 3 and 4.

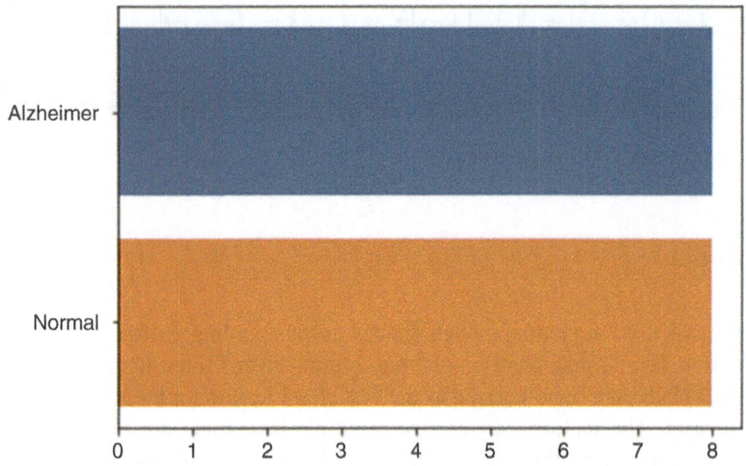

Fig. 2 Number of training data per category

Fig. 3 Sample data in data/train\Normal

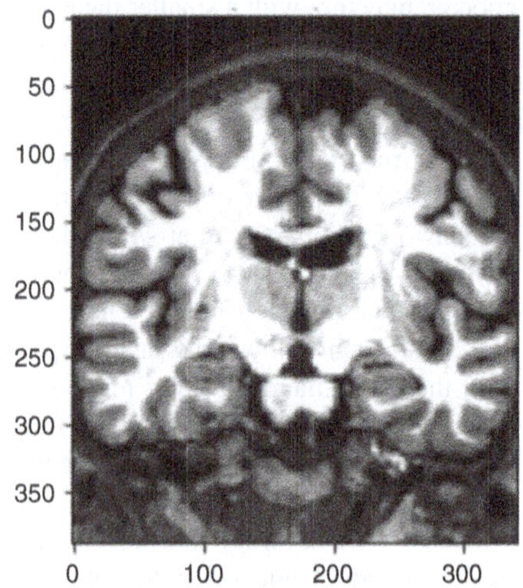

4.2 *Transfer Learning (TL)*

We use a prearranged significant learning model (VGG16) as the explanation behind our image classifier model, and a while later, retrain the model on our own data, for instance, move learning.

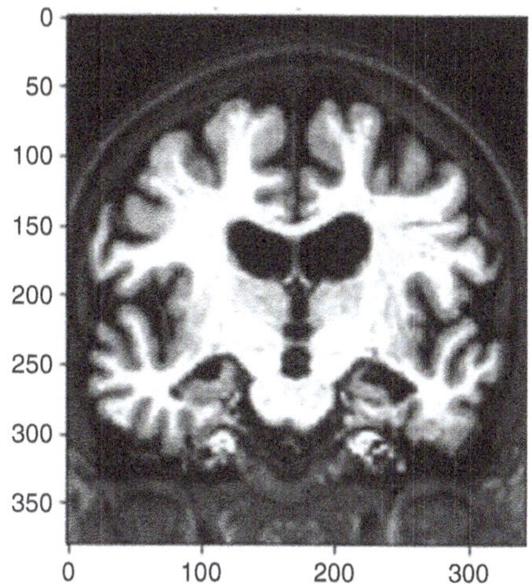

Fig. 4 Sample data in data/train\Alzheimer

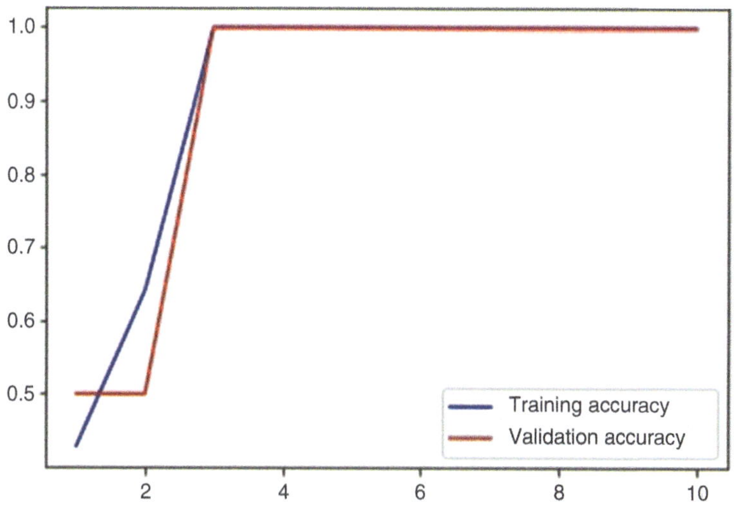

Fig. 5 Training and validation accuracy

4.3 Assess Model Accuracy

Subsequent to performing move learning, the model is assessed on train and test information as indicated by the exactness and misfortune measures. In the event that the exactness at one point ranges to one and misfortune to zero, then it is considered an exceptionally precise model as it appears in Figs. 5 and 6.

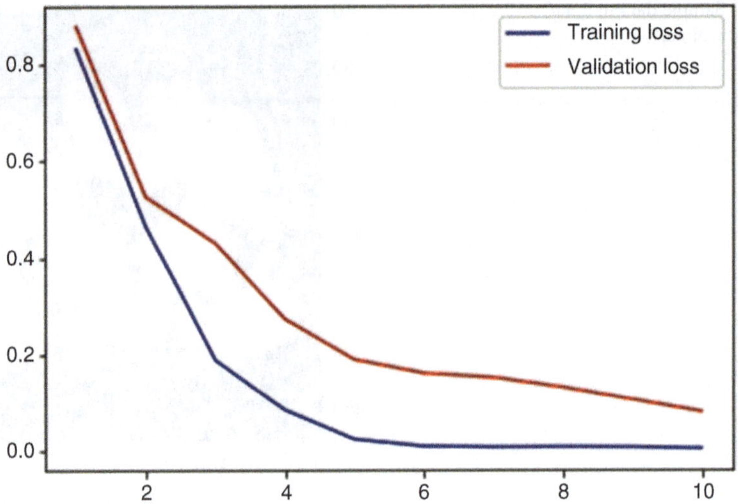

Fig. 6 Training and validation loss

4.4 Plot Confusion Matrix

A confusion matrix is a table that is often used to describe the performance of a classification model on a set of test data for which the true values are known. It allows the visualization of the performance of an algorithm. In this section, we plot the confusion matrix for identifying the correct actual and predicted values in Fig. 7, as shown below.

4.5 Model Prediction

The predicted and actual result of brain images shows the classification between normal and AD brain images in Figs. 8 and 9.

5 Future Scope

These days, a term super-goal is developing in the field of clinical science. It is a method where high goal picture is built from a grouping of inspecting restricted low goal perceptions. Scientists have built a super-goal nanoscope that gives a 3D perspective on cerebrum particles with multiple times more noteworthy detail than ordinary magnifying lens. With the assistance of nanoscope, the organic changes in

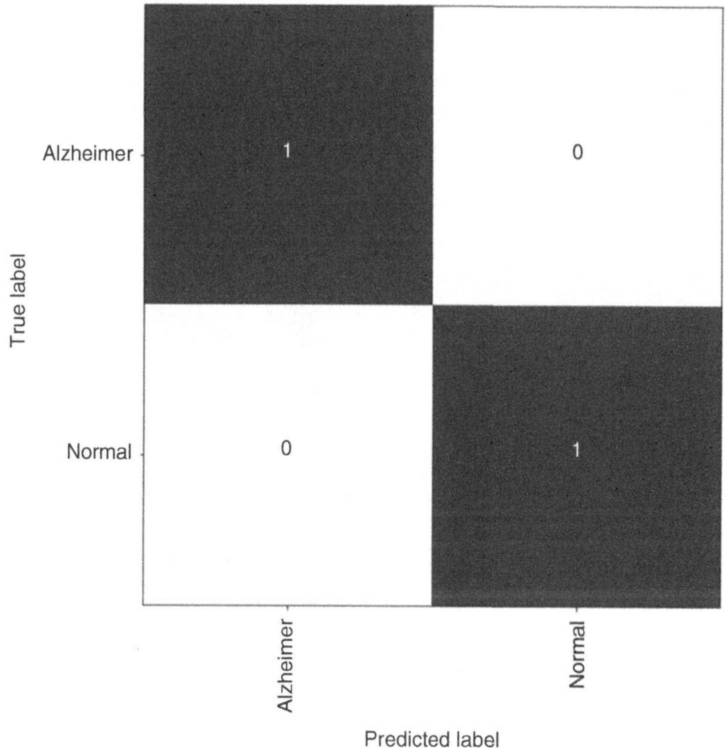

Fig. 7 Confusion matrix

Fig. 8 The actual and predicted result is positive for Alzheimer

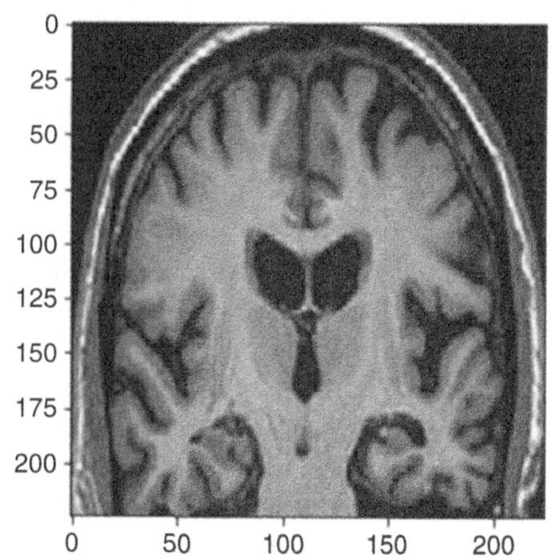

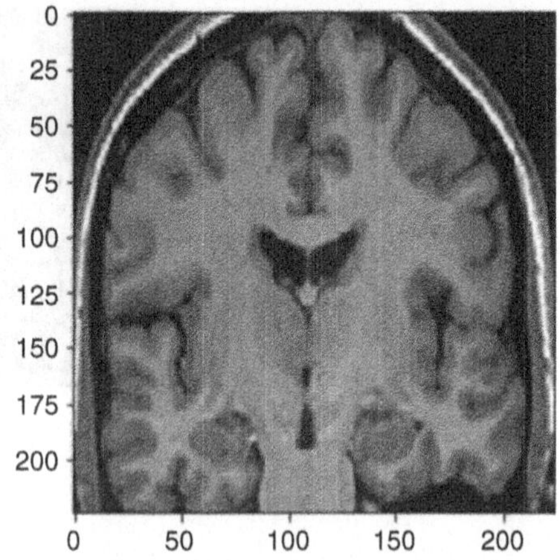

Fig. 9 The actual and predicted outcome is normal

the cerebrum can be seen and the arrangement of harming structure of mind can be halted. Our assignment is to build a system with the assistance of super-goal and profoundly figuring out how to distinguish Alzheimer disease (AD) with more precise methods.

6 Conclusion

Super-objective imaging has diverse steady applications, for instance, clinical imaging, video discernment, and stargazing. In clinical imaging, pictures are picked up for clinical deliberation purposes and for giving data about the presence of systems, and the physiologic and metabolic exercises of the volume underneath the skin. Clinical imaging is a significant end instrument to pick the closeness of express afflictions. Therefore, broadening the image goal ought to basically improve as far as possible regarding restorative treatment. Moreover, an unmatched goal may broadly improve altered ID and picture division results. As the end on 3D convolutional networks, it was seen that 3D convolutional networks perform well on requesting MRI cerebrum channels. Prior to rescaling, we describe ability to produce a convolutional neural network, so gathering should be conceivable between Alzheimer disease and normal person. Thereafter, we will describe the activation work in the CNN and plot the cerebrum 2D and 3D picture. In any case, it is as yet hard to state whether a 3D convolutional network is for each situation better; therefore, further assessment is expected to demonstrate this hypothesis. This is ordinary because a delicate scholarly weakness will consistently incite an Alzheimer's finding and there is no indisputable line isolating them. It is in this manner hard to viably mastermind an individual precisely. Further assessment super-resolution imaging should be done to smoothen out this model significantly further, specifically without time and resource objectives, since adjusting significant neural associations requires different runs of comparable model with different limits. Investigation should in like manner be conceivable in joining distinctively specific models. For example, the blend of a model that can decisively perceive clinically customary (CN) and Alzheimer's infection (AD) alongside a model that can exactly determine the end results of the Alzheimer's sickness (AD). Super objective imaging is steady for better examination, and for earlier acknowledgment of Alzheimer's disease concerning region-based division, edge-based division, and hierarchical division.

References

1. Tiwari, A. (2020). Multidimensional medical imaging analysis Alzheimer's disease via super-resolution imaging and machine learning. In *International Conference on Innovative Computing and Communication (ICICC 2020)*. Elsevier SSRN. https://ssrn.com/abstract=3564459.
2. Du, J., Wang, L., Liu, Y., Zhou, Z., He, Z., & Jia, Y. (2020). Brain MRI super-resolution using 3D dilated convolutional encoder–decoder network. *IEEE Access, 8*, 18938–18950. https://doi.org/10.1109/ACCESS.2020.2968395.
3. Razmjooy, N., Ashourian, M., Karimifard, M., Estrela, V. V., Loschi, H. J., do Nascimento, D., et al. (2020). Computer-aided diagnosis of skin cancer: A review. *Current Medical Imaging, 16* (7), 781–793.
4. Razmjooy, N., Estrela, V. V., & Loschi, H. J. (2020). Entropy-based breast cancer detection in digital mammograms using world cup optimization algorithm. *International Journal of Swarm Intelligence Research (IJSIR), 11*(3), 1–18.
5. Dixit, M., Tiwari, A., Pathak, H., & Astya, R. (2018). An overview of deep learning architectures, libraries and its applications areas. In *International Conference on Advances in Computing, Communication Control and Networking (ICACCCN-2018)*. IEEE Xplorer, pp. 293–297. ISBN: 978-1-5386-4119-4/18.
6. de Jesus, M. A., et al. (2020, April). Using transmedia approaches in STEM. In *2020 IEEE Global Engineering Education Conference (EDUCON)*. IEEE, pp. 1013–1016.
7. Estrela, V. V., et al. (2019). Why software-defined radio (SDR) matters in healthcare? *Medical Technologies Journal, 3*(3), 421–429.
8. Tiwari, A., & Gupta, K. K. (2015). An effective approach of digital image watermarking for copyright protection. *International Journal of Big Data Security Intelligence, 2*(1), 7–17. https://doi.org/10.14257/ijbdsi.2015.2.1.02. ISSN: 2383-7047 SERSC.
9. Misra, I., Gambhir, R. K., Manthira Moorthi, S., Dhar, D., & Ramakrishnan, R. (2012). An efficient algorithm for automatic fusion of RISAT-1 SAR data and Resourcesat-2 optical images. In *2012 4th International Conference on Intelligent Human Computer Interaction (IHCI)*. IEEE, pp. 1–6.
10. Tiwari, A., Jain, N. K., & Tomar, D. (2014). Chakrabortya, D., Thakurb, S., Jeyarama, A., Murthyc, Y. K., & Dadhwalc, V. K. (2012). Texture analysis for classification of RISAT-II images. *International Archives of the Photogrammetry, Remote Sensing and Spatial Information Sciences, 39*(B3), 461–466.
11. Mitchell, A. J., & Shiri-Feshki, M. (2009). Rate of progression of mild cognitive impairment to dementia – meta-analysis of 41 robust inception cohort studies. *Acta Psychiatrica Scandinavica, 119*(4), 252265.
12. Alzheimer's Association. (2015). 2015 Alzheimer's disease facts and figures: Includes a special report on disclosing a diagnosis of Alzheimer's disease. *Alzheimer's and Dementia, 11*(3), 332–384.
13. Maqsood, M., Nazir, F., Khan, U., & Aadil, F. (2019). Transfer learning assisted classification and detection of Alzheimer's disease stages using 3D MRI scans. *Sensors, 19*, 2645. https://doi.org/10.3390/s19112645.

14. Anitha, R., Prakash, & Jyothi, S. (2016). A segmentation technique to detect the Alzheimer's disease using super-resolution imaging. In *International Conference on Electrical, Electronics, and Optimization Techniques (ICEEOT)*.
15. Dessouky, M. M., & Elrashidy, M. A. (2016). Feature extraction of the Alzheimer's disease images using different optimization algorithms. *Journal of Alzheimers Disease & Parkinsonism, 6*(2), 1000230. https://doi.org/10.4172/2161-0460.1000230.
16. Patro, S., & Nisha, V. M. (2019). Early detection of Alzheimer's disease using super-resolution imaging. *International Journal of Engineering Research & Technology (IJERT), 8*(5), 1–4.
17. Islam, J., & Zhang, Y. (2018). Brain MRI analysis for Alzheimer's disease diagnosis using an ensemble system of deep convolutional neural networks. *Brain Informatics, 5*, 2. https://doi.org/10.1186/s40708-018-0080-3.
18. Aroma, R. J., et al. (2020). Multispectral vs. hyperspectral imaging for unmanned aerial vehicles: Current and prospective state of affairs. In V. V. Estrela, J. Hemanth, O. Saotome, G. Nikolakopoulos, & R. Sabatini (Eds.), *Imaging and sensing for unmanned aircraft systems* (Vol. 2, 7, pp. 133–156). London: IET. https://doi.org/10.1049/PBCE120G_ch7.
19. Deshpande, A., et al. (2020). Deep learning as an alternative to super-resolution imaging in UAV systems. In V. V. Estrela, J. Hemanth, O. Saotome, G. Nikolakopoulos, & R. Sabatini (Eds.), *Imaging and sensing for unmanned aircraft systems* (Vol. 2, 9, pp. 177–212). London: IET. https://doi.org/10.1049/PBCE120G_ch9.
20. Du, J., Wang, L., et al. (2020). Brain MRI super-resolution using 3D dilated convolutional encoder–decoder network. *IEEE Access, 8*, 18938–18950. https://doi.org/10.1109/ACCESS.2020.2968395.
21. Pham, C.-H., Tor-Díez, C., & Rousseau, F. (2019). Multiscale brain MRI super-resolution using deep 3D convolutional networks. *Journal of the Computerized Medical Imaging Society, 77*, 101647. https://doi.org/10.1016/j.compmedimag.2019.101647.
22. Ebrahimi-Ghahnavieh, A., Luo, S., & Chiong, R. (2019). Transfer learning for Alzheimer's disease detection on MRI images. In *2019 IEEE International Conference on Industry 4.0, Artificial Intelligence, and Communications Technology (IAICT)*, BALI, Indonesia, pp. 133–138. https://doi.org/10.1109/ICIAICT.2019.8784845.
23. Ucuzal, H., Arslan, A. K., & Çolak, C. (2019). Deep learning based-classification of dementia in magnetic resonance imaging scans. In *2019 International Artificial Intelligence and Data Processing Symposium (IDAP)*, Malatya, Turkey, pp. 1–6. https://doi.org/10.1109/IDAP.2019.8875961.
24. Biju, K. S., Alfa, S. S., Lal, K., Antony, A., & Akhil, M. K. (2017). Alzheimer's detection based on segmentation of MRI image. *Procedia Computer Science*, Elsevier B.V, *115*, 474–481. https://doi.org/10.1016/j.procs.2017.09.088.
25. Hong, X., et al. (2019). Predicting Alzheimer's disease using LSTM. *IEEE Access, 7*, 80893–80901. https://doi.org/10.1109/ACCESS.2019.2919385.
26. Herrera, L. J., Rojas, I., Pomares, H., Guillén, A., Valenzuela, O., & Baños, O. (2013). Classification of MRI images for Alzheimer's disease detection. In *2013 International Conference on Social Computing*, Alexandria, VA, pp. 846–851. https://doi.org/10.1109/SocialCom.2013.127.
27. Lin, W., Tong, T., Gao, Q., Guo, D., Du, X., Yang, Y., et al. (2018). Convolutional neural networks-based MRI image analysis for the Alzheimer's disease prediction from mild cognitive impairment. *Frontiers in Neuroscience, 12*, 777. https://doi.org/10.3389/fnins.2018.00777.
28. Suk, H.-I., & Shen, D. (2013). Deep learning-based feature representation for AD/MCI classification. In *Proceedings of the medical image computing and computer-assisted intervention MICCAI 2013* (pp. 583–590). Berlin Heidelberg: Springer.
29. Ott, A., Breteler, M. M., Bruyne, M. C., Van Harskamp, F., Grobbee, D. E., & Hofman, A. (1997). Atrial fibrillation and dementia in a population-based study: The Rotterdam study. *Stroke, 28*, 316–321.
30. Liu, S., et al. (2015). Multimodal neuroimaging feature learning for multiclass diagnosis of Alzheimer's disease. *IEEE Transactions on Biomedical Engineering, 62*(4), 1132–1140.

Image Enhancement Using Nonlocal Prior and Gradient Residual Minimization for Improved Visualization of Deep Underwater Image

Rahul Khoond, Bhawna Goyal, and Ayush Dogra

1 Introduction

Underwater imaging is a crucial problem in ocean dynamics. These underwater images (UIs) bear information about ocean nature, its flora, and fauna. However, due to scattering and absorption of light, these images undergo a color change, lower contrast, blurring effect, and nonlinear illumination [1]. The wavelength attenuation is responsible for inconsistent colorcasts. In contrast, scattering fetches distance-dependent variables into a clear image. This results in a hazy vision, making it blurred and visually unpleasant. Such low underwater images are less suitable for underwater object detection, underwater robot inspection, and night fishing. Deep underwater images suffer from blue-green illuminations. This phenomenon comes into play when a shorter wavelength, that is, blue, is absorbed last. Hence, that affects the image and makes it appear blue [2–4]. A large number of methods have been proposed in the literature to reform the low visibility of UIs. Like the underwater image, the image captured on a hazy day also undergoes light absorption and scattering. Thus, UI enhancement is an inverse problem akin to image dehazing. This chapter proposes intuitively that the dehazing algorithm brings considerable response while implementing it on the underwater image.

Super-resolution (SR) techniques can also reconstruct blurred images. Image SR is an emerging latest imaging technology. It is the process in which the high resolution (HR) of an image (or video) is restored from its low-resolution (LR) image (or video) [5]. In the LR image, the pixel density is less, whereas in HR, an image's pixel density is substantially higher, which produces a more detailed image [6–8]. SR enhances the image resolution in such a manner that it presents

R. Khoond · B. Goyal (✉)
Department of ECE, Chandigarh University, Mohali, India

A. Dogra
Ronin Institute, Montclair, NJ, USA

A. Deshpande et al. (eds.), *Computational Intelligence Methods for Super-Resolution in Image Processing Applications*, https://doi.org/10.1007/978-3-030-67921-7_14

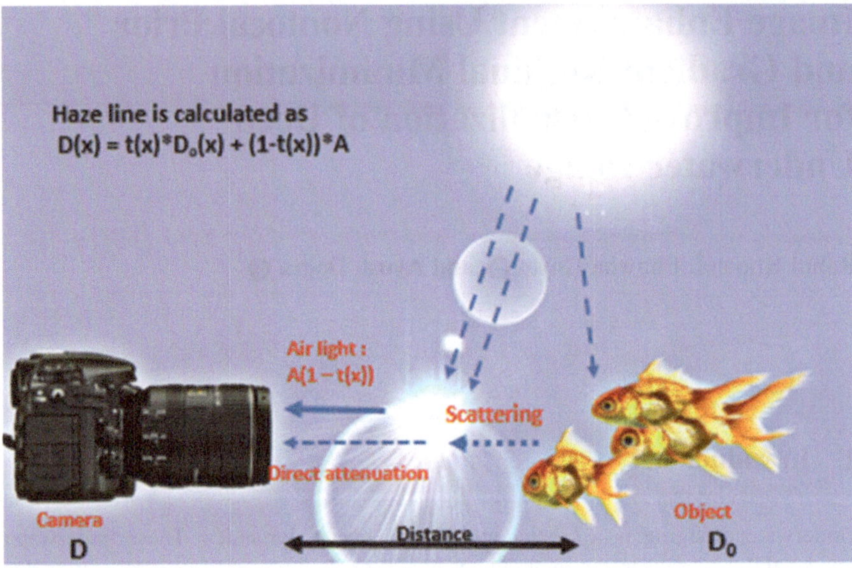

Fig. 1 Image dehazing

better information for humans and for machines. Based on the number of inputs, SR can be categorized into single-image SR (SISR) or multi-image SR (MISR). SISR is an ill-posed problem due to the unavailability of ground truth from LR images. The interpolation-based, reconstruction-based, and deep learning (DL)-based methods are the main techniques which deal with SR problems. The main advantage of the SR approach is that it can obtain HR images even without the presence of LR images. SR has enhanced resolving power, which overcomes the study of the molecular structure of cells [9]. Video information enhancement, surveillance, medical diagnosis, remote sensing, astronomical observation, and biometric information identification are its main applications [10].

In image dehazing, the following are some of the major works: He et al. [11] introduced the valid dark channel prior (DCP) for single-image dehazing. The haze-free image contains at least one channel, which has nearly zero pixel intensity in its channel. The method from [12] used an optimization method to conclude that contextual regulation compels the transmission hypothesis. In [13], Zhu et al. estimated the object's sensed depth to the camera for attenuation in color. According to them, haze-free images are prettily elective by some separate colors that introduce rigid clusters in the RGB space (Fig. 1).

In [14], Dana proposed a nonlocal prior, which rectifies the disadvantages of [15] by enhancing the hazed image's visibility. Wang et al. [16] dehazed an image using a physical model and improved the brightness by using a multiscale retinex restoration algorithm [17]. Seiichi [18] introduced image dehazing with a joint trilateral filter, which overcomes the extra smoothing caused by the bilateral filter. In [19], Chen proposed the gradient residual minimization (GRM) approach, which estimates the haze in the first step, and, in the second step, it minimizes the possible visual artifacts. Berman and Avidan [20] proposed image dehazing with the introduction of surround filter and DCP; here, the transmission is estimated by filtering the input image in three color spaces, that is, RGB, Lab, and HSV. This chapter presents a robust method that aims to increase underwater images' visibility and enhance the overall image features and region description. The NLP-GRM method delivers a hybridization of nonlocal prior and GRM. With this, one can harness the advantages of both of these methods while overcoming limitations of each. Firstly, the hazed image is processed with a nonlocal prior that estimates the haze line and identifies per pixel transformation. The obtained image is referred to as a source image to GRM, which overcomes the disadvantages of nonlocal image dehazing and gives a refined and enhanced dehazed image [19, 20]. In this study, the hazy images contaminated with fog-like films are targeted to be enriched. This study designs a hybrid image dehazing method. The current problem statement can also be targeted with the help of super-resolution methods. The present problem considers a deep-learning-based, very deep super-resolution (VDSR) method for underwater image enhancement.

This chapter is organized as follows: Section 2 illustrates the different methods of image dehazing. Section 3 discusses image enhancement using super resolution. Section 4 describes the NLP-GRM methodology in detail. Section 5 elaborates on results and discussion. Conclusion is provided in Sect. 6.

2 Image Dehazing

2.1 *Gradient Residual Minimization (GRM)*

GRM is a two-step mechanism of image dehazing. It first calculates the transmission map and recovers the dehazed image with gradient residual minimization. The guided total generalized variation (TGV) refines the transmission map [21]. Besides, the remaining results from taking the minimum residual of the gradient. If the initial transmission is t′ and the guided image is $\boldsymbol{I}$, optimization becomes

$$\mathbf{min}\,(t,w)\{\alpha_1 \int |\boldsymbol{H}^{\mathbf{0.5}}\Big(\nabla t - w\,|\,dx + \alpha_0 \int |\,\nabla(w)\,|\,dx + \int |\,t - t'\,|\,dx\}, \quad (1)$$

where $H^{0.5}$ is an anisotropic diffusion tensor [8].

The steps involved in the algorithm of guided TGV are summarized in Algorithm 1.

Algorithm 1

Initialization: $to = \widetilde{t}, w_0, t'_0, w'_0, p_0, q_0 = 0, \sigma_p > 0, \sigma_q > 0, \tau_t > 0, \tau_w > 0$

For i = 0:max-iteration do

$pk{+}1 = \boldsymbol{P}[pk + \sigma.{*}p.{*}\ \alpha 1.{*}(\boldsymbol{H}^{1/2}\ (\nabla t'^{k} - w'^{k}))]$

$qk+1 = \boldsymbol{P}[q\ k + \sigma.{*}q.{*}\alpha 0.{*}\nabla w'^{k}]$
$t^{k+1} = threshold\ \tau\ (t^{k} + \tau u \alpha_1 \nabla^T \boldsymbol{H}^{1/2}\ pk+1)$
$w^{k+1} = w^{k} + \tau_w(\alpha_o \nabla^T\ q\ k+1 + \alpha_1 \boldsymbol{H}^{1/2}\ pk+1)$
$t'^{k+1} = t^{k+1} + \boldsymbol{\theta}(t^{k+1} - t'^{k})$
$w'^{k+1} = w^{k+1} + \boldsymbol{\theta}(w^{k+1} - w'^{k})$
end

Once the transmission map is refined the scene radiance, *Do* is recovered as [19]

$$\min{(Do)}\left\{\frac{1}{2}\int\left\|(Dot - (D - A + At)\right\|_2^2 dx + \eta \int \|\nabla D - \nabla Do\|_2^2\right. \tag{2}$$

The summarized algorithm describes the overall recovery of an image as follows:

Algorithm 2
Initialize $\boldsymbol{E}^0 = \boldsymbol{0}, \boldsymbol{D}_0^0 = (\boldsymbol{D} - \boldsymbol{A})/\boldsymbol{t} + \boldsymbol{A}$
For $k = 0$ to max-iteration do
$Z_b = \boldsymbol{D} - \boldsymbol{E}^k - \boldsymbol{A} + \boldsymbol{At} - \boldsymbol{Dt}$
$\boldsymbol{Z} = \boldsymbol{arg\ min}\left\{\frac{1}{2}\int \|Zt - Zb\|_2^2 dx + \eta \int \|\nabla Z\|_1 dx\right\}$
$\boldsymbol{D}_0^{k+1} = \boldsymbol{I} + \boldsymbol{Z}$
$\boldsymbol{E}^{k+1} = threshold\ (\boldsymbol{D} - \boldsymbol{D}_0^{k+1}\boldsymbol{t} - (\boldsymbol{1} - \boldsymbol{t})\boldsymbol{A})$
end

2.2 Dark Channel Prior (DCP)

In the dark channel prior state, the low intensity of color at some pixel causes haze in an image. For an image *D*, we have

$$\boldsymbol{Do}^{\mathbf{dark}}(\boldsymbol{x}) = \min{(\min{(y \in \boldsymbol{\Omega}(\boldsymbol{x}))}(\boldsymbol{Do}^c(\boldsymbol{y})))} \tag{3}$$

where Do^c represents the color channel of *D*, and $\Omega(x)$ is a local patch at (x). If *D* is the haze-free outdoor image, in that case, D^{dark} should be lower or equal to zero. In order to remove the haze, atmospheric light A must be approximated. The highest intensity pixel is referred to as atmospheric light A to determine its dark channel, choosing the top 0.1% bright pixels beside among these pixels. After approximation of *A*, $\Omega(x)$ is considered constant, and if $t'\ (x)$ is patch transmission beside applying these parameters from Eq. (3) yields

$$\min{(y \in \boldsymbol{\Omega}(\boldsymbol{x}))}(\boldsymbol{D}^c(\boldsymbol{y})) = \boldsymbol{t}'(\boldsymbol{x}) \min{(y \in \boldsymbol{\Omega}(\boldsymbol{x}))}(\boldsymbol{Do}^c(\boldsymbol{y})) + (\boldsymbol{1} - \boldsymbol{t}'(\boldsymbol{x}))\boldsymbol{A}^c. \tag{4}$$

Performing minimum operation on three channels independently, the equation becomes

$$\min\left(y \in \Omega(x)\right)\left(\frac{D^c(y)}{A^c}\right) = t'(x) \min\left(y \in \Omega(x)\right)\left(\frac{D^c(y)}{A^c}\right) + (1 - t'(x)), \tag{5}$$

Then, min operation on three channels is obtained on the above equation and obtained

$$\min\left(y \in \Omega(x)\right)\left(\frac{D^c(y)}{A^c}\right) = t'(x) \min\left(\left(\min\left(y \in \Omega(x)\right)\left(\frac{D^c(y)}{A^c}\right) + (1 - t'(x)).\right.\right. \tag{6}$$

As discussed earlier, the dark channel for a haze-free image should be equal to zero.
Therefore,

$$Do^{\text{dark}} = \min(c)\left(\min\left(y \in \Omega(x)\right)(Do^c(y))\right) = 0. \tag{7}$$

$$\text{For all values of } A^c, \quad \min(c)\left(\min\left(y \in \Omega(x)\right)\frac{(Do^c(y))}{A^c}\right) = 0. \tag{8}$$

Using Eqs. (8) and (6), the transmission t' turns out to be

$$t'(x) = 1 - \min(c)\left(\min\left(y \in \Omega(x)\right)\frac{(Do^c(y))}{A^c}\right), \tag{9}$$

with the transmission map, the scene radiance is estimated from

$$D(x) = t(x).Do(x) + (1 - t(x)).A. \tag{10}$$

Replacing $t(x)$ with $t'(x)$, we have the original dehazed image recovered as

$$Do(x) - \left[\frac{D(x) - A}{\max\left(t(x),\ to\right)}\right] + A. \tag{11}$$

2.3 Nonlocal Prior

A nonlocal algorithm consists of four steps:

1. Finding haze line
2. Estimated the initial transmission
3. Regularization
4. Dehazing

2.3.1 Haze Line

The haze line of an image is calculated by estimating the value of atmospheric light (A). For that, pixels are converted into a spherical coordinate system. Hence, A lies in spherical coordinate, where r is the radius of the sphere, θ and ϕ are longitude and latitude serially, respectively. Pixel colors are presented around the air light in spherical coordinates. The color represents the number of pixels, which are lining in each direction. The image varies only in terms of t, whereas in spherical coordinate, t only effects $v(x)$ without disturbing θ or ϕ [20]. Therefore, the pixel would be in a uniform haze line if their θ and ϕ values are equal; besides, pixel on every haze line has similar value in its dehazed image. To determine the same haze line, we have grouped the pixel on their $[\phi(x), \theta(x)]$ values based on the nearer sample point on the plane [22]. This implementation could be enhanced by using the KD tree [23] from predefining patchwork and querying it for every pixel, the haze line is estimated and calculated.

2.3.2 Estimating Transmission

For an obtained haze line and atmospheric light estimation, the distance between object and camera is essential, and it must be known for the calculation of transmission.

2.3.3 Regularization

It refines the transmission, which gives an inferior range of the transmission. The transmission estimation is rooted in the depredate version of the transmission line.

2.3.4 Dehazing

The dehazed image is calculated with the known value of the transmission $t(x)$.

Algorithm 3 summarizes the Guided TGV steps (Fig. 2).

Algorithm 3
[Input, output] = $[\boldsymbol{D(x), A, D_0(X), t'(X)}]$
1: $\boldsymbol{DA(x) = D(x) - A}$
2: Conversion of Da into spherical coordinates and obtain $[\mathbf{r(x), \varphi(x), \theta(x)}]$
3: pixel clustering based on $[\boldsymbol{\varphi}(\mathbf{x}), \boldsymbol{\theta}(\mathbf{x})]$. Each cluster is a haze-line **(H)**.
4: **for** cluster **Hdo**
5: Estimate max radius: $r'max(x) = max(x \in H) \{r(x)\}$
6: **for** pixel x **do**
7: transmission estimation: $t'(x) = r(x)\ r'max$
8: refining $t'(x)$ which minimize transmission
9: Calculate the dehazed image

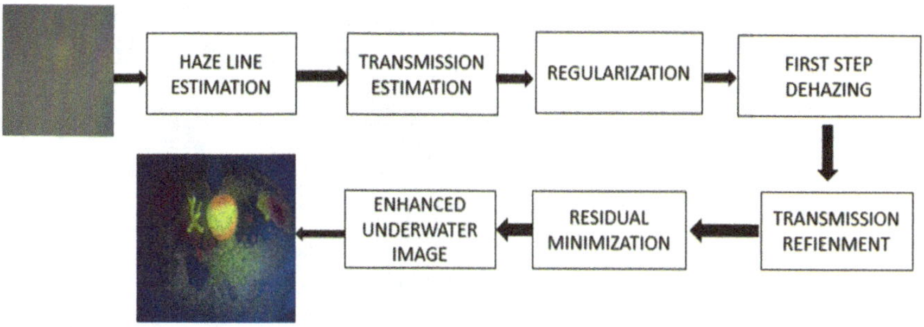

Fig. 2 Methodology for image dehazing

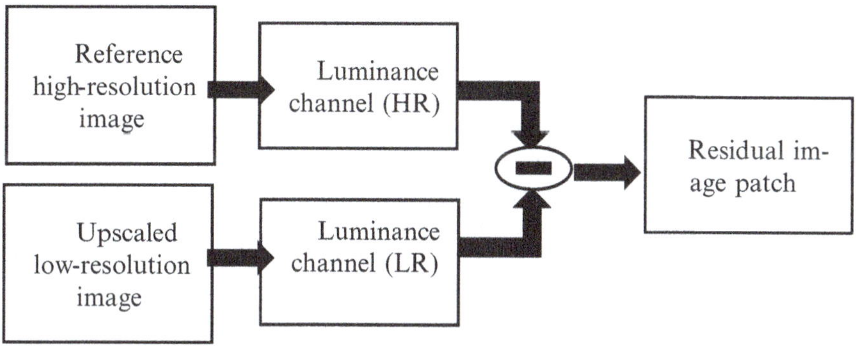

Fig. 3 Very deep super-resolution methodology

3 Image Quality Enhancement Using Super Resolution

Very deep SR (VDSR) is a convolution neural network (CNN)-based architecture for single-image SR [24]. The VDSR network finds out the texture mapping of LR and HR images. LR and HR images have identical image details and different high-frequency details, which makes the mapping compatible. VDSR makes use of the residual learning approach, which learns to estimate the residual image. The difference between a high-resolution reference image and the LR image has been upscaled using bicubic interpolation in SR residual image. A residual image consists of the HR details of an image (Fig. 3).

The VDSR network obtains a residual image from the luminance channel, describing each pixel's brightness of the LR and HR images. VDSR is trained using luminance channels of reference HR image and upscaled LR image. After the VDSR learns to obtain a residual image, the final HR image is calculated by adding the obtained residual image to the up sampled LR image (Fig. 4).

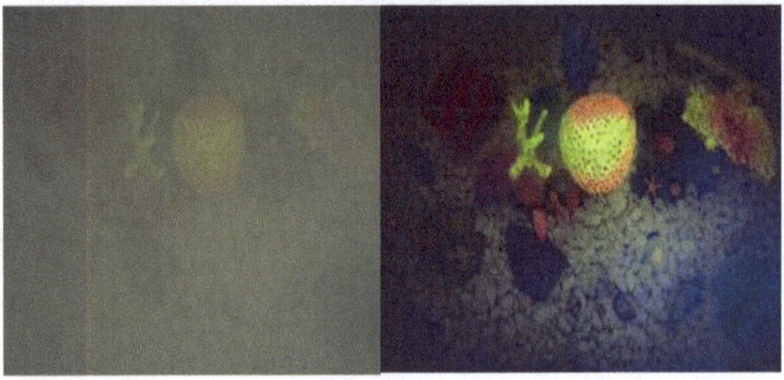

Fig. 4 Dehazed image using image super resolution

4 Integrating Nonlocal Prior to GRM

Dehazing of an image is the process of removing haze formed on an image by atmospheric effects, which cause it to appear gray or blue [25]. As we have already discussed, haze is due to the scattering or absorption of particles such as smoke, moisture, and dust vapor. Our main contribution is to make the underwater image more enhanced and noise free. The haze mathematical model, as given by Koschmieder, states that

$$\boldsymbol{D}(x,y) = \boldsymbol{t}(x,y) * \boldsymbol{Do}(x,y) + (\mathbf{1} - \boldsymbol{t}(x,y)) * \boldsymbol{A}, \tag{12}$$

where $\boldsymbol{D}(x, y)$ is the intensity of the pixel at (x, y), $\boldsymbol{Do}(x, y)$ is the pixel intensity of the obtained dehazed image at $\boldsymbol{t}(x, y)$ (which is the part of the light the camera fails to capture), and $\boldsymbol{A}$ is the air-light color represented in 3-D RGB color space. $\boldsymbol{t}(x,y)$ * $\boldsymbol{Po}(x,y)$ is the direct transmission beside $(\mathbf{1} - \boldsymbol{t}(x,y)) * \boldsymbol{A}$ is the additional scene added in the image due to scattering or absorption. In homogeneous weather, one can write $\boldsymbol{t}$ as

$$\mathbf{t(x,y)} = \mathbf{exp}^{-\boldsymbol{\beta}\mathbf{L(x,y)}}, \tag{13}$$

where $\boldsymbol{t}(x, y)$ is the visibility of each pixel in homogeneous weather, β is the scattering coefficient, and $\boldsymbol{L}$ is scene depth.

Thus, with the known values of $\boldsymbol{t}$ and $\boldsymbol{A}$, we can generate the dehazed image $\boldsymbol{Do}$ from the obtained Eq. (12).

Considering these parameters, the authors introduced nonlocal with GRM under tensor parameters in such a way that we have received enhanced dehazed underwater image. The present work first gave the input image to a nonlocal prior [20]. Here, the nonlocal output is fed to the GRM, and at last, the enhanced dehazed underwater image can be modeled by Eqs. (12) and (13).

Here, β depends on the wavelength and t is different for each pixel of an image. Non-local Image Dehazing algorithm consists of four steps:

1. Finding haze line
2. Estimated the initial transmission
3. Regularization and
4. Dehazing

Haze line: It is calculated by estimating the value of A. Therefore, modifying (12) as

$$\boldsymbol{D_A}(\boldsymbol{x}) = \boldsymbol{D}(\boldsymbol{x}) - \boldsymbol{A}, \tag{14}$$

$$\Rightarrow D_A(x) = t(x) - [Do(x) - A]. \tag{15}$$

Converting the pixels of an image into a spherical coordinate system, hence $D_A(x)$ lies in spherical coordinate, r is the radius of the sphere (i.e., $\|\boldsymbol{D} - \boldsymbol{A}\|$), and θ and ϕ are longitude and latitude serially. Pixel colors are presented around the air light in spherical coordinates. The color represents the number of pixels, which are lining in each direction. From Eq. (12), the image varies only in terms of $\boldsymbol{t}$, whereas in spherical coordinate, t only effects $\mathbf{v(x)}$ without disturbing θ or ϕ [1].

$$D(x) \approx D(y) \Rightarrow \phi(x) \approx \phi(y), \text{and } \theta(x) \approx \theta(y). \tag{16}$$

Therefore, the pixel would be in a uniform haze line if their θ and ϕ values are equal; besides, pixel on every haze line has similar value in its dehazed image $\boldsymbol{Do}$. To determine the same haze line, we have grouped the pixel on their $[\phi(\boldsymbol{x}), \theta(\boldsymbol{x})]$ values based on the nearer sample point on the plane [22]. This implementation could be enhanced by using the KD tree [23], from predefining patchwork, and querying it for every pixel. Therefore, the haze line estimate is calculated as

$$\boldsymbol{D_1} - \boldsymbol{A} = \boldsymbol{\alpha}(\boldsymbol{D_2} - \boldsymbol{A}) \Rightarrow \boldsymbol{D_1} = (\boldsymbol{1} - \boldsymbol{\alpha})\boldsymbol{A} + \boldsymbol{\alpha D_2}. \tag{17}$$

Estimating transmission: For an obtained haze line $\boldsymbol{Do}$ and $\boldsymbol{A}$, $\boldsymbol{v}(\boldsymbol{x})$ depends on the distance between object and camera.

$$\boldsymbol{v}(\boldsymbol{x}) = \boldsymbol{t}(\boldsymbol{x})|\boldsymbol{D}(\boldsymbol{x}) - \boldsymbol{A}|, \quad \boldsymbol{0} \le \boldsymbol{t}(\boldsymbol{x}) \le \boldsymbol{1}, \tag{18}$$

where $t = 1$ for the major radial coordinate; therefore, $v_{\max}$ is calculated as

$$v_{\max} = |\boldsymbol{J} - \boldsymbol{A}|. \tag{19}$$

By combining Eqs. (18) and (19), the transmission map $t(x)$ becomes

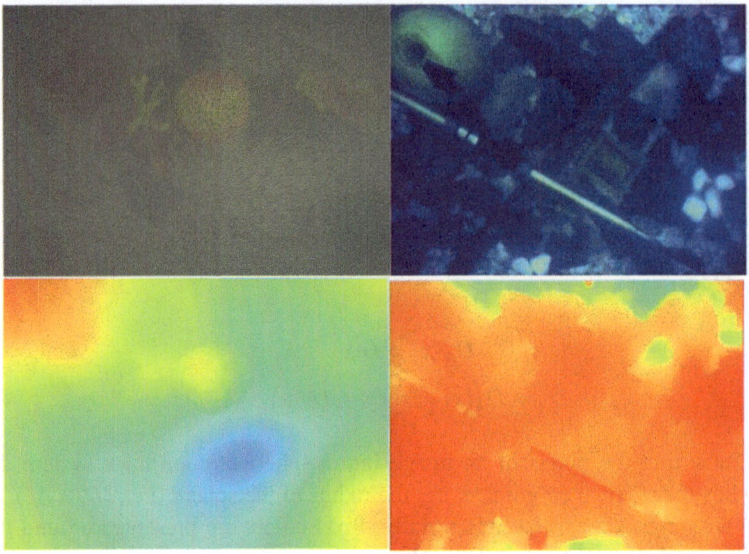

Fig. 5 Source images (L-R, I column): Estimated transmission images with nonlocal (L-R, II column)

$$t(x) = \frac{v(x)}{v_{\max}}, \tag{20}$$

with $r_{\max}$ as the maximum if the haze line consists of haze-free pixels

$$v'_{\max}[x] = \max [r(x)], x \in H. \tag{21}$$

The estimation consists of solving the two previous equations (Fig. 5). Hence, transmission becomes

$$t(x) = \frac{v(x)}{v'_{\max}}. \tag{22}$$

Regularization

If Do is positive ($Do \geq 0$), which gives an inferior range for the transmission

$$t_{\mathrm{Le}} = 1 - \min \left\{ \frac{I_c(x)}{A_c} \right\} . t_{\mathrm{Le}}. \tag{23}$$

The transmission estimation is rooted in the depredate version of t'_{Le}

$$t'_{\mathbf{Le}}(x) = \max\{t'(x), t_{\mathbf{Le}}(x)\}, \tag{24}$$

Mathematically, we can minimize the function to

$$\sum\nolimits_x\left\{\left[t'(x) - t'_{\mathrm{Le}}(x)\right]^{2/}\sigma^2(x)\right\} + \lambda\sum\nolimits_x\sum\nolimits_{y\in Nx}\left\{\left[t'(x) - t'(y)\right]^{2/}\|D(x) - D(y)\|^2\right\}, \tag{25}$$

where λ controls the trade-off between source data and smoothness term, $N_{(x)}$ is a neighbor of x, and $\sigma(x)$ represents the standard deviation [20].

Dehazing with the known value of $t(x)$ dehazed image is calculated by

$$D'_0(x) = \{D(x) - [1 - t'(x)].A\}/t'(x). \tag{26}$$

Once the NLP stage receives the dehazed underwater image, the authors observed an extra unwanted enhancement in the image, which should be decreased for the output. Furthermore, it works well only for the images captured from a long distance, whereas residual minimization observed output image is useful only for the desired distance. Thus, using this disadvantage, the authors again passed the observed output as an input to GRM [19].

GRM works on a two-step mechanism in which at first transmission is refined and in second minimum residual of the image is calculated. Transmission refinement is calculated without recovering the 3D scene [26].

Once this is calculated, the remaining image recovery is estimated by residual minimization [19]. Here, we minimize the residual of the gradient between output and input image. By combining GRM with the haze model, the optimization becomes

$$\min(D, E)\left\{\frac{1}{2}\int \left\|Dot - (D - E - A + At\right\|_2^2 dx + \lambda\int \|E\|_0 dx + \eta\int \|\nabla D - \nabla Do\|_0 dx\right\}, \tag{27}$$

where Do is the number of zeros, and η is the weighting parameter (Fig. 6).

The above equation finds out the total number of nonzero gradients of D, which should be at the position of I although their magnitude could not be the same [27]. Due to air particles' presence in the input image or output image of nonlocal that may cause large error E in required output, even we assume E is the rarebit. To recuperate the obscure image, we solved the optimization problem

$$\min(D, E)\left\{\frac{1}{2}\int \left\|Do - (D - E - A + At\right\|_2^2 dx + \lambda\int \|E\|_1 dx + \eta\int \|\nabla D - \nabla I\|_1 dx\right\}. \tag{28}$$

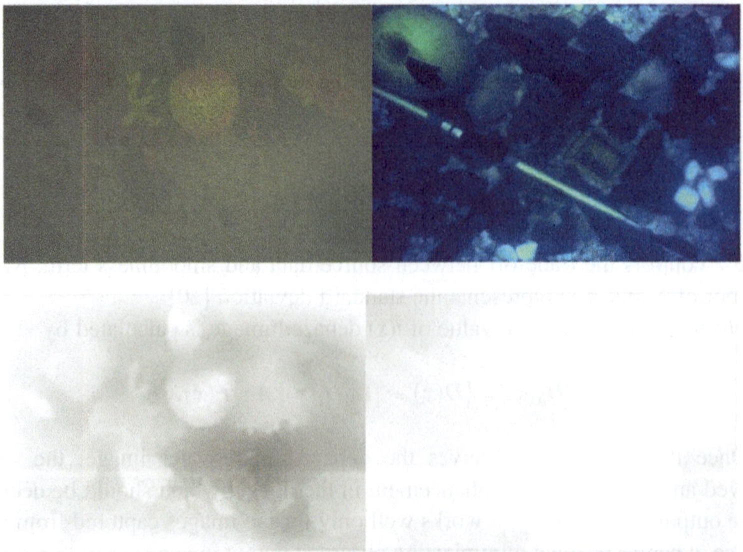

Fig. 6 Source images (L-R, I column): Estimated transmission images with NLP-GRM (L-R, II column)

Here, λ is the regularization parameter, subtracting E from D we are remaining with $D - E$, D_0, and transmission map A, which completes the haze model from Eq. (30) (Fig. 7).

$$\mathbf{min}\,(z)\left\{\frac{\mathbf{1}}{\mathbf{2}}\int\left\|(\boldsymbol{Z}+\boldsymbol{D})-\left(\boldsymbol{D}-\boldsymbol{E}-\boldsymbol{A}+\boldsymbol{A}\boldsymbol{t}\right\|_2^2\boldsymbol{dx}+\eta\int\|\nabla\boldsymbol{Z}\|_1\boldsymbol{dx}\right\}. \tag{29}$$

We alternatively solved this by minimizing the energy function once Z is known. Besides, J could easily be obtained with $D_0 = D + Z$, which has a soft-thresholding solution. Hence, it can be rewritten as follows:

$$\mathbf{min}\,(z)\left\{\frac{\mathbf{1}}{\mathbf{2}}\int\left\|\boldsymbol{Dot}-(\boldsymbol{D}-\boldsymbol{E}-\boldsymbol{A}+\boldsymbol{A}\boldsymbol{t}\right\|_2^2\boldsymbol{dx}+\lambda\int\|\boldsymbol{E}\|_1\boldsymbol{dx}\right. \tag{30}$$

The final output image can be obtained with the help of above equations.

5 Experimental Setup, Results, and Discussions

The method proposed (NLP-GRM) in this chapter is being evaluated using visual analysis and objective metrics. For objective evaluation, we have taken entropy, PSNR, and MSE, which are expressed as

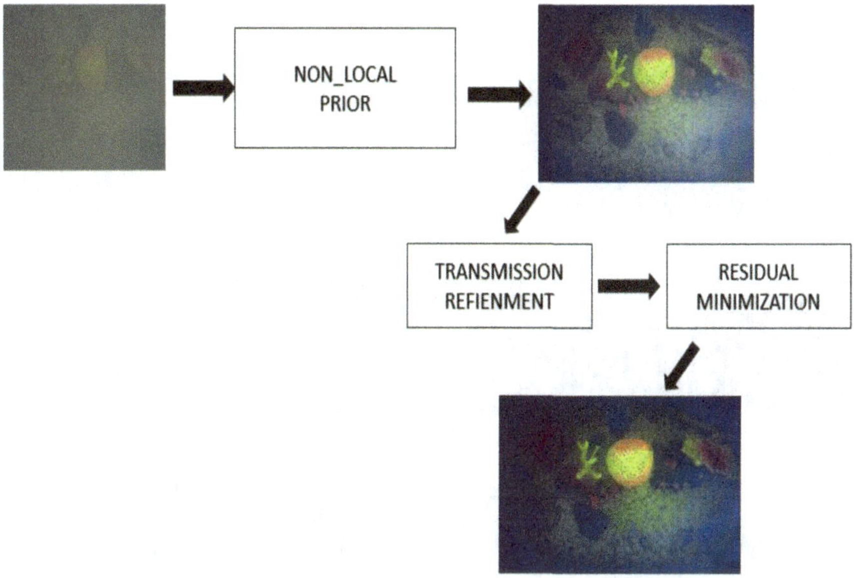

Fig. 7 NLP-GRM methodology

$$\text{MSE} = \frac{1}{ab}\sum_{i=0}^{a-1}\sum_{j=0}^{b-1}[D(i,j) - K(i,j)], \tag{31}$$

$$\text{PSNR} = 10.\log_{10}\left(\frac{\max * \max}{\text{MSE}}\right) = 20.\log_{10}\left(\frac{\max}{M}\right), \tag{32}$$

and

$$\text{Entropy} = -\sum_{i=0}^{a-1} Mi * \log_b Mi. \tag{33}$$

where $M = \sqrt{\text{MSE}}$, a is number of gray levels (256 for 8-bit images), and Mi is the pixel probability. For the evaluation of the NLP-GRM method, we tested it on the turbid dataset. We analyzed various underwater images which are captured from various distances. We have taken three kinds of images in which one is captured from a long distance (with blue hue over the image due to reflection of light and color transformation), one is captured from a moderate underwater, and the other one is captured from a short distance. For experimentation, underwater images of size 1024×768, 1027×768, 1024×695, respectively, have been tested on a machine with Intel® core ™ i5 with 4 GB RAM MATLAB 2014. The NLP-GRM technique and the other state-of-the-art methods have been compared via objective metrics to establish the versatility of this work's method. For the analyses, we have taken parameter values as $\gamma = 1.5$, $A_1 = 0.53$, $A_2 = 0.53$, $A_3 = 0.53$, $\alpha = 0.5$, $\alpha_1 = 0.5$. Under this parameter set, we have obtained haze-free and enhanced underwater images. Besides, the image's information appears well, the visual quality is

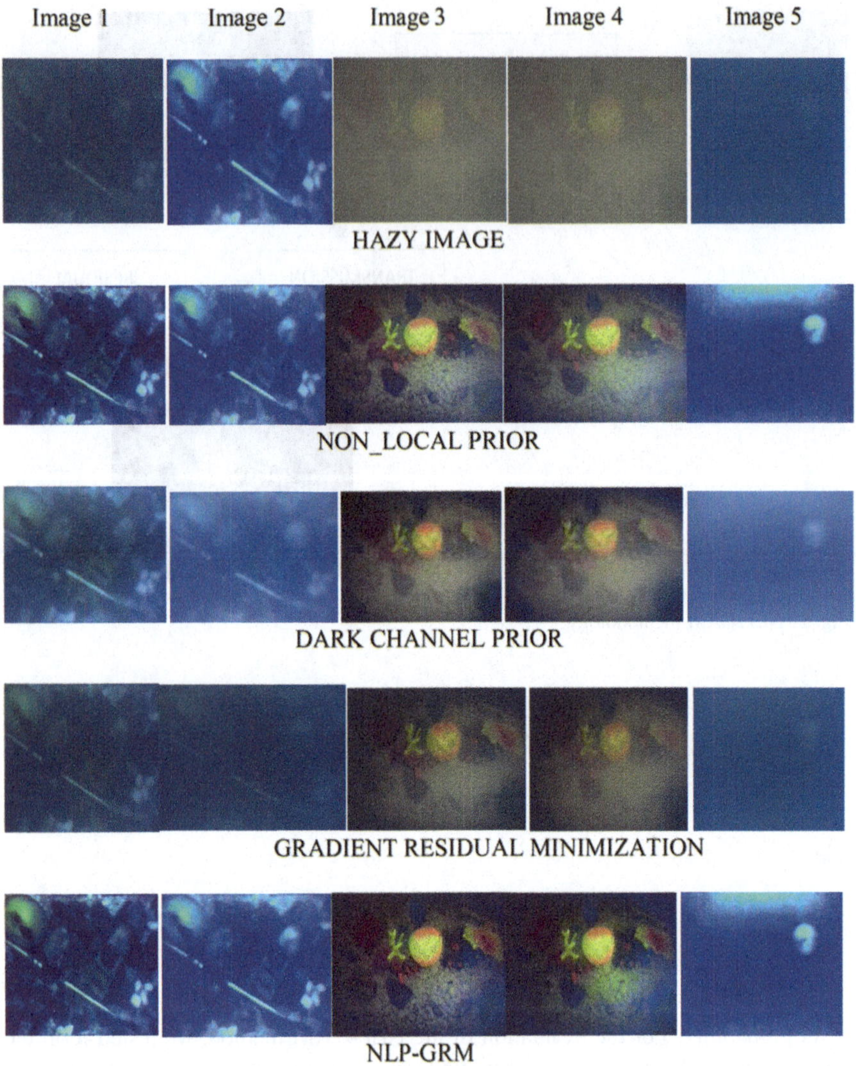

Fig. 8 Visual results for NLP-GRM and state-of-the-art methods

enhanced, edge features are well estimated, and noise is removed. Figure 8 shows the NLP-GRM method and techniques from other authors whose comparison relies on MMSE, PSNR, and entropy. The results of these parameters are discussed in Table 1 [28].

Robust image enhancement using GRM provides optimum results for the images taken from a certain distance. Otherwise, it will decrease the visual aspects of an image. Nonlocal prior visual elements are enhanced, and pixels are over smoothened beside edges are not refined. The DCP, which is environmentally dependent, does not increase the underwater image's quality and ameliorates quantitative results.

Table 1 Objective evaluation

Original images	Methods	MSE	Entropy	PSNR (dB)
Image 1	Nonlocal	9.38	7.25	12.648
	DCP	**1.00**	7.27	12.875
	GRM	3.78	6.59	**15.342**
	NLP-GRM	3.60	**4.33**	15.134
Image 2	Nonlocal	2.68	7.45	11.411
	DCP	4.78	7.05	11.205
	GRM	7.69	5.85	16.141
	NLP-GRM	**2.40**	**4.42**	**15.585**
Image 3	Nonlocal	1.76	7.21	11.947
	DCP	2.97	7.08	11.835
	GRM	1.27	6.78	12.780
	NLP-GRM	**0.68**	**4.70**	**13.352**
Image 4	Nonlocal	1.63	7.43	12.380
	DCP	2.85	7.13	11.889
	GRM	1.78	6.69	12.430
	NLP-GRM	**1.33**	**4.71**	**12.727**
Image 5	Nonlocal	2.65	7.30	11.423
	DCP	4.18	6.30	11.392
	GRM	1.77	5.48	16.767
	NLP-GRM	**1.31**	**4.37**	**18.375**

Therefore, we have combined nonlocal and GRM together so that the obtained optimum results are outstanding in quality and quantity. In our discussion, we have compared our NLP-GRM method with DCP, nonlocal, and robust minimization and analyzed them through objective metrics. Figure 8 shows the visual quality of each technique.

The obtained results refer that our approach works exactly for underwater image enhancement. In Fig. 9, we have considered two different input underwater image datasets captured from two different distances. We have seen optimum results in terms of visual quality and qualitative metrics. Due to the large scattering of light, the image captured from a longer distance has turned blue. The NLP-GRM method successfully dehazed all the images captured from different distances. While comparing the values of objective metrics, that is, MSE, entropy, and PSNR, it can be stated that the NLP-GRM method performs better.

6 Conclusion

This chapter proposes an efficient image dehazing method enabling high visual perception of the objects underwater. Due to the absorption and scattering of light, images are gravely hazed and degraded, which results in low contrast of the

Fig. 9 Source images (L-R, I column): Dehazed images with the NLP-GRM method (L-R, II column)

underwater image. The NLP-GRM method aims to design an efficient algorithm that carries out superior results under a different environmental condition in terms of visual analysis and objective evaluation. Our approach integrates the gradient residual minimization (GRM) and the nonlocal prior to a new method called NLP-GRM. With the help of a nonlocal dehazing algorithm, the haze lines and further the image textures and features are preserved using GRM. As demonstrated in the results, the NLP-GRM method outperforms state-of-the-art methods in terms of edge and feature details and objective metrics as well. The scattering of underwater light results in the formation of haze lines and blurred vision. The hybridized NLP-GRM technique removed the haze lines while preserving the feature details. Later, the authors will improve the NLP-GRM approach performance by tuning parameters (γ, A_1, A_2, A_3, α, α_1). Additional tests using a wide range of datasets are necessary.

The described algorithm also contains some disadvantages, such as high execution time, low response to estimate transmission light for very deep underwater images, and the quality evaluation may be improper for deep UR images. In future, data fusion could be used to remove the artifacts. Histogram equalization can be modified with new algorithms for color distortion improvements.

References

1. Serikawa, S., & Lu, H. (2014). Underwater image dehazing using joint trilateral filter. *Computers & Electrical Engineering, 40*(1), 41–50.
2. Wang, J., Lu, K., Xue, J., He, N., & Shao, L. (2018). Single image dehazing based on the physical model and MSRCR algorithm. *IEEE Transactions on Circuits and Systems for Video Technology, 28*(9), 2190–2199.
3. Goyal, B., Dogra, A., Agrawal, S., Sohi, B. S., & Sharma, A. (2020). Image denoising review: From classical to state-of-the-art approaches. *Information Fusion, 55*, 220–244.
4. Dogra, A., Goyal, B., & Agrawal, S. (2017). From multi-scale decomposition to non-multi-scale decomposition methods: A comprehensive survey of image fusion techniques and its applications. *IEEE Access, 5*, 16040–16067.
5. Yang, J., Wright, J., Huang, T. S., & Ma, Y. (2010). Image super-resolution via sparse representation. *IEEE Transactions on Image Processing, 19*(11), 2861–2873.
6. Deshpande, A., & Patavardhan, P. (2017). Multiframe super-resolution for long range captured Iris polar image. *IET Biometrics, 6*(2), 108–116.
7. Deshpande, A., Patavardhan, P., Estrela, V. V., & Razmjooy, N. (2020). Deep learning as an alternative to super-resolution imaging in UAV systems. In V. V. Estrela, J. Hemanth, O. Saotome, G. Nikolakopoulos, & R. Sabatini (Eds.), *Imaging and sensing for unmanned aircraft systems* (Vol. 2, 9, pp. 177–212). London: IET. https://doi.org/10.1049/PBCE120G_ch9.
8. Estrela, V. V., Hemanth, J., Loschi, H. J., Nascimento, D. A., Iano, Y., & Razmjooy, N. (2020). Computer vision and data storage in UAVs. In V. V. Estrela, J. Hemanth, O. Saotome, G. Nikolakopoulos, & R. Sabatini (Eds.), *Imaging and sensing for unmanned aircraft systems* (Vol. 1, 2, pp. 23–46). London: IET. https://doi.org/10.1049/PBCE120F_ch2.
9. Mahapatra, D., Bozorgtabar, B., & Garnavi, R. (2019). Image super-resolution using progressive generative adversarial networks for medical image analysis. *Computerized Medical Imaging and Graphics, 71*, 30–39.
10. Yue, L., Shen, H., Li, J., Yuan, Q., Zhang, H., & Zhang, L. (2016). Image super-resolution: The techniques, applications, and future. *Signal Processing, 128*, 389–408.
11. Zhu, Q., Mai, J., & Shao, L. (2015). A fast single image haze removal algorithm using color attenuation prior. *IEEE Transactions on Image Processing, 24*(11), 3522–3533.
12. Berman, D., & Avidan, S. (2016). Nonlocal image dehazing. In *Proceedings of the IEEE international conference on computer vision and pattern recognition (CVPR)* (pp. 1674–1682). Piscataway: IEEE.
13. Ancuti, C., Ancuti, C. O., Haber, T., & Bekaert, P. (2012). Enhancing underwater images and videos by fusion. In *2012 IEEE conference on computer vision and pattern recognition* (pp. 81–88). Piscataway: IEEE.
14. Yang, M., Hu, K., Du, Y., Wei, Z., Sheng, Z., & Hu, J. (2020). Underwater image enhancement based on conditional generative adversarial network. *Signal Processing: Image Communication, 81*, 115723.
15. Meng, G., Wang, Y., Duan, J., Xiang, S., & Pan, C. (2013). Efficient image dehazing with boundary constraint and contextual regularization. In *Proceedings of the IEEE international conference on computer vision*, pp. 617–624.
16. Hou, W., Gray, D. J., Weidemann, A. D., Fournier, G. R., & Forand, J. L. (2007). Automated underwater image restoration and retrieval of related optical properties. In *2007 IEEE international geoscience and remote sensing symposium* (pp. 1889–1892). Piscataway: IEEE.
17. Galdran, A., Alvarez-Gila, A., Bria, A., Vazquez-Corral, J., & Bertalmío, M. (2018). On the duality between retinex and image dehazing. In *Proceedings of the IEEE conference on computer vision and pattern recognition*, pp. 8212–8221.
18. Li, C., Guo, C., Ren, W., Cong, R., Hou, J., Kwong, S., et al. (2019). An underwater image enhancement benchmark dataset and beyond. arXiv preprint arXiv:1901.05495.

19. Chen, C., Do, M. N., & Wang, J. (2016). Robust image and video dehazing with visual artifact suppression via gradient residual minimization. In *European conference on computer vision* (pp. 576–591). Cham: Springer.
20. Berman, D., & Avidan, S. (2016). Nonlocal image dehazing. In *Proceedings of the IEEE conference on computer vision and pattern recognition*, pp. 1674–1682.
21. Werlberger, M., Trobin, W., Pock, T., Wedel, A., Cremers, D., & Bischof, H. (2009). Anisotropic Huber-L1 optical flow. *BMVC, 1*(2), 3.
22. Schechner, Y. Y., & Averbuch, Y. (2007). Regularized image recovery in scattering media. *IEEE Transactions on Pattern Analysis and Machine Intelligence, 29*(9), 1655–1660.
23. Otair, D. (2013). Approximate k-nearest neighbour based spatial clustering using KD tree. arXiv preprint arXiv:1303.1951.
24. Kim, J., Kwon Lee, J., & Mu Lee, K. (2016). Accurate image super-resolution using very deep convolutional networks. In *Proceedings of the IEEE conference on computer vision and pattern recognition*, pp. 1646–1654.
25. Meng, G., Wang, Y., Duan, J., Xiang, S., & Pan, C. (2013). Efficient image dehazing with boundary constraint and contextual regularization. In *Proceedings of the IEEE international conference on computer vision (ICCV)*, pp. 617–624.
26. Yang, M., Hu, J., Li, C., Rohde, G., Du, Y., & Hu, K. (2019). An in-depth survey of underwater image enhancement and restoration. *IEEE Access, 7*, 123638–123657.
27. Anwar, S., & Li, C. (2019). Diving deeper into underwater image enhancement: A survey. arXiv preprint arXiv:1907.07863.
28. Treibitz, T., & Schechner, Y. Y. (2012). Turbid scene enhancement using multi-directional illumination fusion. *IEEE Transactions on Image Processing, 21*(11), 4662–4667.

Relative Global Optimum-Based Measure for Fusion Technique in Shearlet Transform Domain for Prognosis of Alzheimer's Disease

Suranjana Mukherjee and Arpita Das

1 Introduction

The task of merging multimodal images in the same scene is known as fusion. In medical diagnosis, a single-modality image is not enough to provide all the information alone [44, 45]. For example, magnetic resonance imaging (MRI) provides the anatomical information related to soft-tissue regions in special high resolution. The most popularly used MRI sequences are T1-weighted and T2-weighted scans. The MR-T1 imaging technique produces the detailed anatomical structure, while MR-T2 prominently highlights the differences between the normal and pathological structures of tissues [46]. Hence, the anatomical features like shrinking of gray matter and enlargement of ventricles due to the higher deposition rate of cerebrospinal fluid are well visualized from MRI, which is very significant in analyzing AD [1–3].

From positron emitted computed tomography (PET) and single-photon emission computed tomography (SPECT), functional information like blood flow, food activity, and metabolism of the affected organ is obtained in low spatial resolution [7, 8, 41–43]. The red, yellow, and dark blue colors in the PET scan image of the brain signify the healthy blood flow, a gradual decrease in metabolism due to shrinking of blood vessels, and no proper flow, respectively [1–6]. Analysis of the huge amount of data is time-consuming [47]. Under this situation, the fused image of different modalities in one composite frame saves valuable time for a doctor in the diagnostic procedure as well as reduces the storage cost of preserving multiple images.

The fusion algorithm design focuses on preserving relevant details and spatial features for increasing the clinical applicability of medical images for accurate assessment of affected organs or tissues. The various fusion approaches are broadly

---

S. Mukherjee (✉) · A. Das
Department of Radio Physics and Electronics, University of Calcutta, Kolkata, West Bengal, India

A. Deshpande et al. (eds.), *Computational Intelligence Methods for Super-Resolution in Image Processing Applications*, https://doi.org/10.1007/978-3-030-67921-7_15

classified as pixel level, feature level, and decision level [1, 7, 9]. At the pixel level, the fusion schemes are performed directly on the image pixels. There are two aspects of pixel-level methods: advanced image decomposition schemes and advanced image fusion rules. Advanced image decomposition schemes are classified into three categories: single-scale methods, multiscale transform, and two-scale methods [1, 6, 7, 11–25]. The activity level measure of input images is there to combine the respective image coefficient. The procedures like principal component analysis (PCA) and general intensity hue saturation transform (IHST) are referred to as single-scale methods [10–13].

A linear combination of vectors is constructed from the source images to form principal components in a PCA-based transform [10, 11]. IHST suffers from spectral distortion during arithmetic combination [12, 13]. Hence, original color details of the source images are degraded in the fused images. It also experiences a lack of visual clarity as the intensity plane is average of red (R), green (G), and blue (B) plane. The multiscale transform (MST) source image is decomposed to one approximate and many directional subbands [14].

In MST, the wavelet transform (WT) approaches as the discrete wavelet transform, stationary wavelet transforms, and dual-tree complex wavelet transforms capture information from three directions – horizontal, vertical, and diagonal [15–17]. Hence, contourlet transform (CT) aims to overcome WT's directional shortcoming [18].

In comparison to WT, CT is multidirectional and anisotropic. Moreover, CT can render edges and capture other singularities along the curvature efficiently. But it suffers from the effect of pseudo-Gibbs phenomena during up- and down-sampling processes [19]. In this view, the nonsubsampled contourlet transform (NSCT) [20] and nonsubsampled shearlet transform (NSST) [21] proposed by Cunha et al. and Easley et al., respectively, were source images and subimages of the same size. As a result, the pseudo-Gibbs phenomenon is suppressed appropriately. In comparison to NSCT, shearlets do not have any restrictions for shearing. Moreover, NSST is computationally efficient, as the inverse NSST only requires a summation of shearing filters rather than the inverse directional filter bank of NSCT.

In recent years, the NSCT- and NSST-based fusion approaches are prevalent [22–30]. For example, Bhatnagar et al. proposed a multimodal medical image fusion scheme in the NSCT domain where low- and high-frequency subbands are combined using phase congruency and directive contract enhancement scheme, respectively [22]. In another approach, Bhatnagar et al. proposed the normalized Shannon entropy measure and directive contrast enhancement-based fusion rule for the low- and high-frequency subband, respectively [23]. Yang et al. presented a fuzzy type-2 fusion scheme in the NSCT domain [24].

In some of the recent works, the pulse-coupled neural network (PCNN) model has been utilized. For example, Das et al. proposed a neuro-fuzzy fusion model in the NSCT domain [25–27]. Yin et al. presented an adaptive-PCNN fusion model in the NSST domain [26]. Gupta proposed an approach to fuse MRI and CT images in the NSST domain employing low- and high-frequency computing of the local energy and a simplified PCNN model, respectively [27]. To capture the high-frequency components' salient features, a sum-modified Laplacian operator is present in the feeding input of the PCNN model. Liu et al. suggested a moving frame decomposition framework followed by NSST decomposition to fuse MRI and CT [28]. In their approach, lower and higher subband coefficients are merged employing a simple averaging process and a 3×3 window-based sum-modified Laplacian operator to highlight every fine detail in the fused images. Ullah et al. presented a local feature-based fuzzy fusion scheme in the NSST domain for integrating CT and MRI to extract information of interest [29]. Many recent research works utilized the NSST. In one of our previous works, a vague set theory-based segmented fusion technique seems to integrate meaningful information related to investigate tissues in the NSST domain, and fusion results are very encouraging [30]. Hence, NSST is considered in the present work as well.

The process of improving the resolution is called super resolution (SR). SR has a significant role in improving the resolution for better representation. In some of the research works [31–34], the source images' low resolution (LR) is also improved with the fusion scheme. Yin et al. presented a fusion framework where the LR source images are first interpolated to enhance the resolution and then fused by sparse representation method [31]. In another work by Aymaz et al., interpolation has also been used to resize with the source images with a PCA-based fusion rule [32]. Li et al. recommended a scheme for the simultaneous image fusion, SR, edge enhancement, and noise suppression using a fractional differential and variation model the source images for extracting detail features [33]. The geometry of the source images is extracted by structure tensor. The edges are enhanced optimally by the utilization of the Euler-Lagrange differential operator. Then, for SR, the down-sampling operator is there to resize with the original one, and the fused image provides better visualization in terms of resolution and clarity. Zhong et al. presented a mixed SR-fusion model in MST, where a convolution neural network (CNN) model is employed to enhance the resolution of the directional subbands [34]. In one of our earlier works, a feature-based simultaneous SR-fusion scheme has been proposed [35]. In that work, before the PCA-based fusion rule, the resolution of each of the red (R), green (G), and blue (B) color plane is enhanced appropriately by applying the fuzzy c-means algorithm.

The high-frequency subband images (HFSIs) and low-frequency subband images (LFSIs) are integrated based on higher informative features and relative global optimum found from a particle swarm optimizer (PSO). In the present work, the maximum intensity-based selection is utilized for *LFSI*. Besides this, the *PSO* model proposed by Kennedy et al. [36] is explored to represent well both the input images with relatively good visualization and optimal edge enhancement. The detailed components of LR functional information are getting important simultaneously with high-resolution (HR) anatomical information. A relative factor is multiplied with each of the detail subband components to improve the resolution.

The new concept of the weighted combination of HFSIs attempts to visualize both modalities most refined textures better. The maximum of global optimum obtains the weight factors in *PSO*. The PSO model's fitness function is set as the static objective evaluation indices like *Ent*, *SD*, *AG*, and *SF* of the source images.

A suitable fitness function promotes the particles to move quickly to their best optimal location. This scheme is utilized to highlight every finer detail of the HFSI.

This chapter is organized as follows. Section 2 describes the proposed methodology. Experimental results and its analysis are given in Sect. 3, and finally, some conclusions have been drawn in Sect. 4.

2 Methodology

The source images are decomposed in the NSST domain. The NSST analysis utilizes the "maxflat" pyramidal filter, and images can be handled in three levels. A nonsubsampled pyramidal filter bank avoids the pseudo-Gibbs phenomenon. The basic framework of the proposed fusion scheme is described in Fig. 1.

Figure 1 describes the necessary steps of the proposed fusion scheme as follows:

1. Extract the (R, G, B) components of the functional image ***B***.
2. Decompose the anatomical image ***A*** and each (R, G, B) component of image-B by employing three-level NSST with nonsubsampled Laplacian pyramidal filter bank.
3. Combine the high- and low-frequency coefficients according to the proposed fusion rules.
4. Reconstruct the fused image by applying inverse NSST.

The details of every salient block are described in the following sections.

2.1 Nonsubsampled Shearlet Transform (NSST)

NSST overcomes the lack of directionality of conventional WT. Two essential steps of NSST are multiscale partitioning and directional localization [21, 26]. A nonsubsampled pyramidal filter achieves the shift-invariance insensitivity in the

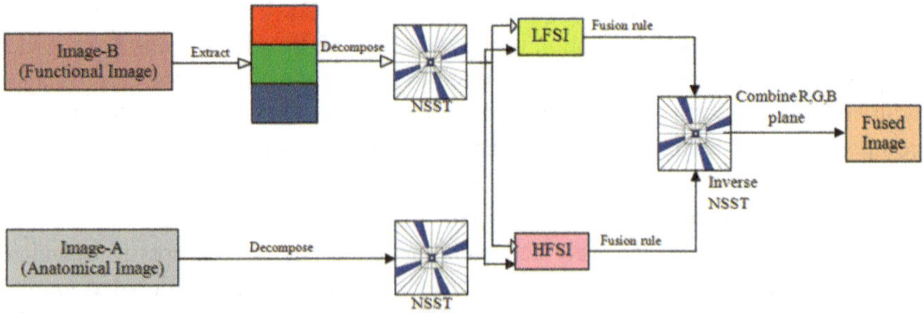

Fig. 1 Block diagram of the proposed approach

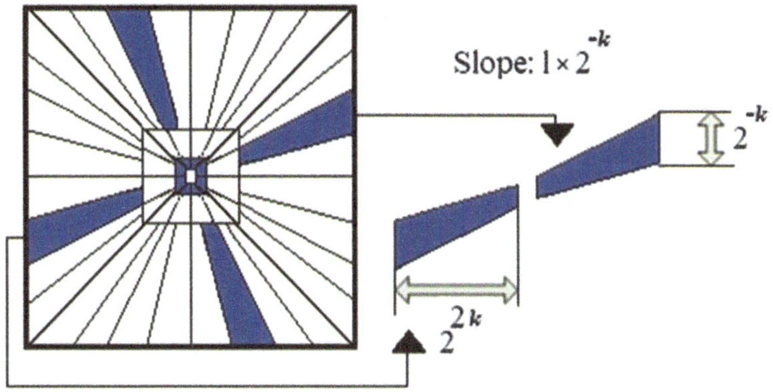

Fig. 2 Representation of shearlets

multiscale partitioning. This scheme suppresses the Gibbs phenomenon adequately as convolutions replace downsamplers, thus preserving the source images' edge/color information within the fused images. In the directional localization, the frequency plane is decomposed into an LFSI and several trapezoidal HFSI utilizing a shift-invariant shearing filter. For the decomposition k-th level, a pair of trapezoids known as shearlets of size $2^{2k} \times 2k$ (approx) is produced and oriented along with the line slope $l \times 2^{-k}$. l is an integer value as in Fig. 2.

NSST is computationally efficient because the summation of the shearlets restores fused coefficients.

2.2 Combination of Low-Frequency Information

The *LFSIs* are merged by utilizing the maximum absolute intensity-based selection rule, which is very simple and computationally efficient. The decision mapping is as follows:

$$\text{LFSI}_{F,k,l} = \begin{cases} A_{A,k} \text{ for } A_{A,k} \geq A_{B,k}\,(i,j) \\ A_{B,k} \text{ for } A_{A,k} < A_{B,k}\,(i,j) \end{cases} \tag{1}$$

where $A_{A,k}$ and $A_{B,k}$ are the absolute intensity of the LFSI of images ***A*** and ***B*** with the k-th decomposition level at the (i,j) spatial location.

2.3 Combination of High-Frequency Information

To highlight every salient feature of *HFSI*, the maximum of the global best in *PSO* is estimated for both images. Then, the relative global best becomes

$$Q_A = \frac{\max(\mathrm{gbest}_A)}{\max(\mathrm{gbest}_A) + \max(\mathrm{gbest}_B)}, \quad \text{and}$$

$$Q_B = \frac{\max(\mathrm{gbest}_B)}{\max(\mathrm{gbest}_A) + \max(\mathrm{gbest}_B)}$$

Now, the HFSI of images *A* and *B* can merge as follows:

$$\mathrm{HFSI}_{F,k} = Q_A * \mathrm{HFSI}_{A,k} + Q_B * \mathrm{HFSI}_{B,k} \tag{2}$$

The maximum of *gbest* is obtained from *PSO*. The details of *PSO* are described as follows. The *PSO* is a simple, naturally inspired stochastic optimization technique inspired by the food searching behavior of a flock of birds or a school of fish [2, 36–38]. A school of fish or a flock of birds is called the swarm, and the individual swam members are called particles. It is the iterative process to find the global optimum. The parameter setting of PSO is done, according to Xu et al. [38]. If the number of particles is significantly less, the PSO model may have the problem of local convergence. On another side, for a large number of particles, the convergence time is more. If the number of particles lies in the range from 20 to 50, it is then appropriate. In this study, there are 20 particles. The maximum number of iteration is 100, and fitness function *f* is estimated by the static properties like *Ent*, *SD*, *AG*, and *SF* of the input *HFSI*.

In *PSO*'s best position, each particle is called the individual best or "*pbest.*" In contrast, the best place found by the whole population is known as the global best or "*gbest.*" Global best is well known to all and updated instantly when any particle finds a new best position. The maximum of "*gbest*" is *d*. *PSO* is iterative and is governed in the *t*-th iteration, position $[\boldsymbol{x}_n(t)]$, and velocity $[\boldsymbol{v}_n(t)]$ update equations of the *n*-th particle of the swarm. It is described as follows:

$$v_{n+1}(t) = c_0 v_n(t) + c_1\,\mathrm{rand}()[p_{\mathrm{best}} - x_n(t)] + c_2\,\mathrm{rand}()[p_{\mathrm{best}} - x_n(t)], \tag{3}$$

and

$$x_n(t+1) = x_n(t) + v_n(t), \tag{4}$$

where c_0 is the inertia coefficient, c_1 and c_2 are acceleration coefficients, and random numbers are denoted as *rand()* generated uniformly in the range [0, 1].

On the other hand, C_1 and C_2 are responsible for tuning the cognition and the social terms, respectively, updating the velocity in Eq. (3). The larger values of C_1 and C_2 encourage exploration and exploitation, respectively.

Initially, the algorithm is explored in diverse search spaces to broaden the range of all possible solutions. Satisfactory simulation hinges on two propositions: nearest-neighbor velocity matching and craziness determined by the inertia coefficient. This craziness introduces some random changes in the particle's velocity, and the system varies significantly. As a result, a thought-provoking and life-like appearance happen in the simulation process. The lower value of C_0 reduces the velocity of particles

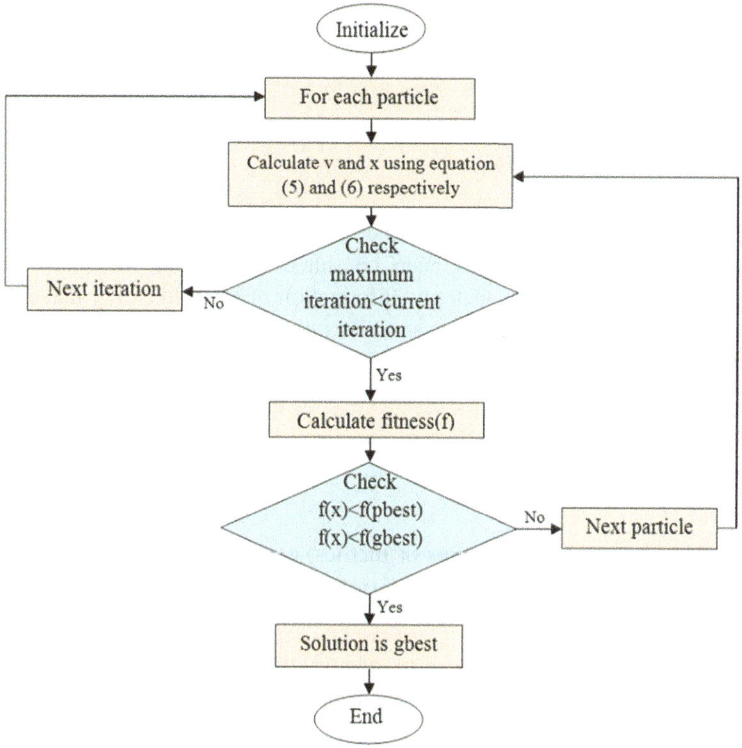

Fig. 3 Flow diagram of PSO

to zero to reach the global best. The selection of coefficients is updated in the velocity equation, which determines the swarm's convergence ability to find the optimum. The proposed PSO flowchart appears in Fig. 3.

3 Experimental Results

3.1 Dataset Preparation

The proposed fusion scheme is implemented on different sagittal slices of MRI and PET images to integrate anatomical and functional information in a single frame. The PET-MRI brain images are of a 70-year-old man suffering from memory loss about 9 months before imaging and having a history of atrial fibrillation/rapid heart rate as described by Harvard University [39]. MRI and PET images reveal the gray matter detailing as well as widened hemispheric sulci and abnormality in regional cerebral metabolism, correspondingly. Experimental results obtained by this fusion method provide the complete information to study further investigation and progression of AD.

There are two performance evaluation aspects for the assessment of the proposed fusion algorithm: objective and subjective.

3.2 Subjective Analysis

The subjective evaluation relies on the comprehensive ability of humans. In this view, an expert radiologist has been consulted to evaluate the fused images. According to him, the entire anatomic information of MRI, like gray matter details and surrounding edges of ventricles as well as the color information of PET related to metabolism, are well integrated into fused images.

3.3 Objective Analysis

Mathematical parameters (measures or metrics) evaluate the salient features of the fused images. The *Ent*, *SD*, *AG*, and *SF* of the images are evaluated to analyze the recommended approach [38, 40]. These parameters are explained as follows.

- *Ent:* This parameter is employed to quantify the information content present in the image and its texture distribution. The more excellent the entropy value means better incorporation of information in the fused image. The mathematical description is as follows.

$$\text{Ent} = \sum_{i=0}^{255} p(z_i) \log_2 p(z_i), \tag{5}$$

  where $p(z_i)$ denotes the probability of randomly variable intensity z_i.
- *SD:* It is evaluated to estimate the clarity and visual quality of the image information. A higher value of *SD* corresponds to the superior visual clarity of the image. The mathematical representation is as follows:

$$\text{SD} = \sum_{i=0}^{255} (p(z_i) - m)^2 p(z_i), \tag{6}$$

  where m represents the average intensity of the image.
- *AG:* The average of gradient values in the X and Y directions is to estimate the sharpness related to edges, curves, notches, and region boundaries of images. A greater value of AG signifies better integration of fine texture information. Its expression is

$$\text{AG} = \frac{1}{M \times N} \sum_{i=0}^{M-1} \sum_{j=0}^{N-1} \sqrt{\frac{1}{2}\left(\left(\frac{\partial G_{x,i,j}}{\partial x_{i,j}}\right)^2 + \left(\frac{\partial G_{y,i,j}}{\partial y_{i,j}}\right)^2\right)}, \tag{7}$$

where $\frac{\partial G_{x,i,j}}{\partial x_{i,j}}$ and $\frac{\partial G_{y,i,j}}{\partial y_{i,j}}$ represent the gradient of the image in the X and Y directions, respectively.

- *SF:* It is used to indicate the clarity and overall activity level of the images. The higher value of SF denotes more information integrated into the fused image with better contrast. Its formula is

$$\mathrm{SF} = \sqrt{\mathrm{RF}^2 + \mathrm{CF}^2}, \tag{8}$$

where RF and CF represent the row and column frequency of fused image $\boldsymbol{F}$. The RF and CF are calculated as follows. The objective assesment is present in the Table1.

$$\mathrm{RF} = \frac{1}{M \times N}\sqrt{\sum_{i=1}^{M}\sum_{j=2}^{N}\left[F(i,j) - F(i,j-1)\right]^2}, \tag{9}$$

and

$$\mathrm{CF} = \frac{1}{M \times N}\sqrt{\sum_{i=2}^{M}\sum_{j=1}^{N}\left[F(i,j) - F(i-1,j)\right]^2} \tag{10}$$

The parametric evaluation shows that the proposed fusion scheme has successfully preserved all the relevant information from the fused images with good visual clarity and contrast.

Table 1 Objective evaluation of parameters of the proposed fusion techniques

Images	Evaluation parameters	Functional image	Anatomical image	Fused image
Set 1 of Fig. 4	Ent	4.1731	4.1393	**5.1228**
	SD	0.2810	0.2308	**0.3420**
	AG	0.0203	0.0295	**0.0299**
	SF	0.0904	0.893	**0.1065**
Set 2 of Fig. 4	Ent	3.9221	4.3188	**5.1992**
	SD	0.3065	0.2511	**0.3578**
	AG	0.0100	0.230	**0.0266**
	SF	0.0946	0.975	**0.1108**
Set 3 of Fig. 4	Ent	3.9779	4.3873	**5.3077**
	SD	0.3117	0.2255	**0.3491**
	AG	0.0104	0.0354	**0.0291**
	SF	0.0963	0.933	**0.1176**
Set 4 of Fig. 4	Ent	3.7715	4.5346	**5.3870**
	SD	0.2723	0.2315	**0.3330**
	AG	0.0090	0.0270	**0.0312**
	SF	0.0842	0.956	**0.1204**

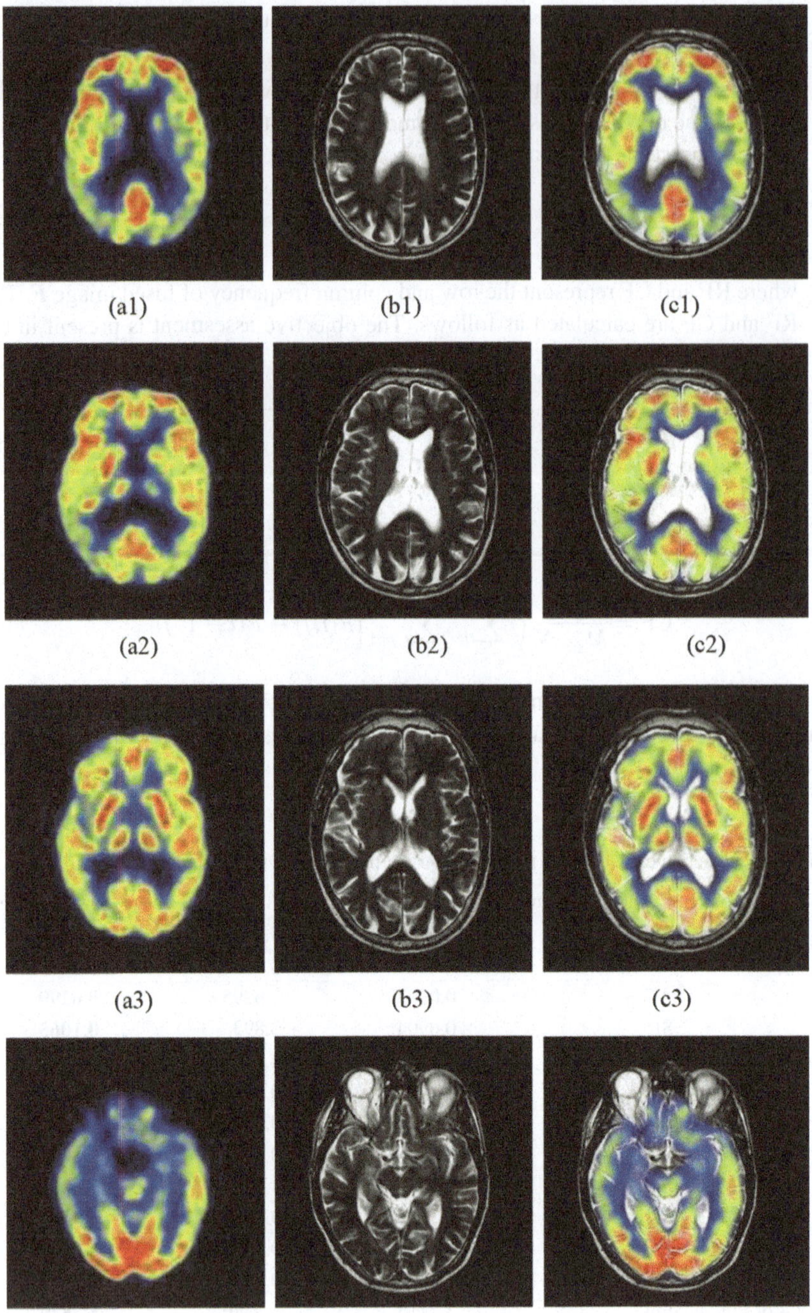

Fig. 4 Fusion of multimodal images: (a1–a4) PET; and (b1–b4) MRI (c1–c4) fused images

4 Conclusion

This chapter focuses on designing an effortless and computationally efficient fusion approach to combine anatomical and functional information. The NSST-based decomposition can capture the salient directional information and overcomes the directional limitation of WT approaches. The maximum intensity-based selection of approximate images is useful in preserving gross data with enhanced clarity and activity. The relative global optimum can determine the proper ratio to acquire multimodal information. The proposed fusion rule for merging the approximate and detail information is good enough to combine the source images' informative features. The proposed approach can restore every salient texture related to edges, curvatures, and notches with enhanced information restoration, activity level preservation, and clarity. But the proposed fusion is more straightforward than the advanced neural network-based methods. The integrated images are good enough to assist in accurate diagnosis as well as those that may reduce the storage space of preserving multiple images. In the future, we may try to develop another fusion scheme that will be computationally efficient and inexpensive as the present one. Then, the present scheme will be compared with that approach.

Acknowledgments This work was supported by the Center of Excellence (CoE) and funded by the World Bank, MHRD India. The authors are grateful to Dr. Pratip Nandi of R.G. Kar, Medical College and Hospital, Kolkata, and Dr. Soumitra Halder, MD, in Radio Diagnosis of Metiabruz Super Speciality Hospital for providing valuable comments on the subjective evaluation of the proposed fusion scheme.

References

1. Du, J., Li, W., Lu, K., & Xiao, B. (2016). An overview of multimodal medical image fusion. *Neurocomputing, 215*, 3–20.
2. Das, A., & Bhattacharya, M. (2017). Study on neurodegeneration at different stages using MR images: Computational approach to registration process with optimization techniques. *Computer Methods in Biomechanics and Biomedical Engineering, Imaging & Visualization, 5*(3), 165–182.
3. Braunwald, E., Fauci, A. S., Kasper, D. L., Hauser, S. L., Longo, D. L., Jameson, J. L., et al. (2001). Alzheimer's disease and other dementias. *Harrison's Principles of Internal Medicine*, McGraw Hill, *17*(2), 2393–2394.
4. Chang, D. J., Zubal, I. G., Gottschalk, C., Necochea, A., Stokking, R., Studholme, C., et al. (2002). Comparison of statistical parametric mapping and SPECT difference imaging in patients with temporal lobe epilepsy. *Epilepsia, 43*(1), 68–74.
5. Horn, J. F., Habert, M. O., Kas, A., Malek, Z., Maksud, P., Lacomblez, L., et al. (2009). Differential automatic diagnosis between Alzheimer's disease and frontotemporal dementia based on perfusion SPECT images. *Artificial Intelligence in Medicine, 47*(2), 147–158.
6. Gigengack, G., Ruthotto, R., Burger, B., Wolters, C. H., Xiaoyi Jiang, X., & Schafers, K. P. (2012). Motion correction in dual gated cardiac PET using mass-preserving image registration. *IEEE Transactions on Medical Imaging, 31*(3), 698–712.

7. James, A. P., & Dasarathy, B. V. (2014). Medical image fusion: A survey of the state of the art. *Information Fusion, 19*, 4–19.
8. Shen, R., Cheng, I., & Basu, A. (2013). Cross-scale coefficient selection for volumetric medical image fusion. *IEEE Transactions on Biomedical Engineering, 60*(4), 1069–1079.
9. Petrovic, V. S., & Xydeas, C. S. (2004). Gradient-based multiresolution image fusion. *IEEE Transactions on Image Processing, 13*(2), 228–237.
10. Desale, R. P., & Verma, S. V. (2013). Study and analysis of PCA, DCT & DWT based image fusion techniques. In *International conference on signal processing, image processing & pattern recognition* (pp. 66–69). Piscataway: IEEE.
11. Shahdoosti, H. R., & Ghassemian, H. (2016). Combining the spectral PCA and spatial PCA fusion methods by an optimal filter. *Information Fusion, 27*(1), 150–160.
12. Ming, S., Shun-Chi, S., Chyun, H., & Ping, S. H. (2001). A new look at IHS-like image fusion methods. *Information Fusion, 2*(3), 177–186.
13. Daneshvar, S., & Ghassemian, H. (2010). MRI and PET image fusion by combining IHS and retina-inspired models. *Information Fusion, 11*(2), 114–123.
14. Liu, Y., Liu, S., & Wang, Z. (2015). A general framework for image fusion based on multiscale transform and sparse representation. *Information Fusion, 24*(1), 147–164.
15. Chibani, Y., & Houacine, A. (2003). Redundant versus orthogonal wavelet decomposition for multisensor image fusion. *Pattern Recognition, 36*(4), 879–887.
16. Pajares, G., & Cruz, J. M. (2004). A wavelet-based image fusion tutorial. *Pattern Recognition, 37*(9), 1855–1872.
17. Singh, R., & Khare, A. (2014). Redundant discrete wavelet transform based medical image fusion. *Advances in Signal Processing and Intelligent Recognition Systems, 264*(1), 505–515.
18. Do, M. N., & Martin, V. (2005). The contourlet transform: An efficient directional multiresolution image representation. *IEEE Transactions on Image Processing, 14*(12), 2091–2106.
19. Yang, L., Guo, B. L., & Ni, W. (2008). Multimodality medical image fusion based on multiscale geometric analysis of contourlet transform. *Neurocomputing, 72*(1), 203–211.
20. Cunha, A. L. D., Zhou, J., & Do, M. N. (2006). The nonsubsampled contourlet transform: Theory, design, and applications. *IEEE Transactions on Image Processing, 15*(10), 3089–3101.
21. Easley, G., Labate, D., & Lim, W. Q. (2008). Sparse directional image representations using the discrete shearlet transform. *Applied and Computational Harmonic Analysis, 25*(1), 25–46.
22. Bhatnagar, G., Wu, Q. M. J., & Liu, Z. (2013). Directive contrast based multimodal medical image fusion in NSCT domain. *IEEE Transactions on Multimedia, 15*(5), 1014–1024.
23. Bhatnagar, G., Wu, Q. M. J., & Liu, Z. (2015). A new contrast based multimodal medical image fusion framework. *Neurocomputing, 157*, 143–152.
24. Yang, Y., Que, Y., Huang, S., & Lin, P. (2016). Multimodal sensor medical image fusion based on type-2 fuzzy logic in NSCT domain. *IEEE Sensors Journal, 16*(10), 3735–3745.
25. Das, S., & Kundu, M. K. (2013). A neuro-fuzzy approach for medical image fusion. *IEEE Transactions on Biomedical Engineering, 60*(12), 3347–3353.
26. Yin, M., Liu, X., Liu, Y., & Chen, X. (2018). Medical image fusion with parameter-adaptive pulse coupled neural network in nonsubsampled shearlet transform domain. *IEEE Transactions on Instrumentation and Measurement, 68*(1), 49–64.
27. Gupta, D. (2018). Nonsubsampled shearlet domain fusion techniques for CT–MR neurological images using improved biological inspired neural model. *Biocybernetics and Biomedical Engineering, 38*(2), 262–274.
28. Liu, X., Mei, W., & Du, H. (2018). Multimodality medical image fusion based on image decomposition framework and nonsubsampled shearlet transform. *Biomedical Signal Processing and Control, 40*, 343–350.
29. Ullah, H., Ullah, B., Wu, L., Abdalla, F. Y., Ren, G., & Zhao, Y. (2020). Multi-modality medical images fusion based on local-features fuzzy sets and novel sum-modified-Laplacian in non-subsampled shearlet transform domain. *Biomedical Signal Processing and Control, 57*, 101724.

30. Mukherjee, S., & Das, A. (2020). Vague set theory based segmented image fusion technique for analysis of anatomical and functional images. *Expert Systems with Applications, 159*, 113592.
31. Yin, H., Li, S., & Fang, L. (2013). Simultaneous image fusion and super-resolution using sparse representation. *Information Fusion, 14*(3), 229–240.
32. Aymaz, S., & Köse, C. (2019). A novel image decomposition-based hybrid technique with super-resolution method for multi-focus image fusion. *Information Fusion, 45*, 113–127.
33. Li, H., Zhengtao, Y., & Mao, C. (2016). Fractional differential and variational method for image fusion and super-resolution. *Neurocomputing, 171*, 138–148.
34. Zhong, J., Yang, B., Li, Y., Zhong, F., & Chen, Z. (2016). Image fusion and super-resolution with convolutional neural network. In *Pattern recognition, communications in computer and information science* (Vol. 663). Singapore: Springer.
35. Mukherjee, S., & Das, A. (2020). Effective fusion technique using FCM based segmentation approach to analyze Alzheimer's disease. In *Smart healthcare analytics in IoT enabled environment* (pp. 91–107). Cham: Springer.
36. Kennedy, J., & Eberhart, R. (1995). Particle swarm optimization. In *IEEE international conference on neural networks, Perth, WA* (pp. 1942–1948). Piscataway: IEEE.
37. Das, A., & Bhattacharya, M. (2011). Affine-based registration of CT and MR modality images of human brain using multiresolution approaches: Comparative study on genetic algorithm and particle swarm optimization. *Neural Computing and Applications, 20*(2), 223–237.
38. Xu, X., Shan, D., Wang, G., & Jiang, X. (2016). Multimodal medical image fusion using PCNN optimized by the QPSO algorithm. *Applied Soft Computing, 46*, 588–595.
39. Johnson, K. A., & Becker, J. A. (2001). *The whole brain atlas*. Available online at http://www.med.harvard.edu/aanlib/home.html.
40. Gonzalez, R. C., Woods, R. E., & Eddins, S. L. (2004). *Digital image processing using MATLAB*. Pearson-Prentice-Hall: Upper Saddle River.
41. Arshaghi, A., et al. (2020). Image transmission in UAV MIMO UWB-OSTBC system over Rayleigh channel using multiple description coding (MDC). In *Imaging and sensing for unmanned aircraft systems: Volume 2: Deployment and applications* (pp. 67–90). Stevenage: IET.
42. Razmjooy, N., Estrela, V. V., & Loschi, H. J. (2019). A study on metaheuristic-based neural networks for image segmentation purposes. Data Science Theory, Analysis and Applications, Taylor and Francis.
43. de Jesus, M. A., et al. (2020). Using transmedia approaches in STEM. In *2020 IEEE global engineering education conference (EDUCON)* (pp. 1013–1016). Piscataway: IEEE.
44. Estrela, V. V., et al. (2019). Why software-defined radio (SDR) matters in healthcare? *Medical Technologies Journal, 3*(3), 421–429.
45. Aroma, R. J., et al. (2020). Multispectral vs. hyperspectral imaging for unmanned aerial vehicles: Current and prospective state of affairs. In V. V. Estrela, J. Hemanth, O. Saotome, G. Nikolakopoulos, & R. Sabatini (Eds.), *Imaging and sensing for unmanned aircraft systems* (Vol. 2, 7, pp. 133–156). London: IET. https://doi.org/10.1049/PBCE120G_ch7.
46. Razmjooy, N., Ashourian, M., Karimifard, M., Estrela, V. V., Loschi, H. J., do Nascimento, D., et al. (2020). Computer-aided diagnosis of skin cancer: A review. *Current Medical Imaging, 16* (7), 781–793.
47. Deshpande, A., et al. (2020). Deep learning as an alternative to super-resolution imaging in UAV systems. In V. V. Estrela, J. Hemanth, O. Saotome, G. Nikolakopoulos, & R. Sabatini (Eds.), *Imaging and sensing for unmanned aircraft systems* (Vol. 2, 9, pp. 177–212). London: IET. https://doi.org/10.1049/PBCE120G_ch9.

Conclusion

This book is motivated by the aspiration to highlight the evolution of super-resolution (SR) and computational intelligence (CI) and it consists of various chapters with applications revolving around these subjects. The content offers to help to gain insight into the knowledge evolved and practiced. CI usage has been rising worldwide with the inclusion of big data, artificial intelligence, and so on. Since data are the new oil of the current era, there are high demands for CI-based possibilities and analyses. Computer vision fueled by super-resolution advances carries a step forward in obtaining unblurred high-resolution images through algorithms in today's world. CI arranges for a swift and satisfying output prompted by intelligent algorithms.

It is an often desire or motivation for the eyes to witness something better than a blurred portrait of an image. The editors realized and acknowledged that the importance and the need for high resolution in images are two crucial aspects of the application, that is, for human information and machine perception. For the interpretation of humans, visual information should be improved and, in some cases, interpolated. On the other hand, the machine needs an improvised image for perception to make decisions. It is an underlying fact, though the more the resolution, the more the image is detailed with high pixel density. Thus, this image rendering advancement helps computer vision applications across many domains such as surveillance at airports or borders for security, medical imaging, forensic, and satellite imaging. Whatever vision task is at play, enhanced imageries engendered by super-resolution are needed to improve the image for further processing. Since the new holy grail entails understanding and/or explaining visual data, processing raw multimodality images and videos to obtain rich, high-resolution versions of them is a task on the rise.

The gathering of like minds in penning down the thoughts and knowledge to establish this book was also positively intimidating. It is overwhelming to find the authors with the possibilities of them directing their energy in such a specific research goal and then pouring interest in writing a book. With authors and

A. Deshpande et al. (eds.), *Computational Intelligence Methods for Super-Resolution in Image Processing Applications*, https://doi.org/10.1007/978-3-030-67921-7

contributors from around the globe, it heightens the anxiousness and enthusiasm. The aligning of knowledge and all the contributors' thoughts to successfully pursue this book to perfection is an attractive task. The more in-depth research we get leads to more diverse detailed results and the possibility of performing different data exploration types. The editors believe that this tome's research and publishing shall expand minds and benefit humanity while aligning other professionals' likes.

Index

A. Deshpande et al. (eds.), *Computational Intelligence Methods for Super-Resolution in Image Processing Applications*, https://doi.org/10.1007/978-3-030-67921-7

P

Q

R

S

The manufacturer's authorised representative in the EU is Springer Nature Customer Service Centre GmbH, Europaplatz 3, 69115 Heidelberg, Germany. If you have any concerns regarding our products, please contact ProductSafety@springernature.com

Printed and bound by CPI Group (UK) Ltd, Croydon, CR0 4YY

28/09/2026

02224091-0015